Tellurite Glass Smart Materials

Raouf El-Mallawany

Editor

Tellurite Glass Smart Materials

Applications in Optics and Beyond

 Springer

Editor
Raouf El-Mallawany
Physics Department
Faculty of Science
Menoufia University
Shebin ElKoum, Egypt

ISBN 978-3-030-09524-6   ISBN 978-3-319-76568-6   (eBook)
https://doi.org/10.1007/978-3-319-76568-6

Printed on acid-free paper

This Springer imprint is published by the registered company Springer International Publishing AG part of Springer Nature.
The registered company address is: Gewerbestrasse 11, 6330 Cham, Switzerland

Contents

Contributors

Raja J. Amjad Department of Physics, COMSATS Institute of Information Technology, Lahore, Pakistan

Luiz Carlos Barbosa Depto de Eletrônica Quântica, Instituto de Física Gleb Wataghin, Universidade estadual de Campinas, Campinas, SP, Brazil

P. R. Biju School of Pure & Applied Physics, Mahatma Gandhi University, Kottayam, India

Enver Fernandez Chillcce BrPhotonics- CPqD, São Paulo, Brazil

Raouf El-Mallawany Physics Department, Faculty of Science, Menoufia University, Shebin ElKomm, Egypt

M. H. Fernandes Laboratory for Bone Metabolism and Regeneration, Faculty of Dental Medicine, University of Porto, Porto, Portugal

REQUIMTE/LAQV – University of Porto, Porto, Portugal

Cicero Omegna Filho Luxtec Optical System, Campinas, SP, Brazil

Gejo George School of Pure & Applied Physics, Mahatma Gandhi University, Kottayam, India

P. S. Gomes Laboratory for Bone Metabolism and Regeneration, Faculty of Dental Medicine, University of Porto, Porto, Portugal

REQUIMTE/LAQV – University of Porto, Porto, Portugal

L. A. Gómez-Malagón Escola Politécnica de Pernambuco, Universidade de Pernambuco, Recife, PE, Brazil

L. Grenho Laboratory for Bone Metabolism and Regeneration, Faculty of Dental Medicine, University of Porto, Porto, Portugal

REQUIMTE/LAQV – University of Porto, Porto, Portugal

G. Hungerford HORIBA Jobin Yvon IBH Ltd, Glasgow, UK

Carlos Jacinto Grupo de Nano-Fotônica e Imagens, Instituto de Física, Universidade Federal de Alagoas, Maceió, AL, Brazil

Jayasankar C. K. Department of Physics, Sri Venkateswara University, Tirupati, India

Cyriac Joseph School of Pure & Applied Physics, Mahatma Gandhi University, Kottayam, India

Luciana R. P. Kassab Faculdade de Tecnologia de São Paulo/CEETEPS, São Paulo, SP, Brazil

Saman Q. Mawlud University of Salahaddin/College of Education/Physics Department, Iraq-Kurdistan, Erbil, Iraq

S. H. Nandyala School of Metallurgy and Materials, University of Birmingham, Edgbaston, Birmingham, UK

V. P. Prakashan School of Pure & Applied Physics, Mahatma Gandhi University, Kottayam, India

M. Reza Dousti Grupo de Nano-Fotônica e Imagens, Instituto de Física, Universidade Federal de Alagoas, Maceió, AL, Brazil

M. S. Sajna School of Pure & Applied Physics, Mahatma Gandhi University, Kottayam, India

Weslley Q. Santos Grupo de Nano-Fotônica e Imagens, Instituto de Física, Universidade Federal de Alagoas, Maceió, AL, Brazil

M. S. Sanu School of Pure & Applied Physics, Mahatma Gandhi University, Kottayam, India

Dariush Souri Department of Physics, Faculty of Science, Malayer University, Malayer, Iran

A. Stamboulis School of Metallurgy and Materials, University of Birmingham, Edgbaston, Birmingham, UK

P. Syam Prasad Department of Physics, National Institute of Technology Warangal, Warangal, Telangana, India

N. V. Unnikrishnan School of Pure & Applied Physics, Mahatma Gandhi University, Kottayam, India

M. J. Valenzuela Bell Departamento de Física, Universidade Federal de Juiz de Fora, Juiz de Fora, MG, Brazil

Venkata Krishnaiah K Laser Applications Research Group, Ton Duc Thang University, Ho Chi Minh City, Vietnam

Faculty of Applied Sciences, Ton Duc Thang University, Ho Chi Minh City, Vietnam

Department of Physics, RGM College of Engineering and Technology, Nandyal, India

P. Venkateswara Rao Department of Physics, The University of the West Indies, Kingston, Jamaica

Venkatramu V Department of Physics, Yogi Vemana University, Kadapa, India

Chapter 1
Some Physical Properties of Tellurite Glasses

Raouf El-Mallawany

Abstract The present chapter summarizes some physical properties of tellurite glasses when modified with transition metal oxides (TMO) and with rare-earth oxides (REO). The physical properties were thermal, optical, electrical, and mechanical. Also, some recent applications of tellurite glasses have been collected.

1.1 Introduction

Solid-state physics is the study of rigid matter, through methods such as quantum mechanics, crystallography, electromagnetism, and metallurgy, and it is the largest branch of condensed matter physics. Solid-state physics studies how the large-scale properties of solid material result from their atomic-scale properties and form a theoretical basis of materials science. Processing of new noncrystalline solid, tellurite glass has been achieved in the last 60 years. The research work on the smart tellurite glasses have been published in earlier articles [1–4] but remained virtually unknown to materials and device engineers until 1990s when linear, nonlinear, and dispersion properties of modified tellurite glasses were reported [5, 6]. Tellurite glasses have been mushroomed in the field known academically as the "Physics of Noncrystalline Solids." Smart materials are designed materials to have one or more properties that can be significantly changed in a controlled fashion by external stimuli. Moreover, research effort is seen as providing the fundamental base for finding new smart noncrystalline solids prepared by nano-processing to get new physical data. Artificial neural network (ANN) technique has been used to simulate and predict any physical property for applications in therapy, radiation shielding, and renewable energy as the efficiency enhancements in solar cells using photon downconversion in Tb/Yb-doped tellurite glass have been reported [7–11]. Also, many books [12–14] have been published and five video seminar (loaded in the Internet) for international educational purposes entitled "An Introduction to Tellurite Glasses" [15]. Prior to article [4] (1984), tellurite glasses were considered

R. El-Mallawany (✉)
Physics Department, Faculty of Science, Menoufia University, Shebin ElKomm, Egypt

1

as intermediate glass former and no international attention on it. Pure TeO_2 and binary TeO_2-WO_3 glass have been prepared. Since 1984 and up till now, the research activities were focused on developing new materials by modifying TeO_2 glass with rare-earth oxides (REO) or transition metal ions (TMI) to get unique optical, electrical, thermal, elastic, vibrational properties and its applications and have opened a new horizon in new physical aspects of semiconducting tellurite glasses research [16–40]. Figure 1.1 shows some samples of $80TeO_2$-$5TiO_2$-$(15\text{-}x)$ WO_3-x A_nO_m where A_nO_m is Nb_2O_5, Nd_2O_3, and Er_2O_3; $x = 0.01$, 1, 3, and 5 mol% for Nb_2O_5; and $x = 0.01$, 0.1, 1, 3, 5, and 7 mol% for Nd_2O_3 and Er_2O_3.

Recently there has been significant progress in tellurite (TeO_2)-based glasses based on their unique and interesting physical properties when modified with transition metal or rare-earth oxides [41–50]. The interesting physical properties were low phonon energy, low melting temperature, high refractive index, large transparency window, electrical shielding, and high thermal and chemical stability which enable the glasses for optical applications. Luminescence and energy transfer studies on Sm^{3+}/Tb^{3+} co doped telluroborate glasses for WLED applications, production, and characterization of femtosecond laser-written double line waveguides in heavy metal oxide glasses, sensitization of Ho^{3+} on the 2.7 μm emission of Er^{3+} in $(Y0.9La0.1)_2O_3$ transparent ceramics, spectroscopy of Yb- doped tungsten-tellurite glass have been reported [41–50]. Also, assessment of its lasing properties, and spectroscopic studies of Dy^{3+} ion doped tellurite glasses for solid state lasers, white LEDs and yellow to orange-reddish glass phosphors: Sm^{3+}, Tb^{3+} and Sm^{3+}/Tb^{3+} in zinc tellurite-germanate glasses have been reported [41–50]. The two glass series (ordinary glass, glass with nanoparticles) were successfully prepared by using melt quenching method [7]. The effect of erbium and erbium nanoparticles has been studied by:

- The average size of nanoparticles in glass with nanoparticles was found in the range $\approx$23.53 nm,
- The density of ordinary glass is less than glass with nanoparticles due to the increasing compactness in the glass system.
- The refractive index of glass with nanoparticles is higher than ordinary glass system which is caused by the change in density.
- The absorption peaks of the ordinary glass are two times intense than glass with nanoparticles which correspond to the restriction of electrons in nanoparticles.

Hundreds of scientific and few review articles have been reported on tellurite glasses modified with transition metals and also with rare earths. Figure 1.2 shows the statistical distribution of tellurite glasses articles in 1951–2018.

Fig. 1.1 Samples of $80TeO_2$-$5TiO_2$-$(15\text{-}x)$ WO_3-xA_nO_m where A_nO_m is Nb_2O_5, Nd_2O_3, and Er_2O_3; $x = 0.01$, 1, 3, and 5 mol% for Nb_2O_5; and $x = 0.01$, 0.1, 1, 3, 5, and 7 mol% for Nd_2O_3 and Er_2O_3

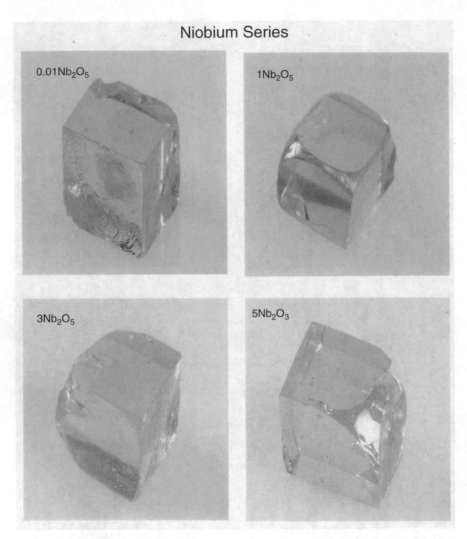

Fig. 1.1 (continued)

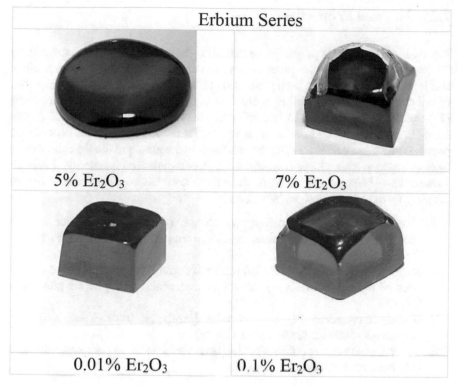

Fig. 1.1 (continued)

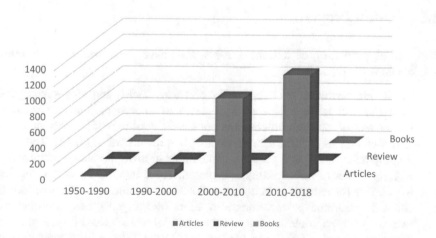

Fig. 1.2 Tellurite glasses scientific articles, review articles, and books by Scopus research engine 1951–January 2018

1.2   Thermal Properties

Processing and thermal properties of tellurite glasses have been investigated by different techniques for the pure, binary, ternary, and quaternary noncrystalline solids; tellurite glasses have been processed [12, 13, 25]. Selected thermal properties of solids like thermal expansion coefficient (α), glass transformation temperature (T_g), specific heat capacity (C_p), and difference in specific heat capacities (ΔC_p) from solid (C_{pg}) to liquid (C_{pt}) have been measured. Theoretical correlations between the previous measured values for different modifiers in the glass and the structure factors like: average stretching force constant, average crosslink density, energy of crystallization have been achieved. Also, theoretical correlations have been achieved between thermal expansion coefficient and:

  I. Vibrational properties like long wavelength (Gruniesen parameter at high enough temperature greater than the Debye temperature (estimated from ultrasonic measurements).

  II. Thermal expansion coefficient (α) has been correlated with the second derivatives of the potential energy of a diatomic chain as explained previously [12, 13].

  III. Gruenisen parameter has been correlated with the Born exponent for the amorphous solids for the first time [12, 13].

  IV. Thermal expansion coefficient (α) has been correlated with compressibility and both have been discussed structurally.

1.3   Optical Properties

The optical properties of tellurite glasses task have been focused into three subdirections [5, 12, 13]:

  I. Linear and nonlinear refractive indices: in early 1990s and after solving the theoretical difficulties with optical properties of amorphous solids by calculating number of ions/unit volume and polarizability of ions [5]. Analysis of the refractive indices of these semiconducting amorphous solids [5, 51]. The relationship between refractive indices and number of ion/unit volume (N/V) along with the values of polarizability is explored. These two factors were primarily responsible for reductions in both the dielectric constant and refractive index, although electronic polarizations also affect optical properties, coordination number (CN), and field intensity of the modifier which were all gathered and discussed. Based on that and started from 1992 [5], the nonlinear optical properties of these glasses have measured, analyzed, and attracted the researchers worldwide. Also, the halogen effect on the optical properties has been archived along with thermal luminescence and spectroscopic properties like luminescence properties of $Er3^+$ ions in tellurite glasses. Also, erbium nano particles for photonic applications and determination of nano-size particles was performed by electron energy scattering, EDX have been achieved [7].

II. Ultraviolet absorption (UV): experimental procedures to measure the UV of bulk and thin film, oxide, or halide along with theoretical concepts to find the energy gap and band tail width. Direct and indirect transition of the electrons has been judged. Also, derivative absorption spectrum fitting (DASF) method and absorption spectra fitting (ASF) have been used to calculate energy gap, $(E_{DASF}^{opt.})$, $(E_{ASF}^{opt.})$ [52, 53].

III. Infrared absorption (IR) and Raman spectroscopy: the two complementary nondestructive characterization techniques; both provide extensive information about the structure and vibrational properties of these new noncrystalline solids in pure and modified forms either by oxide or oxyhalide. The basis for quantitative interpretation of the absorption bands in the IR spectra has been provided by using values of stretching force and constant and reduced mass of the vibrating cation-anion. Such interpretations showed that the CN determines the primary forms of these spectra. The deconvolution of the FTIR data was made by origin 8 using Gaussian-type function to help us to identify all bands appear in the spectra. The structural peculiarities and Raman spectra of tellurite based glasses have been achieved [31]. A fresh look at the problem has been achieved. Also, the TeO_2–WO_3 glass system as simulated by ab initio calculations has been achieved [31].

1.4 Electrical Properties

Electrical properties (conduction and insulation) of tellurite glasses have been collected. The AC, DC, and dielectric constant (ε) of the present amorphous solids have been measured by using different frequencies and wide temperature ranges [12, 13, 24, 27–29, 54]:

I. ITheoretical analysis of the electrical conduction based on hopping mechanism at high, room, and low temperatures by calculating hopping activation energy, electron-phonon coupling. Also, the candidate explores the dependence of the semiconducting behavior of tellurite glasses on the ratio of low to high valence states in the modifiers by using ESR at low temperature.

II. Dielectric constant and loss factor which have been found to vary inversely with frequency and directly with temperature, the rates of change ($\partial \varepsilon / \partial f$) and ($\partial \varepsilon / \partial T$) have been measured. The candidate has analyzed the experimental data according to the value of the polarizability which depend on the types and percentage of the modifier that present in these semiconductors. Data on the electrical modulus and relaxation behavior of semiconducting tellurite glasses were discussed in terms of the polarizable atoms per unit volume.

1.5 Mechanical Properties

Fourth task was the mechanical properties of tellurite glasses. By using ultrasonic equipment, hydrostatic and uniaxial pressure, and elastic moduli (SOEC), their derivatives (TOEC) have been measured. Elastic properties provide considerable information about the structure and interatomic potential of solids. Gruneisen parameter which expresses the volume dependence of the normal mode frequency has been calculated. The role of the modifier in the elastic properties of tellurite glasses has been measured and theoretically analyzed [4, 30]. Also, application of ultrasonic has been used for:

I. Detection of micro phase separation in amorphous solids has been achieved.
II. Correlations between the acoustic Debye temperature and the theoretically calculated optical Debye temperature from IR absorption has been achieved, while the acoustic Debye temperature has been discussed in details according to the number of vibrating atoms per unit volume (N/V) when modified with REO or TMO. Also, internal frication (Q^1) at low temperature has been measured at low temperature by using ultrasonic equipments in order to find the relaxation processes in this amorphous solid material. Radiation effect on (Q^1) has been measured. Simulation of acoustic properties of some tellurite glasses and prediction of ultrasonic parameters at low temperatures for tellurite glasses using artificial neural network (ANN) technique have been achieved.

1.6 Applications

Applications of tellurite glasses have been increased every day. The effects of Yb^{3+}-Er^{3+} co-doped TeO_2-PbF_2 oxyfluoride tellurite glasses with different Yb^{3+} concentrations and characterized their upconversion properties. Intense emission bands at 527, 544, and 657 nm corresponded to the Er^{3+} transitions, and the maximum was obtained at an Yb^{3+}-to-Er^{3+} molar ratio of 3. When this glass was applied at the back of amorphous silicon solar cells in combination with a rear reflector, a 0.45% improvement in efficiency was obtained under co-excitation of AM1.5 and 400 mW 980 nm laser radiation. Maximum external quantum efficiency and luminescence quantum efficiency of 0.27% and 1.35%, respectively, were achieved at 300 mW excitation [11].

Based on the increase in dc electrical conductivity of tellurite glasses with increasing γ-irradiation doses in the temperature range of 300–573 k [54], the observed increase in conductivity is attributed to the relative increase in the charge carriers that produced by γ-rays. In 2013, effect of pre-readout annealing treatments on thermoluminescence (TL) mechanism in tellurite glasses at therapeutic radiation doses level has been established [9]. The higher sensitization for thermal annealing on thermoluminescence, TL mechanism in the region 550–600 °C for $80(TeO_2)$–5 (TiO_2)–$(15-x)$ (WO_3)–$(x)A_nO_m$ where $A_nO_m = Nb_2O_5$, Nd_2O_3, Er_2O_3, and $x = 5$ mol

% has been measured. The behavior of trap centers and luminescence centers has been investigated for tellurite glasses doped with rare-earth oxides irradiated at 0.5 up to 2 Gy and annealed at different temperatures in the range 350–700 °C. The behavior of the three types of tellurite glasses is analyzed regarding to their kinetic parameters and luminescence emission which enhance the claim of tellurite glasses for use as thermoluminescence detector and TLD material at therapeutic radiation doses. The higher sensitization for thermal annealing on TL mechanism in the region 550–600 °C for $80(TeO_2)–5(TiO_2)–(15 - x) (WO_3)–(x) A_nO_m$ where $A_nO_m = Nb_2O_5, Nd_2O_3, Er_2O_3$, and $x = 5$ mol% has been measured.

Moreover, in 2017, the photonic applications of tellurite glasses, lasers utilizing tellurite glass-based gain media, tellurite glasses for optical amplifiers, broadband emission in tellurite glasses, tellurite glass fibers for mid-infrared nonlinear applications, tellurite thin films produced by RF sputtering for optical waveguides and memory device applications, laser writing in tellurite glasses and supercontinuum generation in tellurite optical fibers, and tellurite glasses for plasmonics have been published [14]. Beside the above the present book will summarize the some physical properties of tellurite glasses and then cover radiation shielding by simulation, tellurite glass materials for energy conversion technology and laser devices, structural and luminescence properties for laser applications, and optothermal, optical properties in the presence of gold nanoparticles and lanthanide-doped zinc oxyfluorotellurite glasses as a new smart material. Additional chapters address the properties and uses of tellurite glasses in optical sensing, significance of near-infrared (NIR) emissions, solar cell, laser, luminescent display applications, solar energy harvesting, and development of bioactive-based tellurite-lanthanide (Te-Ln)-doped hydroxyapatite composites for biomedical applications. Figure 1.3 represented the sketch diagram of the present tellurite glass smart materials book.

Chapter 2 summarizes the radiation shielding properties for tellurite glasses to show their superior properties for potential applications. The shielding properties of the present glasses cover such properties as density and shielding parameters: mass attenuation coefficient, μ/ρ; line attenuation coefficient, μ; effective atomic numbers, Z_{eff}; half value layers, HVL; mean free path, MFP; and exposure buildup factors, EBF. Values of mass attenuation coefficient have been computed using WinXCOM program and shielding rate of each glass shielding material tested by ^{60}Co, ^{137}Cs, and ^{125}I gamma ray sources. Chapter 3 presented tellurite glass materials for potential applications for solar energy technology and laser devices. It is because these materials present very efficient optical and physical properties. Tellurite glasses doped with PbTe, CdTe, and rare-earth materials are considerable, due to its practical importance in technological applications such as integrated optics, optoelectronics, lasers, broadband optical amplifiers, and solar energy conversion. The challenges in the solar energy technology research are to increase the conversion efficiency compared with that of silicon solar cells and consequently to make them more cost effective for commercial applications. Tellurite glasses doped with rare-earths were demonstrated as very broadband optical amplifiers and laser devices.

In addition, Chap. 4 discussed in details both structural and luminescence properties of heavy metal oxide-based TeO_2 glasses incorporated by some rare-earth

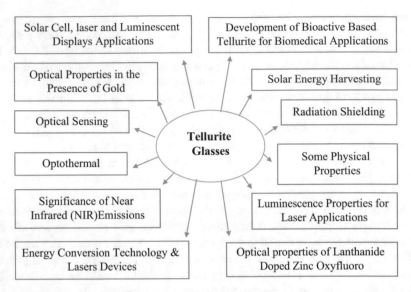

Fig. 1.3 Sketch of the physics of new noncrystalline solid tellurite glass smart materials and their latest applications

ions. The glasses were developed by conventional melt quenching method, and the structural analysis was done by XRD, FTIR, and Raman. The XRD patterns confirm the amorphous nature of the samples, and the FTIR characterization showed the formation of more non-bridging oxygen atoms in the glass network with the inclusion of rare-earth ions. Spectroscopic characterizations such as optical absorption, photoluminescence, and decay profile measurements were performed on the glasses. The Judd-Ofelt theory has been employed on optical absorption spectra to evaluate the Judd-Ofelt (J-O) intensity parameters Ω_λ ($\lambda = 2, 4, 6$). The measured J-O intensity parameters were used to determine the emission transition probability (A_R), stimulated emission cross section ($\sigma(\lambda_p)$), branching ratios (β_R), and radiative lifetimes (τ_R) for various emission transitions from the excited levels of rare-earth ions in the host glass network. The obtained results showed the use of the glasses for potential applications in the field of laser technology. Moreover, Chap. 5 showed that transition metal oxide containing glasses (TMOGs) have special and unique optothermal properties of some vanadate-tellurate oxide glasses. One can verify how the optical, thermal, and thermoelectric properties vary with composition. In other words, optical and thermal characterization of such glassy systems can be investigated versus the composition with the aim of finding the more potential candidates in optical applications. Moreover, beside thermal stability, different parameters such as elastic moduli, optical bandgap, molar volume (V_m), oxygen molar volume (V_O^*), oxygen packing density (OPD), molar refraction (R_m), metallization criterion (M), and the concentration of non-bridging oxygen ions (NBOs) can be evaluated and discussed as the most important factors on the properties and applications of a material. In brief, optical applications (such as active material in optical fibers) of oxide glasses need the high thermal stable glasses with narrower bandgap. It should be noted that any suggestion for optical applications needs precise determination of optical properties such as energy bandgap, which effect on the evaluation of other related optical parameters and so in optical device manufacturing and applications; in the case of bandgap determination, derivative absorption spectrum fitting method (abbreviated as DASF) has been recently proposed; this method is briefly introduced in this work. Also, such thermal stable glasses with good thermoelectric properties are promising materials which can be used in solar cells and photovoltaic (PV) panels as heat pumps to elevate the PV efficiency. An attempt has been made to discuss these subjects, giving more light in the field of optothermal aspects.

Chapter 6 presented the effect of gold nanoparticles (Au NPs) in optical properties in the presence of gold nanoparticles. The homogeneous distribution and growth of spherical and non-spherical Au NPs (average size ~3.36 ± 0.076 nm to ~10.62 ± 0.303 nm) in the glassy matrix were evidenced from the transmission electron microscopy (TEM) analysis. The UV-Vis-NIR absorption spectra showed nine bands corresponding to transition bands from ground state $^6H_{5/2}$ to excited states $^6P_{3/2}$, $^4I_{11/2}$, $^6F_{11/2}$, $^6F_{9/2}$, $^6F_{7/2}$, $^6F_{5/2}$, $^6F_{3/2}$, $^6H_{15/2}$, and $^6F_{1/2}$ in which the most intense bands were $^6F_{9/2}$, $^6F_{7/2}$, $^6F_{5/2}$, and $^6F_{3/2}$. Authors assert that these tellurite glass nanocomposites can be used for developments of the solid-state lasers and nanophotonics applications. Also, Chap. 7 discussed the optical properties of zinc oxyfluorotellurite glasses doped with rare-earth ions and current challenges faced in this field. Zinc-tellurite glasses are among the most important heavy metal glass compositions with a wide range of excellent structural, thermal, chemical, and optical properties. When doped with rare-earth ions, zinc-tellurite glasses show superior properties than those of other glass compositions, such as wide broadband luminescence and efficient upconversion emissions of Er^{3+} ions, as well as high rare-earth solubility, which facilitate the incorporation of sensitizers such as Ce^{3+} and Yb^{3+} ions. When modified with some fluoride components, the optical and thermal stability of rare-earth-doped zinc-tellurite glasses do not change drastically, while the average phonon energy stays in a low-range energy, and the excited state lifetime of the rare-earth ions increases due to the different site symmetries provided by F^{-1} ions, rather than O^{-2} ions. The recent developments in the oxyfluorotellurite glasses doped with different lanthanides are given in this chapter, which are compared to those achieved with the zinc-tellurite oxide glasses. Moreover, Chap. 8 is a review of the fundamentals of thermometry by luminescence spectroscopy, the theoretical background to calculate the thermal sensibility of rare-earth ion-doped materials, and some examples of the thermometry done on tellurite glasses as well as some other host matrices. For example, Er^{3+}-doped tellurite glasses are given as thermal sensors applied in the visible spectral region, where the fluorescent intensity ratio of two green emission bands plays the role to optically determine the temperature. On the other hand, Nd^{3+} ion-doped glasses could be used to measure the temperature in the near-infrared region, using the intensity ratio variations of principal emissions in the 800–1400 nm spectral range. Thermal sensibility of each case is compared to various glass host compositions.

Chapter 9 reviewed the NIR emission properties of RE^{3+} ions in multicomponent tellurite glasses. The chapter includes some important tellurium-based glasses as potential host materials for RE^{3+} ions having near-infrared (NIR) emissions. The influences of the composition on the spectral as well as laser parameters of certain rare-earth transitions investigated by several researchers are also detailed so as to apply them in a wide variety of practical applications. It also covers some basic theories necessary to explain the spectroscopic features of interest, the required experimental evidences, and the representative data related to the topic from the previous reports. The recent developments in the intensification in NIR lumines-cence of lanthanide-embedded tellurite-based hosts due to the co-doping of the metal nanoparticles are also addressed.

Chapter 10 presented tellurite glasses, solar cell, and laser and luminescent display applications to meet the increasing energy demand, respectively. This chapter reviewed recent results of the management of the solar spectrum on a solar cell using rare-earth ion-doped TeO_2-ZnO glasses, with and without metallic nanoparticles, as a cover slip. Transparent rare-earth-doped materials as glasses can absorb light at shorter wavelength and emit light at longer wavelengths, the well-known downconversion process; besides they have the advantage of easy preparation and high doping concentration of rare-earth ions. In this context tellurite glasses appear as potential candidates because of their wide transmission window (400–5000 nm), low phonon energy (800 cm^{-1}) when compared to silicate, and thermal and chemical stability. Few vitreous hosts have been investigated to be used as cover slip to enhance the performance of conventional solar cell; so the lack of studies using rare-earth-doped glasses on the top of standard solar cells has moti-vated the recent reports that are reviewed in this chapter. The role of the downconversion process to increase the solar cell efficiency has been achieved. It is shown that the management of Tb^{3+} and Yb^{3+} ion concentration can be optimized to modify the solar spectrum and consequently increase the solar cell efficiency. It was demonstrated that plasmon-assisted efficiency enhancement could be obtained for commercial Si and GaP solar cells, respectively, covered with Eu^{3+}-doped TeO_2-ZnO glasses with silver nanoparticles. Tellurite glasses have also proven to be adequate hosts for rare-earth ions and for the nucleation of metallic nanoparticles (NPs). Also this chapter reviewed results of the modification introduced by different Nd_2O_3 concentrations on the laser operation of TeO_2-ZnO glasses. The control and improvement of the photoluminescence efficiency due to the nucleation of gold NPs in Yb^{3+}/Er^{3+}-doped TeO_2-PbO-GeO_2 glasses are also reviewed. It is shown that the nucleation of silver NPs in Tb^{3+}-doped TeO_2-ZnO-Na_2O-PbO glass contributes for the large enhancement in the blue-red spectrum.

Moreover, Chap. 11 presented lanthanide-doped tellurite glasses for solar energy harvesting. Lanthanide-doped materials exhibit the high photoluminescence quan-tum yield (PLQY) in the visible and near-infrared regions. Recently, these materials can be integrated with Si solar cells to create an additional electron hole pairs through optical conversion (upconversion and downconversion) processes. How-ever, it is crucial to identify the suitable materials which convert light energy into electrical energy. Lanthanide-doped tellurite glasses have the advantages over other low phonon energy materials (i.e., fluoride glasses) that exhibit the properties including low phonon energy, wide transmission (ranging from visible to infrared region), and high refractive index. Low phonon energy of tellurite glasses favors in the enhancement of photoluminescence quantum yield for optical conversion. Moreover, high PLQY glasses could be employed on the top and bottom of PV cells to improve the photocurrent further. In addition, TiO_2-modified lanthanide tellurite glasses may also enrich the photocatalytic activity in the visible region of electromagnetic spectrum.

Finally, Chap. 12 showed the development of bioactive-based tellurite-lanthanide (Te-Ln)-doped hydroxyapatite composites for biomedical applications. The research on lanthanides materials for medical applications is promising. Consequently, investigations into the role of Te-Ln-HA host scaffold materials for bone repair is a relatively new approach that deserves a special attention. Human skeletal bone loss is a major health concern in the twenty-first century, with massive socioeconomic implications. The objective is to develop bioactive-based tellurite glasses for biomedical applications. In this study, tellurium oxide (TeO_2) and lanthanide (Ln^{3+})-doped borate host systems have been developed and incorporated in the hydroxyapatite (HA) matrix in vitro biological studies. In the proposed work, the following scientific questions will be addressed:

- Whether the tellurite-lanthanides (Te-Ln) host glasses reinforced hydroxyapatite (HA) ceramic materials can influence the cell behavior, such as proliferation, differentiation, and apoptosis?
- Does the proposed system show any toxicity on cells?
- Impact of Te-Ln on cell/tissue surface architectural integrity and how the materials will react with bacterial adhesion property and monitoring the growth of the major bacteria (e.g., *E. coli*, *Staphylococcus aureus*, *Staphylococcus epidermidis*, and *Pseudomonas aeruginosa*) that are known to be involved in orthopedic implant-associated infections.
- Does this Te-Ln material will show any luminescence response?

References

1. J.E. Stanworth, Tellurite glasses. Nature **169**, 581–582 (1952)
2. A. Yakhkind, Tellurite Glasses. J. Am. Ceramic Soc. **49**, 670–675 (1966). https://doi.org/10.1111/j.1151-2916.1966.tb13197.x
3. B. Flynn, A. Owen and J. Robertson, Proc. of the "7-th inter. conf. on amorphous and liquid semi", UK, 678 (1977)
4. E.F. Lambson, G.A. Saunders, B. Bridge, R.A. El-Mallawany, The elastic behaviour of TeO_2 glass under uniaxial and hydrostatic pressure. J. Non-Cryst. Solids **69**(1), 117–133 (1984)
5. R.El-Mallawany, The optical properties of tellurite glasses. J. Appl. Phys. **72**(5), 1774–1777 (1992)
6. G. Ghosh, Sellmeier coefficients and chromatic dispersions for some Tellurite glasses. **78**, 2828–2830 (1995). https://doi.org/10.1111/j.1151-2916.1995.tb08060.x

7. M.N. Azlan, M.K. Halimah, R. El-Mallawany, M.F. Faznny, C. Eevon, Optical properties of zinc borotellurite glass system doped with erbium and erbium nanoparticles for photonic applications. J. Mater. Sci. Mater. Electron. **28**, 4318–4327 (2017). https://doi.org/10.1007/s10854-016-6056-2

8. R. El-Mallawany, M.S. Gaafar, Y.A. Azzam, Prediction of ultrasonic parameters at low temperatures for tellurite glasses using ANN. Chalcogenide Lett. **11**(5), 227–232 (2014)

9. R. El-Mallawany, H.M. Diab, Effect of pre-readout annealing treatments on TL mechanism in tellurite glasses at therapeutic radiation doses level. Measurement **46**, 1722–1725 (2013)

10. R. El-Mallawany, M.I. Sayyed, M.G. Dong, Comparative shielding properties of some Tellurite glasses: Part 2. J. Non-Crystalline Solids **474**, 16–23 (2017)

11. F. Yang, C. Liu, D. Wei, Y. Chen, L. Jingxiao, S.-e. Yang, Er^{3+} –Yb^{3+} co-doped TeO_2–PbF_2 oxyhalide tellurite glasses for amorphous silicon solar cells. Opt. Mater. **36**, 1040–1043 (2014)

12. R. El-Mallawany, Tellurite Glasses Handbook: Physical Properties and Data, CRC Press, USA, 1st (2002), ISBN: 0849303680 9780849303685, 540 Page, (www.crcpress.com, engineering / Chemical,T)

13. R. El-Mallawany, Tellurite Glasses Handbook: Physical Properties and Data (CRC Press, USA) 2nd ed, (2011), ISBN: 9781439849835 1439849838 9781439849842 1439849846, 532 Pages, https://www.crcpress.com/Tellurite-Glasses-Handbook.../978143984

14. R. El-Mallawany, Introduction to Tellurite glasses, springer series in materials science, "Technological Advances in Tellurite Glasses: Properties, Processing and Applications", Edit by V.A.G. Revera and D. Manzani, 254. (2017) 1–13., http://www.springer.com/gp/book/9783319530369

15. R.El-Mallawany, An introduction to Telluirte Glasses, (2005), 5 Video taped lectures provided as an educational resource for the entire international glass community available in video streaming format on IMI website; www.lehigh.edu/imi/resources.htm), http://www.lehigh.edu/~inimif/test_htm/

16. M.M. El-Zaidia, A.A. Ammar, R.A. El-Mallwany, Infra-red spectra, electron spin resonance spectra, and density of (TeO_2) 100− x–(WO_3) x and (TeO_2) 100− x–$(ZnCl_2)$ x glasses. Phys. Status Solidi A **91**(2), 637–642 (1985)

17. A. Abdel-Kader, R. El-Mallawany, M.M. Elkholy, Network structure of tellurite phosphate glasses: Optical absorption and infrared spectra. J. Appl. Phys. **73**(1), 71–74 (1993)

18. I.Z. Hager, R. El-Mallawany, A. Bulou, Luminescence spectra and optical properties of TeO_2–WO_3–Li_2O glasses doped with Nd, Sm and Er rare earth ions. Physica B: Condensed Matter 406 (4), 972–980 (2011) **406**(4), 1844 (2011)

19. I.Z. Hager, R. El-Mallawany, Preparation and structural studies in the (70− x) TeO_2–20WO_3–10Li_2O–xLn_2O_3 glasses. J. Mater. Sci. **45**(4), 897 (2010)

20. N.S. Hussain, G. Hungerford, R. El-Mallawany, M.J.M. Gomes, M.A. Lopes, N. Ali, J.D. Santos, S. Buddhudu, Absorption and emission analysis of RE^{3+} (Sm^{3+} and Dy^{3+}): Lithium Boro Tellurite glasses. J. Nanosci. Nanotechnol. **9**(6), 3672–3677 (2009)

21. R. El-Mallawany, M. Sidkey, A. Khafagy, H. Afifi, Ultrasonic attenuation of tellurite glasses. Mater. Chem. Phys. **37**(2), 197–200 (1994)

22. M.M. Elkholy, R.A. El-Mallawany, Ac conductivity of tellurite glasses. Mater. Chem. Phys. **40**(3), 163–167 (1995)

23. A. El-Adawy, R. El-Mallawany, Elastic modulus of tellurite glasses. J. Mater. Sci. Lett. **15**(23), 2065–2067 (1996)

24. H.M.M. Moawad, H. Jain, R. El-Mallawany, T. Ramadan, M. El-Sharbiny, Electrical conductivity of silver vanadium tellurite glasses. J. Am. Ceram. Soc. **85**(11), 2655–2659 (2002)

25. R. El-Mallawany, Specific heat capacity of semiconducting glasses: Binary vanadium tellurite. Phys. Status Solidi A **177**(2), 439–444 (2000)

26. R. El-Mallawany, A. Abd El-Moneim, Comparison between the elastic moduli of tellurite and phosphate glasses. Phys. Status Solidi A **166**(2), 829–834 (1998)

27. R. El-Mallawany, A.H. El-Sayed, M.M. HA El-Gawad, ESR and electrical conductivity studies of (TeO_2) 0.95 (CeO_2) 0.05 semiconducting glasses. Mater. Chem. Phys. **41**(2), 87–91 (1996)

28. H.M.M. Moawad, H. Jain, R. El-Mallawany, DC conductivity of silver vanadium tellurite glasses. J. Phys. Chem. Solids **70**(1), 224–233 (2009)
29. R. El-Mallawany, Theoretical analysis of the electrical properties of tellurite glasses. Mater. Chem. Phys. **37**(4), 376–381 (1994)
30. R. El-Mallawany, Phase separation ultrasonic detection of microphase separation in Tellurite glasses. Phys. Stat. Soli. (a) **133**, 245 (1992)
31. A. Mirgorodsky, M. Colas, M. Smirnov, T. Merle-Méjean, R. El-Mallawany, Philippe Thomas, structural peculiarities and Raman spectra of TeO_2/WO_3-based glasses: A fresh look at the problem. J. Solid State Chem. **190**, 45–51 (2012)
32. R. El-Mallawany, M. Sidkey, A. Khafagy, H. Afifi, Elastic constants of semiconducting tellurite glasses. Mater. Chem. Phys. **37**(3), 295–298 (1994)
33. M. Sidky, R. El-Mallawany, R. Nakhala, A. El-Moneim, Ultrasonic attenuation at low temperature of TeO_2-V_2O_5 glasses. Phys. Stat. Soli. (a) **159**, 397 (1997)
34. R. El-Mallawany, Theoretical analysis of ultrasonic wave attenuation and elastic moduli of Tellurite glasses. Mater. Chem. Phys. **39**, 161 (1994)
35. R. El-Mallawany, H.M. Diab, Improving dosimetric properties of tellurite glasses. Phys. B Condens. Matter **407**(17), 3580–3585 (2012)
36. M.A. Merzliakov, V.V. Kouhar, G.E. Malashkevich, E.V. Pestryakov, Spectroscopy of Yb-doped tungsten-tellurite glass and assessment of its lasing properties. Opt. Mater. **75**, 142–149 (2018)
37. R. El-Mallawany, M.S. Gaafar, M.A.M. Abdeen, S.Y. Marzouk, Simulation of acoustic properties of some tellurite glasses. Ceram. Int. **40**, 7389–7394 (2014)
38. M.A. Sidky, R.A. El-Mallawany, A.A. Aboushly, Y.B. Saddeek, Relaxation of longitudinal ultrasonic waves in some tellurite glasses. Mater. Chem. Phys. **74**(2), 222–229 (2002)
39. R. El-Mallawany, Devitrification and Vitrification of Tellurite glasses. J. Mater. Sci. Elect. **6** (1) (1995)
40. R.A. El-Mallawany, L.M. Sharaf El-Deen, M.M. Elkholy, Dielectric properties and polarizability of molybdenum tellurite glasses. J. Mater. Sci. **31**(23), 6339–6343 (1996)
41. R. El Allawany, Debye temperature of ternary tellurite glasses at room temperature. Phys. Status Solidi A **130**(1), 103–108 (1992)
42. V. Uma, M. Vijayakumar, K. Marimuthu, G. Muralidharan, Luminescence and energy transfer studies on Sm^{3+}/Tb^{3+} codoped telluroborate glasses for WLED applications. J. Mol. Struct. **1151**, 266–276 (2018)
43. D.S. da Silva, N.U. Wetter, W. de Rossi, L.R.P. Kassab, R.E. Samad, Production and characterization of femtosecond laser-written double line waveguides in heavy metal oxide glasses. Opt. Mater. **75**, 267–273 (2018)
44. Y. Sun, Q. Yang, H. Wang, Y. Shao, Sensitization of Ho^{3+} on the 2.7 μm emission of Er^{3+} in $(Y0.9La0.1)_2O_3$ transparent ceramics. J. Lumin. **194**, 50–55 (2018)
45. R. El-Mallawany, H.A. Afifi, M. El-Gazery, A.A. Ali, Effect of Bi_2O_3 addition on the ultrasonic properties of pentaternary borate glasses. Measurement **116**, 314–317 (2018)
46. V. Himamaheswara Rao, P. Syam Prasad, M. Mohan Babu, P. Venkateswara Rao, T. Satyanarayana, L.F. Santos, N. Veeraiah, Spectroscopic studies of Dy^{3+} ion doped tellurite glasses for solid state lasers and white LEDs. Spectrochim. Acta A Mol. Biomol. Spectrosc. **188**, 516–524 (2018)
47. M.E. Alvarez-Ramos, J. Alvarado-Rivera, M.E. Zayas, U. Caldino, J. Hern_andez-Paredes, Yellow to orange-reddish glass phosphors: Sm^{3+}, Tb^{3+} and Sm^{3+}/Tb^{3+} in zinc tellurite-germanate glasses. Opt. Mater. **75**, 88–93 (2018)
48. S.H. Alazoumia, S.A. Aziza, R. El-Mallawany, U. Sa'ad Aliyud, H.M. Kamari, M.H.M.M. Zaida, K.A. Matoria, A. Ushah, Optical properties of zinc tellurite glasses. Resul. Phys. **9**, 1371–1376 (2018)
49. A. Tubtimtae, S. Phadungdhitidhada, D. Wongratanaphisan, A. Gardchareon, S. Choopun, Tailoring Cu2_xTe quantum-dot-decorated ZnO nanoparticles for potential solar cell applications. Curr. Appl. Phys. **14**, 772–777 (2014)

50. O.A. Zamyatin, M.F. Churbanov, J.A. Medvedeva, S.A. Gavrin, E.V. Zamyatina, A.D. Plekhovich, Glass-forming region and optical properties of the TeO2 – ZnO – NiO system. J. Non-Cryst. Solids **479**, 29–41 (2018)
51. R.El-Mallawany, Evaluation of optical parameters of some tellurite glasses. Optik-Int. J. Light Electron Opt. **125**(20), 6344–6346 (2014)
52. S.H. Elazoumi, H.A.A. Sidek, Y.S. Rammah, R. El-Mallawany, M.K. Halimah, K.A. Matori, M.H.M. Zaid, Effect of PbO on optical properties of tellurite glass. Res. Phys. **8**, 16–25 (2018)
53. R. El-Mallawany, Y.S. Rammah, A. El Adawy, Z. Wassel, Optical and thermal properties of some Tellurite glasses. Am. J. Opt. Photon. **5**(2), 11–18 (2017.) http://www.sciencepublishinggroup.com/j/ajop, doi:10.11648/j.ajop.20170502.11, ISSN: 2330-8486 (Print); ISSN: 2330-8494 (Online)
54. A. Abdel Kader, A. Higazy, R. EL-Mallawany, M. ElKholy, The effect of gamma irradiation on the electrical conductivity of TeO_2 -P_2O_5 and Bi_2O_2 -TeO_2 -P_2O_5 glasses. Radiat. Eff. Defects Solids **124**, 401 (1992)

Chapter 2
Radiation Shielding Properties of Tellurite Glasses

Raouf El-Mallawany

Abstract This chapter summarizes the shielding properties by simulation for tellurite glasses to show their superior properties. The radiation shielding properties of the present glasses such properties as density, shielding parameters: mass attenuation coefficient, μ/ρ, line attenuation coefficient, μ, effective atomic numbers, Z_{eff}, half value layers, HVL, mean free path, MFP and Exposure buildup factors, EBF have been collected. Values of the mass attenuation coefficient have been computed using WinXCOM program.

2.1 Introduction

Glass-forming region, optical spectroscopy, thermal, electrical, mechanical properties, investigation and evaluation of photon attenuation coefficients and evaluation radiation shielding properties of tellurite glasses have reported [1–20]. Now, neutrons, gamma-rays and X-rays are used widely in the world in many applications, for example, food irradiation, environmental protection, manufacture, elemental analysis, medical therapy, etc. However, the exposure for long times to the high penetrating radiation such as gamma rays may cause genetic mutations, cancer, and death. So it is necessary to choose suitable shielding materials to protect people from these harmful rays [21–26]. Concrete shields are used widely for shielding harmful rays [27–30]. However, during the period of using the concrete, the water contained in concrete would be lost. This would be harmful for the structure of concrete. Also, the use of concrete is limited [31]. Glasses would be suitable to make up for the defect of concrete. Besides, glasses are easy to make transparent, have a wide range of composition, and easy to shape. What's more, some good elements for neutron shielding or gamma ray/X-ray shielding could be added into glass materials. Boron element is easy to add into glass and is good for neutron shielding [32].

R. El-Mallawany (✉)
Physics Department, Faculty of Science, Menoufia University, Shebin ElKomm, Egypt

© Springer International Publishing AG, part of Springer Nature 2018
R. El-Mallawany (ed.), *Tellurite Glass Smart Materials*,
https://doi.org/10.1007/978-3-319-76568-6_2

17

Radiation interacts in an absorbing medium to deposit energy which is defined as the absorbed dose, and, when weighted according to the damaging effect of the radiation type, the term effective dose equivalent. A related term is radiation exposure, which applies to air only and is a measure of the amount of ionization produced by X-rays and gamma radiation in air. Radiation dose can be calculated if three things are known [33]:

(a) The mass of the medium being irradiated
(b) The number of "radiations" per unit area (the flux) that impinge on the mass
(c) The amount or rate of energy deposition in the mass specified

For particles, the mass in which the energy is dissipated is just the depth of penetration; for photons, it is necessary to use a unit depth (e.g., 1 cm) due to the probabilistic pattern of interactions represented by the mass energy absorption coefficient (μ_{en}/ρ) (cm^2/g), where μ_{en} is the absorption coefficient (cm^{-1}) and ρ is the density (g/cm^3) of the shielding material. All emitted radiations must be considered in this process, and adjustments should be made for any attenuating medium between the source and the point of interest. Various materials, placed between a source and a receptor, can affect the amount of radiation transmitted from the source to the receptor. Such effects are due to attenuation and absorption of the emitted radiation in the source itself, in material used for encapsulation of the source, or in a shielding barrier. Regardless of how it occurs, shielding is an important aspect of radiation protection since it can be a form of radiation control; therefore, the features of shields and their design, use, and effectiveness warrant specific consideration.

It is worth noting that even though various aspects are straightforward, radiation shielding is a very complex discipline for many radiation sources and the many geometric configurations in which they may occur. Therefore, this presentation focuses on straightforward configurations of point sources, line sources, and area and volume sources which fortunately can be used conservatively to address most of the situations encountered in practical radiation protection. Many shielding problems can be treated in terms of one of these configurations, and in general the exposure will be slightly overestimated such that shield designs would be conservative [33]. For any shielding material, the knowledge of physical parameters such as mass attenuation coefficient, μ/ρ; linear attenuation coefficient, μ; effective atomic numbers, Z_{eff}; half value layers, HVL; mean free path, MFP; and exposure buildup factors, EBF, is essential for understanding the radiation shielding properties.

2.2 Mass Attenuation Coefficient, μ/ρ

Interactions of photon during its passage through a matter will experience attenuation, and its intensity reduces exponentially according to the well-known Beer-Lambert law as given by the next equation:

$$I = I_o \exp[-\mu x] \qquad (2.1)$$

where I and I_o represent the transmitted and incident radiation intensity, respectively, μ represents the linear attenuation coefficient (cm^{-1}), and x is sample mass thickness. The mass attenuation coefficient (μ/ρ) can be used to measure the probability for interactions that happens between the incident photon and matter per unit mass per unit area. The (μ/ρ) values of the present glass systems can be evaluated according to the next formula [34]:

$$\mu/\rho = \sum_i w_i(\mu/\rho)_i \qquad (2.2)$$

w_i, represents the fractional weight of the ith constituent in the glass system and ρ is the density of the shielding medium, and $(\mu/\rho)_i$ of the glass systems were evaluated using WinXCOM program [35]. This software is able to give partial cross sections, total cross sections, and attenuation coefficients for different interaction processes, such as incoherent and coherent scattering, photoelectric absorption, and pair production. By substituting the chemical composition or weight fraction of mixture, the mass attenuation coefficient of the selected glass samples will be generated in the energy range of 1 keV–100 GeV.

Values of mass attenuation coefficient have been reported using WinXCOM program within the energy range 0.01MeV–20 MeV for selected tellurite glass samples in the form of:$(100-x)TeO_2-xMoO_3$, where $x = 20, 30, 45, 50$ mol% [20], $TeO_2-A_nO_m$, $TeO_2-WO_3-B_nO_m$, $TeO_2-WO_3-Er_2O_3-PbO$, where $A_nO_m = La_2O_3$, CeO_2, Sm_2O_3, MnO_2, CoO_3, Nb_2O_5, where $B_nO_m = Er_2O_3$, La_2O_3, Sm_2O_3, CeO_2 [36], $80TeO_2-5TiO_2-(15-x)$ $WO_3-xA_nO_m$, where A_nO_m is $Nb_2O_5 = 0.01$, 5, $Nd_2O_3 = 3$, 5, $Er_2O_3 = 5$ mol% [37], $TeO_2-V_2O_5$, $TeO_2-V_2O_5-TiO_2$, $TeO_2-V_2O_5-CeO_2$, $TeO_2-V_2O_5-ZnO$ [38]. The mass attenuation coefficient (μ/ρ) values of the binary molybdenum tellurite glass as a function of incident photon energy are listed in Table 2.1. The low photon energy and the μ/ρ values are ≈ 1 {1.256–0.037 cm^2 g^{-1} and 1.058 to 0.035 cm^2 g^{-1} for $80TeO_2-20MoO_3$ and $50TeO_2-50MoO_3$}. The (μ/ρ) values are large in the low-energy region ~ 120 cm^2/g for the binary tellurite glasses with 10 mol% La_2O_3, CeO_2, or Sm_2O_3 and in the order of 110 cm^2/g for the other samples. Previously, it has been found that the calculated (μ/ρ) of $80TeO_2-20Bi_2O_3$, $85TeO_2-15Bi_2O_3$ [39], and $80TeO_2-20WO_3$ [40] are higher than $80TeO_2-20V_2O_5$, $80TeO_2-20K_2O$ [41]. It was worth noting that the (μ/ρ) values for photon energy greater than 0.1 MeV. The (μ/ρ) of the selected tellurite glasses is close to $5PbO-45BaO-50B_2O_3$ [42] and lower than $50BaO-50Brosilicate$ [43] and $70PbO-30SiO_2$ [39–44] glasses.

Also, the mass attenuation coefficient μ/ρ of the $80TeO_2-5TiO_2-(15-x)$ $WO_3-xA_nO_m$, where A_nO_m is $Nb_2O_5 = 0.01$, 5; $Nd_2O_3 = 3$, 5; and $Er_2O_3 = 5$ mole % [37], is given in Table 2.1. From Table 2.1, it can be noticed that for the selected glass systems, the mass attenuation coefficient indices increased in the order $5Nb_2O_5 < 0.01Nd_2O_3 < 3Nd_2O_3 < 5Er_2O_3 < 0.01$ Nb_2O_5, and it is clear that by changing the types of rare-earth oxides as well as their concentration, it is possible to change the mass attenuation coefficient values, which indicate the influence of the rare-earth oxides in enhancing the shielding properties of the tellurite glasses.

Table 2.1 Glass composition, density ρ (g/cm^3), and mass attenuation coefficient (μ/ρ) cm^2 g^{-1} in different energies for some tellurite glasses

Glass	Density g cm^{-3}	(μ/ρ) cm^2 g^{-1}	Energy MeV	(μ/ρ) cm^2 g^{-1}	Energy MeV
80TeO$_2$-20MoO$_3$	5.01 [45]	1.256 [20]	0.1	0.037	15
50TeO$_2$-50MoO$_3$	4.6 [45]	1.058 [20]	0.1	0.37	15
90TeO$_2$-10La$_2$O$_3$	5.685 [46]	120 [36]	0.01	0.05	5
90TeO$_2$-10CeO$_2$	5.706 [46]	120 [36]	0.01	0.05	5
90TeO$_2$-10Sm$_2$O$_3$	5.782 [46]	110 [36]	0.01	0.5	5
80TeO$_2$-5TiO$_2$-10WO$_3$-5Nb$_2$O$_5$	5.424 [47]	108.86 [37]	0.01	0.03364	5
80TeO$_2$-5TiO$_2$-10WO$_3$-5Nd$_2$O$_3$	5.750 [47]	115.8 [37]	0.01	0.03385	5
80TeO$_2$-5TiO$_2$-10WO$_3$-5Er$_2$O$_3$	5.896 [47]	119.59 [37]	0.01	0.03392	5
90TeO$_2$-10V$_2$O$_3$	5.213 [48]	114.1 [38]	0.01	0.033	5
50TeO$_2$-50V$_2$O$_3$	4.1 [48]	94.58 [38]	0.01	0.031	5
50TeO$_2$-40V$_2$O$_3$-10TiO$_2$	4.062 [49]	87.79 [38]	0.01	0.029	5

Also, mass attenuation coefficient of TeO$_2$-V$_2$O$_5$ and TeO$_2$-V$_2$O$_5$-TiO$_2$ [38] glasses is given in Table 2.1. The mass attenuation coefficient μ/ρ values are influenced by V$_2$O$_5$ concentration and photon energy, and for all V$_2$O$_5$ concentration, the (μ/ρ) values were decreased exponentially with the increase of energy. Moreover, it is clear that at lower energies, the total interaction of photons with all glasses is high, while at higher energies, it decreases which indicates that the transmission of photon increases through the glass samples. Also, it can be observed that the (μ/ρ) values are decreased with the increase of V$_2$O$_5$ content for all energies. From this figure it is clearly that the (μ/ρ) of all glasses reduce very sharply as the photon energy decreases from 0.01 MeV to 0.02 MeV. The (μ/ρ) of tellurite glass with 10 mol% V$_2$O$_5$ was found to decrease from 114.11 cm^2/g to 17.08 cm^2/g, while in tellurite glass with 50 mol% V$_2$O$_5$, it decreases from 96.01 cm^2/g to 14.55 cm^2/g. Figure 2.1 represented the (μ/ρ) of some selected tellurite glasses in comparison with silicate and borate glasses for 0.01 MeV.

In the low-energy region ($E < 0.01$ MeV), the (μ/ρ) values decrease very sharply due to photoelectric effect. Photoelectric effect predominates at low photon energy as its cross section changes with atomic number ($\sim Z^{4-5}$) and energy ($\sim 1/E^3$). Hence, glasses have highest (μ/ρ) values in this energy region where photoelectric effect dominates and the values of (μ/ρ) decrease as energy increases. This would imply that if there is an increase in the energy of the gamma ray (incident photon energy), there would be a decrease in the attenuation and thus leads to more penetration of the gamma ray in the glass. The variation in (μ/ρ) values with energy can be explained by microscopic photon interactions (photoelectric absorption, Compton scattering, and pair production) as the interaction cross section depends upon atomic number and photon energy. The interaction cross section is directly proportional to (Z^n/E^3) where $n = 4.0$ to 4.5 (Z is atomic number) for photoelectric absorption in low-energy region (region of less than 100 keV); therefore, (μ/ρ) values for the glass samples decreases sharply. In Compton scattering region (photon energy range of

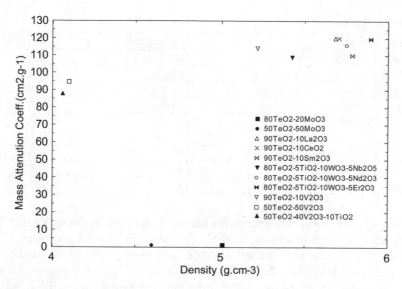

Fig. 2.1 Mass attenuation coefficient (μ/ρ) of some selected tellurite glasses in comparison with silicate and borate glasses for 0.01 MeV

100–800 keV), the interaction cross section is dependent upon (Z/E), whereas cross section is directly proportional to ($Z^2/\ln E$) for pair production (for energy of more than 1 MeV).

2.3   Effective Atomic Numbers, Z_{eff}

Effective atomic number, Z_{eff}, is an appropriate quantity for describing gamma ray interactions and can be defined by relations [50]:

$$Z_{eff} = \frac{\sigma_a}{\sigma_e} \tag{2.3}$$

where σ_a and σ_e are the total atomic cross section and the total electronic cross section, respectively, which can be calculated by using the following relations:

$$\sigma_a = \frac{\mu/\rho}{N_A \sum_i \frac{w_i}{A_i}} \tag{2.4}$$

$$\sigma_e = \frac{1}{N_A} \sum_i \left(\sum_j \frac{f_j A_j}{Z_j} \right) w_i \tag{2.5}$$

Table 2.2 Z_{eff} for some tellurite glasses at 0.1 MeV and 10 MeV

Glass composition mol%	Z_{eff} (0.1 MeV)	Z_{eff} (10 MeV)
$80TeO_2$-$5TiO_2$-$10WO_3$-$5Nb_2O_5$ [36]	48.931	30.360
$80TeO_2$-$5TiO_2$-$10WO_3$-$5Nd_2O_3$ [36]	43.539	33.626
$80TeO_2$-$5TiO_2$-$10WO_3$-$5Er_2O_3$ [36]	39.014	33.263
$80TeO_2$-$20WO_3$ [40]	43.465	33.831
$80TeO_2$-$20V_2O_5$ [38]	53.849	22.127
$80TeO_2$-$20ZnO$ [41]	47.779	28.295
$80TeO_2$-$20MoO_3$ [20]	47.747	29.329
$50TeO_2$-$50MoO_3$ [20]	40.222	28.066
$80TeO_2$-$20V_2O_5$ [38]	38.19	22.12

where A_i, Z_i, and f_i are, respectively, the atomic weight, the atomic number, and the fractional abundance of element i and N_A represents the Avogadro constant.

Recently, Z_{eff} for some tellurite glasses have been published in detail in the energy range 0.1–10 MeV [20, 36–38]. The behavior of Z_{eff} for all glasses is almost identical, and it can be observed that the addition of rare-earth oxides (REO) (Nb_2O_5, Nd_2O_3, and Er_2O_3) leads to the increase of the effective atomic number, which emphasizes our estimation that the addition of REO to the present glasses improves the shielding properties of these glasses (Table 2.2).

The behavior of Z_{eff} has been explained on the basis of dependence of cross section of photoelectric process which varies inversely with the photon energy [38]. Values Z_{eff} reach minimum value in the energy range 1–3 MeV [20, 36–38]. This may be due to dominance of the Compton scattering process, where the interaction cross section is directly proportional to atomic number Z [20, 36–38].

2.4 Half Value Layers, HVL, and Mean Free Path, MEP

Half value layer (HVL) and mean free path (MFP) are convenient parameters to represent gamma ray interactions of a material. The HVL is the thickness of a particular material needed to decrease the intensity of photon to 50% of its initial value, while the MFP is the average distance traveled by a photon in the medium before an interaction occurs [40]. The HVL and MFP of the present glasses were calculated using the next equations:

$$HVL = \frac{0.693}{\mu} \qquad (2.6)$$

$$MFP = \frac{1}{\mu} \qquad (2.7)$$

where μ (cm^{-1}) is the linear attenuation coefficient which is equal to multiplication of mass attenuation coefficient value and density of the glass sample.

Table 2.3 HVL for some tellurite glasses at 0.1 MeV and 10 MeV

Glass composition mol%	HVL (0.1 MeV)	HVL (10 MeV)
80TeO$_2$-5TiO$_2$-10WO$_3$-5Nd$_2$O$_3$ [36]	0.077	3.399
80TeO$_2$-5TiO$_2$-10WO$_3$-5Er$_2$O$_3$ [36]	0.073	3.303
80TeO$_2$-20MoO$_3$ [20]	0.110	4.034
80TeO$_2$-20PbO [27]	0.051	2.942

Both HVL and MFP for tellurite glasses have been published in detail in the energy range 0.1 MeV to 10 MeV [20, 36–38]. It has been shown that HVL are influenced by chemical contents of the glasses and incident photon energy, while at lower photon energy ($E < 0.1$ MeV), the HVL values of the three glass systems are very small. As the incident photon energy increases, the HVL values increase rapidly and reach maximum value at 7 MeV and then start decreasing with further increase in photon energy [20, 36–38]. Some selected values of HVL are tabulated in Table 2.3. The mean free path has opposite behavior for the present noncrystalline solids and tellurite glasses [20, 36–38].

For certain energy, smaller HVL values mean more efficiency of the sample to attenuate the gamma ray photons. Therefore, the lower value of HVL is the better shielding capabilities. From Table 2.3 it can be seen that 80TeO$_2$-5TiO$_2$-10WO$_3$-5Nd$_2$O$_3$, 80TeO$_2$-5TiO$_2$-10WO$_3$-5Er$_2$O$_3$, and 80TeO$_2$- 20PbO [36] glass systems show lower value of HVL which indicated that they provide superior shielding from gamma rays than the other prepared glasses.

2.5   Removal Cross Section for Fast Neutrons (Σ_R)

The removal cross section for fast neutrons Σ_R is the probability of a neutron undergoing certain reaction per unit length of moving through the shielding material and can be given by the expression:

$$\sum_R = \sum_i W_i \left(\sum_R / \rho \right)_i \qquad (2.8)$$

where $\Sigma_{R/\rho}$ (cm^2/g) and W_i represent the mass removal cross section of the ith constituent and the partial density (g/cm^3), respectively. The removal cross sections for fast neutrons (Σ_R) for the selected glass systems can be calculated by using the next equation [51]:

$$\sum_R = W_i \left(\sum_R / \rho \right)_i \qquad (2.9)$$

where W_i is the partial density (g/cm^3) and $\Sigma_{R/\rho}$ (cm^2/g) is the mass removal cross section of the ith constituent obtained from the literature data [52, 53].

Table 2.4 Removal cross section for fast neutrons Σ_R and density for some tellurite glasses

Glass composition mol%	(Σ_R) cm^{-1}	Density
80TeO$_2$-5TiO$_2$-15WO$_3$ [36]	0.09910	
80TeO$_2$-5TiO$_2$–14.99WO$_3$–0.01Nb$_2$O5 [36]	0.101623	5.332
80TeO$_2$-5TiO$_2$-10WO$_3$-5Nb$_2$O$_5$ [36]	0.104939	5.424
80TeO$_2$-5TiO$_2$-12WO$_3$-3Nd$_2$O$_3$ [36]	0.10723	5.633
80TeO$_2$-5TiO$_2$-10WO$_3$-5Nd$_2$O$_3$ [36]	0.109385	5.750
80TeO$_2$-5TiO$_2$-10WO$_3$-5Er$_2$O$_3$ [36]	0.111788	5.896
70Ba-30SiO$_2$ [54]	0.65	

The removal cross section for fast neutrons Σ_R for tellurite glasses has been published [20, 36–38]. Table 2.4 shows values of ΣR for 80TeO$_2$-5TiO$_2$- (15-x)WO$_3$ glass and $x = 0.01$ Nb$_2$O$_5$, 5Nb$_2$O$_5$, 3 Nd$_2$O$_3$, 5 Nd$_2$O$_3$ and 5 Er$_2$O$_3$ mol%, respectively. The highest values of ΣR were found for 5 Nd$_2$O$_3$ and 5 Er$_2$O$_3$ which possess the maximum densities; this indicates that the density of the glass is an important parameter affecting neutron attenuation. Values of ΣR of 70Ba–30SiO$_2$ were 0.065 cm^{-1} [54]. Form this, it has been concluded that the 80TeO$_2$-5TiO$_2$-10WO$_3$-5Nd$_2$O$_3$ and 80TeO$_2$-5TiO$_2$-10WO$_3$-5Er$_2$O$_3$ glasses are the most effective for neutron shielding than the other glasses in this work.

2.6   Exposure Buildup Factors, EBF

The exposure buildup factors, EBF, and the G-P fitting parameters were used to calculate the EBF of glasses as follows:

$$B(E,X) = 1 + \frac{b-1}{K-1}(K^x - 1) \qquad \text{for } K \neq 1$$

$$B(K,X) = 1 + (b-1)x \qquad \text{for } K = 1$$

where,

$$K(E,x) = cx^a + d\,\frac{\tan h\left(\frac{x}{X_K} - 2\right) - \tan h(-2)}{1 - \tan h(-2)} \quad \text{for } x \leq 40 \text{ mfp} \qquad (2.10)$$

where E is the incident photon energy, x is the penetration depth in mfp, and a, b, c, d and X_k are the G-P fitting parameters. The exposure buildup factors, EBF, for tellurite glasses have been published [20, 36–38]. The values of EBF are tabulated in Table 2.5.

Table 2.5 shows as an example of the variation of EBF for some binary molybdenum tellurite glasses [20]. The EBF values increase with increasing penetration

Table 2.5 Exposure buildup factors, EBF, for some tellurite glasses

Energy (MeV)	80TeO$_2$-20MoO$_3$ [20]	50 TeO
1 mfp		
0.015	1.005	1.006
0.15	1.23	1.24
1.5	1.60	1.62
15	1.48	1.38
40 mfp		
0.015	1.014	1.020
0.15	2.04	2.64
1.5	54.81	59.32
15	3443.84	1519.48

depths for glass samples at the selected energies (0.015 and 15 MeV). The EBF values increase as MoO3 content increase in the tellurite glass. At lower energy a significant variation is observed in the EBF between 1 and 40 mfp for the glass with different percentages of 20MoO$_3$ mol% and 50MoO$_3$ mol% which is explained by the fact that photoelectric effect is dominant in the low-energy regions and the probability of Compton scattering is nearly negligible.

References

1. O.A. Zamyatin, M.F. Churbanov, J.A. Medvedeva, S.A. Gavrin, E.V. Zamyatina, A.D. Plekhovich, Glass-forming region and optical properties of the TeO$_2$ –ZnO–NiO system. J. Non-Cryst. Solids **479**, 29–41 (2018)
2. G. Lakshminarayanaa, S.O. Baki, M.I. Sayyed, M.G. Dong, A. Lira, A.S.M. Noor, I.V. Kityk, M.A. Mahdi, Vibrational, thermal features, and photon attenuation coefficients evaluation for TeO$_2$-B$_2$O$_3$-BaO-ZnO-Na$_2$O-Er$_2$O$_3$-Pr$_6$O$_{11}$ glasses as gamma arrays shielding materials. J. -Non-Cryst. Solids **481**, 568–578 (2018)
3. M.A. Merzliakov, V.V. Kouhar, G.E. Malashkevich, E.V. Pestryakov, Spectroscopy of Yb-doped tungsten-tellurite glass and assessment of its lasing properties. Opt. Mater. **75**, 142–149 (2018)
4. M.E. Alvarez-Ramos, J. Alvarado-Rivera, M.E. Zayas, U. Caldi-no, J. Hern_andez-Paredes, Yellow to orange-reddish glass phosphors: Sm^{3+}, Tb^{3+} and Sm^{3+}/Tb^{3+} in zinc tellurite-germanate glasses. Opt. Mater. **75**, 88–93 (2018)
5. S.H. Elazoumi, H.A.A. Sidek, Y.S. Rammah, R. El-Mallawany, M.K. Halimah, K.A. Matori, M.H.M. Zaid, Effect of PbO on optical properties of tellurite glass. Res. Phys. **8**, 16–25 (2018)
6. M.I. Sayyed, M. Çelikbilek Ersundu, A.E. Ersundu, G. Lakshminarayana, P. Kostka, Investigation of radiation shielding properties for MeO-PbCl2-TeO$_2$ (MeO =Bi$_2$O$_3$, MoO$_3$, Sb$_2$O$_3$, WO$_3$, ZnO) glasses. Radiat. Phys. Chem. **144**, 419–425 (2018)
7. R. El-Mallawany, Y.S. Rammah, A. El Adawy, Z. Wasses, Optical and thermal properties of some Tellurite glasses. Am. J. Opt. Photon. **5**(2), 11–18 (2017)
8. M.M. El-Zaidia, A.A. Ammar, R.A. El-Mallwany, Infra-red spectra, electron spin resonance spectra, and density of (TeO$_2$) 100− x–(WO$_3$) x and (TeO$_2$) 100− x–(ZnCl$_2$) x glasses. Phys. Status Solidi A **91**(2), 637–642 (1985)

9. I.Z. Hager, R. El-Mallawany, A. Bulou, Luminescence spectra and optical properties of TeO $_2$–WO $_3$–Li $_2$ O glasses doped with Nd, Sm and Er rare earth ions. Physica B: Condensed Matter **406**(4), 972–980 (2011), **406**(4), 1844 (2011)

10. I.Z. Hager, R. El-Mallawany, Preparation and structural studies in the (70− x) TeO$_2$–20WO$_3$–10Li$_2$O–xLn$_2$O$_3$ glasses. J. Mater. Sci. **45**(4), 897 (2010)

11. N.S. Hussain, G. Hungerford, R. El-Mallawany, M.J.M. Gomes, M.A. Lopes, J.D. Nasar Ali, Santos, and S. Buddhudu, absorption and emission analysis of RE^{3+} (Sm^{3+} and Dy^{3+}): Lithium Boro Tellurite glasses. J. Nanosci. Nanotechnol. **9**(6), 3672–3677 (2009)

12. M.M. Elkholy, R.A. El-Mallawany, Ac conductivity of tellurite glasses. Mater. Chem. Phys. **40** (3), 163–167 (1995)

13. A. El-Adawy, R. El-Mallawany, Elastic modulus of tellurite glasses. J. Mater. Sci. Lett. **15**(23), 2065–2067 (1996)

14. R. El-Mallawany, Specific heat capacity of semiconducting glasses: Binary vanadium tellurite. Phys. Status Solidi A **177**(2), 439–444 (2000)

15. R. El-Mallawany, A. Abd El-Moneim, Comparison between the elastic moduli of tellurite and phosphate glasses. Phys. Status Solidi A **166**(2), 829–834 (1998)

16. R. El-Mallawany, A.H. El-Sayed, M.M.H.A. El-Gawad, ESR and electrical conductivity studies of (TeO$_2$) 0.95 (CeO$_2$) 0.05 semiconducting glasses. Mater. Chem. Phys. **41**(2), 87–91 (1996)

17. H.M.M. Moawad, H. Jain, R. El-Mallawany, DC conductivity of silver vanadium tellurite glasses. J. Phys. Chem. Solids **70**(1), 224–233 (2009)

18. R. El-Mallawany, Theoretical analysis of the electrical properties of tellurite glasses. Mater. Chem. Phys. **37**(4), 376–381 (1994)

19. R. El-Mallawany, P. Separation, Ultrasonic detection of microphase separation in Tellurite glasses. Phys. Stat. Sol. (a) **133**, 245 (1992)

20. M.I. Sayyed, R. El-Mallawany, Shielding properties of (100-x)TeO$_2$-(x)MoO$_3$ glasses. Mater. Chem. Phys. **201**, 50e56 (2017)

21. M. Dong, X. Xue, Y. He, et al., A novel comprehensive utilization of vanadium slag: As gamma ray shielding material. J. Hazard. Mater. **318**, 751–757 (2016)

22. Z. Li, X. Xue, S. Liu, et al., Effects of boron number per unit volume on the shielding properties of composites made with boron ores from China. J. Nucl. Sci. Tech. **23**(6), 344–348 (2012)

23. X. Cao, X. Xue, T. Jiang, et al., Mechanical properties of UHMWPE/Sm$_2$O$_3$, composite shielding material. J. Rare Earths **28**(S1, 482–484 (2010)

24. M. Sayed, J.A. Khan, L.A. Shah, et al., Degradation of quinolone antibiotic, norfloxacin, in aqueous solution using gamma-ray irradiation. J. Environ. Sci. Poll. Res. **23**(13), 13155–13168 (2016)

25. L.R. Amparo, G. Elliotpaul, Neutron scattering: A natural tool for food science and technology research. Trends Food Sci. Technol. **20**(11–12), 576–586 (2010)

26. A. Wyszomirska, Iodine-131 for therapy of thyroid diseases. Physical and biological basis. Nucl. Med. Rev. Cent. East. Eur. **15**(2), 120–123 (2012)

27. I. Akkurt, C. Basyigit, S. Kilincarslan, et al., The shielding of γ-rays by concretes produced with barite. Prog. Nucl. Energy **46**(1), 1–11 (2005)

28. I. Akkurt, H. Akyıldırım, B. Mavi, et al., Radiation shielding of concrete containing zeolite. J. Radiat. Measure. **45**(7), 827–830 (2010)

29. I. Akkurt, A.M. El-Khayatt, The effect of barite proportion on neutron and gamma-ray shielding, J. Ann. Nucl. Energy **51**, 5–9 (2013)

30. B. Oto, A. Gür, M.R. Kaçal, et al., Photon attenuation properties of some concretes containing barite and colemanite in different rates. J. Ann. Nucl. Ener. **51**, 120–124 (2013)

31. C.M. Lee, Y.H. Lee, K.J. Lee, Cracking effect on gamma-ray shielding performance in concrete structure. J. Prog. Nucl. Energy **49**(4), 303–312 (2007)

32. J.C. Khong, D. Daisenberger, G. Burca, et al., Design and characterization of metallic glassy alloys of high neutron shielding capability. J. Sci. Rep. **6**, 36998 (2016)

33. J.E. Martin, Physics for Radiation Protection, 3rd[rd] Edition, (WILEY-VCH Verlag GmbH & Co. KGaA, Weinheim, ISBN:978-3-527-41176-4, 670 pages, 2013)
34. M.F. Kaplan, *Concrete Radiation Shielding* (John, New York, 1989)
35. L. Gerward, N. Guilbert, K.B. Jensen, H. Levring, WinXCom—A program for calculating X-ray attenuation coefficients. Radiat. Phys. Chem. **71**, 653–654 (2004)
36. M.G. Dong, R. El-Mallawany, M.I. Sayyed, H.O. Tekin, Shielding properties of $80TeO_2$–$5TiO_2$–$(15-x)$ WO_3–xAnOm glasses using WinXCom and MCNP5 code. Radiat. Phys. Chem. **141**, 172–178 (2017)
37. R. El-Mallawany, M.I. Sayyed, Comparative shielding properties of some tellurite glasses: Part 1. Physica B, **539c**, 133–140 (2018)
38. R. El-Mallawany, M.I. Sayyed, Comparative shielding properties of some tellurite glasses: Part 2. J. Non-Crys. Sol. **474**, 16–23 (2017)
39. M.I. Sayyed, Bismuth modified shielding properties of zinc boro-tellurite glasses. J. Alloy. Compd. **688**, 111–117 (2016)
40. M.I. Sayyed, S.I. Qashou, Z.Y. Khattari, Radiation shielding competence of newly developed TeO2-WO3 glasses. J. Alloy. Compd. **696**, 632–638 (2017)
41. M.I. Sayyed, Investigations of gamma ray and fast neutron shielding properties of tellurite glasses with different oxide compositions. Can. J. Phys. **94**, 1133–1139 (2016)
42. S.R. Manohara, S.M. Hanagodimath, K.S. Thind, L. Gerward, On the effective atomic number and electron density: A comprehensive set of formulas for all types of materials and energies above 1 keV. Nucl. Instrum. Methods Phys. Res. B **266**, 3906–3912 (2008)
43. R. Bagheri et al., Gamma ray shielding study of barium bismuth borosilicate glasses as transparent shielding materials using MCNP-4C code, XCOM program, and available experimental data. Nucl. Eng. Technol. (2016). https://doi.org/10.1016/j.net.2016.08.013
44. K.J. Singh, N. Singh, R.S. Kaundal, K. Singh, Gamma-ray shielding and structural properties of $PbO-SiO_2$ glasses. Nucl. Instrum. Methods Phys. Res. B **266**, 944–948 (2008)
45. R. El-Mallawany, M. Sidkey, A. Khafagy, H. Afifi, Elastic constants of semiconducting tellurite glasses. Mater. Chem. Phys. **37**, 295–298 (1994)
46. R.A. El-Mallawany, G.A. Saunders, Elastic properties of binary, ternary and quaternary rare earth tellurite glasses. J. Mater. Sci. Lett. **7**(8), 870–874 (1988)
47. R. El-Mallawany, I.A. Ahmed, Thermal properties of multicomponent tellurite glass. J. Mater. Sci. **43**(15), 5131–5138 (2008)
48. M.A. Sidkey, R. El Mallawany, R.I. Nakhla, A.A. El-Moneim, Ultrasonic studies of (TeO_2) 1-x-(V_2O_5) x glasses. J. Non-Cryst. Solids **215**(1), 75–82 (1997)
49. R. El-Mallawany, N. El-Khoshkhany, H. Afifi, Ultrasonic studies of (TeO_2) 50–(V_2O_5) 50– x (TiO_2) x glasses. Mater. Chem. Phys. **95**(2), 321–327 (2006)
50. S.R. Manohara, S.M. Hanagodimath, K.S. Thind, On the effective atomic number and electron density: A comprehensive set of formulas for all types of materials for all types of materials and energies above 1 keV. Nucl. Instrum. Methods Phys. Res. B **266**(18), 3906–3912 (2008)
51. Y. Elmahroug, B. Tellili, C. Souga, Determination of shielding parameters for different types of resins. Ann. Nucl. Energy **63**, 619–623 (2014)
52. M.F. Kaplan, *Concrete Radiation Shielding* (Longman Scientific and Technology, Longman Group UK, Limited, Essex, 1989)
53. A.B. Chilten, J.K. Shultis, R.E. Faw, *Principle of Radiation Shielding* (Prentice-Hall, Englewood Cliffs, 1984)
54. V.P. Singh, N.M. Badiger, J. Kaewkhao, J. Non-Cryst. Solids **404**, 167 (2014.) https://doi.org/10.1016/j.jnoncrysol.2014.08.003

Chapter 3
Tellurite Glass Materials for Energy Conversion Technology and Lasers Devices

Luiz Carlos Barbosa, Cicero Omegna Filho, and Enver Fernandez Chillcce

Abstract Tellurite glass materials present greater potential applications for solar energy technology and laser devices, it is, because these materials present very efficient optical and physical properties. Tellurite glasses doped with PbTe, CdTe, and rare-earth materials are considerable, due to their practical importance in technological applications such as integrated optics, optoelectronics, lasers, broadband optical amplifiers, and solar energy conversion. The challenges in the solar energy technology research are to increase the conversion efficiency compared with that of silicon solar cells and consequently to make them more cost-effective for commercial applications. Tellurite glasses doped with rare earths were demonstrated as very broadband optical amplifiers and laser devices. In this chapter we are involved with study of optical and physical properties of these tellurite glass materials.

3.1 Introduction

Tellurite glass materials present greater potential applications for solar energy technology and laser devices, it is, because these materials present very efficient optical and physical properties. Tellurite glasses doped with PbTe, CdTe, CdTeS, and rare-earth materials are considerable, due to its practical importance in technological applications such as, integrated optics, optoelectronics, lasers, broadband

L. C. Barbosa (✉)
Depto de Eletrônica Quântica, Instituto de Física Gleb Wataghin, Universidade estadual de Campinas, Campinas, SP, Brazil
e-mail: barbosa@ifi.unicamp.br

C. O. Filho
Luxtec Optical System, Campinas, SP, Brazil
e-mail: cicero@luxtec.com.br

E. F. Chillcce
BrPhotonics- CPqD, São Paulo, Brazil
e-mail: efernandez@brphotonics.com

© Springer International Publishing AG, part of Springer Nature 2018
R. El-Mallawany (ed.), *Tellurite Glass Smart Materials*,
https://doi.org/10.1007/978-3-319-76568-6_3

optical amplifiers, and solar energy conversion. The challenges in the solar energy technology research are to increase the conversion efficiency compared with that of silicon solar cells and consequently to make them more cost-effective for commercial applications. Tellurite glasses doped with rare earths were demonstrated as very broadband optical amplifiers and laser devices. In this chapter we are involved with study of optical and physical properties of these tellurite glass materials.

3.2 Formation and Growth of Semiconductor Nanocrystals in Glasses

The concept of "integrated glass substrate" seems to be an interesting approach for a flat glass producer to support the lighting industry and to participate on a fast-growing market in the future.

The current status of important application of glass in the energy conversion technology was photovoltaic industry and role of glass for reducing the cost of solar energy. Photovoltaic conversion of sunlight into electricity is a versatile process that currently contributes to world energy demand moderately. There have been significant research and technological development (RTD) efforts at every stage of the photovoltaic value chain across the world. Over the last two decades, appreciable technological improvements and sustained growth in production volume have resulted in price reduction beyond expectations. This in turn has made photovoltaic power the third most important renewable energy source after hydro and wind powers. Photovoltaic power systems (PVPS) connected to the grid have increased from about 17GWp (2010) to almost 30GWp (2011) in a year. In 2011, the total global installed PVPS reached ~70GWp producing ~ 85TWh electrical energy, which is equal to the production capacity of 12GW nuclear power plants. Over the years, the West has dominated RTD, materials and components production together with system installation activities in the field of PV. However, in recent years while the major parts of RTD and PVPS installations have remained in the West, production has shifted toward the East. The foreseen price decrease currently being in the PV value chain seems to continue over the medium and long terms. Thus, depending on when and where the correct policies are implemented, photovoltaic power could become more competitive with conventional electrical energy production technologies starting from sun-belt countries.

There are currently many large-scale international efforts for improving the consumer acceptance of energy conversion technology (PV modules and systems). At the crystalline silicon or thin-film module level, performance improvement and cost reduction are the primary targets. Over the last decade, efficiencies of the best crystalline silicon cells, CuIn(Ga)S(Se), i.e., CIGS cells, CdTe cells, and thin-film silicon cells, have improved by 5.5%, 4.6%, 5.1%, and 3.9%, respectively. There has also been notable progress in the technology to stabilize crystalline and thin-film modules. In an attempt to reduce cost of per Wp in photovoltaic conversion, an improvement of cell efficiency is just one parameter, and the main cost factors related to all production materials and technologies need to be considered.

Glass is used in crystalline photovoltaic module as a protective and supporting layer, but in thin-film modules, glass also serves as the substrate or superstrate. In a recent evaluation, the relative cost fraction of glass is about 10% in crystalline Si modules and almost 25% in thin-film modules. Recent estimates for the total material cost of a 0.24€/Wp for CIGS modules is dominated by the cost of glass (0.10€/Wp) [1]. Thus research and technological developments in glass for PV play a major role in the future of the photovoltaic industry.

The presentation aimed on the discussion of the status and future prospects of the photovoltaic value chain together with the role of glass and glass RTD for photovoltaic module production. Accordingly, suitable technological developments will be required in production and coatings processes for improved optical, mechanical, and chemical properties of PV glasses parallel to energy conversion technology (RTD activities) in float glass, despite the functional glasses: properties and applications for energy and information.

Despite the growing economy, which is combined with an increasing demand for energy, the chronic overcapacity in the worldwide PV production led to profitless prosperity in the year 2012. Since currently the material cost, PV modules is dominated by flat glass, measures for cost reductions are in the focus and should be considered for future developments too. In particular for the PV industry, cost-effectiveness for solar float glass has to be realized, together with higher transmission, an increased mechanical stability, and higher degree of flatness and homogeneity. Furthermore, an easy application of antireflective coatings, prevention of corrosion layer, a better processability (cutting, grinding, and tempering), and an efficient production of large quantities have to be possible.

By the way, in energy conversion technology, advances and maturation followed by their wider implementation will increase the capture and repurposing of readily available waste-energy sources. A number of technologies and prototypes exist for converting heat energy into electrical energy. Conventional technologies for electricity generation from heat recovery rely on boiling liquids to produce steam that drives turbines. For lower-temperature waste heat, there are variations that use other fluids and their associated gases. In organic generators, working fluids such as propane or toluene are used and have lower boiling points. A number of candidate fluids and applications are discussed in a review by Tchanche et al. [1]. These has cost and complexity advantages over a typical steam cycle, including the avoidance of superheating; this comes at the expense of maximum efficiency of 24%, compared to more than 30% efficiency for its water-based counterpart.3 A number of manufacturers produce ORC systems at costs estimated in the $2–$3 per watt range compared to $1.10–$1.40 per watt.

The most common materials used today are alloys of chalcogenides (materials with a chalcogen or IUPAC group 16 anion). Specifically, these materials are either based on bismuth telluride (Bi_2Te_3) or lead telluride ($PbTe$). Bi_2Te_3 can be alloyed with Bi_2Se_3 to form n-type $Bi_2Te_{3-x}Se_x$ and with Sb_2Te_3 to form p-type $Bi_xSb_{2-x}Te_3$. $PbTe$ can be alloyed with $PbSe$ to form p-type $PbTe_{1-x}Se_x$ and with $SnTe$ to form n-type Pb1-xSnxTe. PbTe has been used successfully by the National Aeronautics and Space Administration (NASA) as radioisotope thermoelectric generators (RTGs) but has been rejected by all current power generation projects because of the lead content and poor mechanical properties during thermal cycling under variable temperature gradients and glasses with high refractive index as tellurite glasses doped with rare-earth compound.

3.3   Formation and Growth of Semiconductor Nanocrystals in Glasses

In the past, it was demonstrated that high nonlinearity optical materials are of great interest for optical communications. Glasses doped with nanocrystals (CdTe, PbTe, CdSe, CdS, and others) show large optical nonlinearities and fast responses (at a femtosecond time scale). The growing of these nanocrystals is now well understood, and good control of the sizes and size dispersions has been demonstrated. PbTe materials, with a bandgap of approximately 3.8 μm, have been used as the well material in quantum wells and confined levels at wavelengths down to 2.6 μm. Figure 3.1a shows a series of absorption spectra for the samples with glass matrix containing SiO_2, ZnO, Al_2O_3, and Na_2O [2] This matrix is doped with 1.5% of PbO and metallic Te. The component materials were heated to 1350 °C and held for 50 min. The melt is then quenched between two stainless steel plates. The

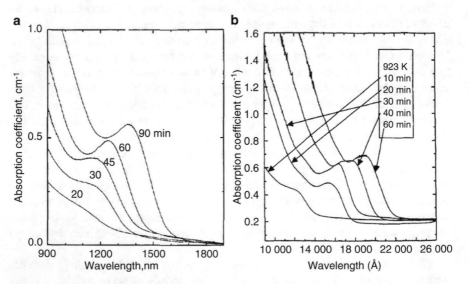

Fig. 3.1 (**a**) Absorption spectra of PbO/Te-doped glass (SiO2, ZnO, Al, O, and Na2O) samples annealed at 525 "C for 20, 30, 45, 60, and 90 min [1]. (**b**) Coefficient of optical absorption of the glass-PbTe composites for ex situ isothermal treatment during the indicated time. (From Craievich et al. [3])

PbTe semiconductor nanocrystals are produced by a subsequent heat treatment, where the sizes are controlled using different temperatures and times. In this series, the nanocrystals were grown in two steps; first, the PbO/Te-doped glass samples was preannealed for 17 h at 450 °C and then were annealed at 525 °C for 20, 30, 45, 60, and 90 min. The two annealing temperatures were chosen to be close to the glass transition temperature (Tg) and softening point temperature (Ts), respectively, as shown in the inset of Fig. 3.1a. Figure 3.1b shows the optical absorption spectra of a transparent lame (about 0.1 mm thick) glass sample [3] . The composition of the glass sample is $52SiO_2-8B_2O_3-20ZnO-20K_2O$ doped with 2 wt% PbO and Te. The glass sample was homogenized at 1623 K during 50 min and quenched to room temperature. The glass sample was held at 923 K for different periods of time, such as shown in Fig. 3.1b.

The experimental results (that correspond to Fig. 3.1b) indicate that nanocrystals are nearly spherical and have an average radius increasing from 16 to 33 Å after 2 h at 923 K. The shape and size distribution of the nanocrystals and the kinetics of their growth were studied by small-angle X-ray scattering (SAXS) during in situ isothermal treatment at 923 K. The kinetics of nanocrystal growth is governed by the classic mechanism of atomic diffusion. The experimental SAXS results suggest that nanocrystals nucleate and grow as a consequence of the diffusion of isolated Pb and Te elements through the glass matrix. Under this assumption, the average crystal radius, R, is expected to have the simple time dependence given by: $R^2 = Kt + R0$, where K is a constant and $R0$ is the initial nanocrystal radius. The experimental values of R for increasing periods of time obey the potential law ($R \propto t^2$), as can be seen in Fig. 3.2. This result is consistent with the proposed mechanism of nanocrystal growth by pure atomic diffusion up to about 1 h of heat treatment. As the concentration of Pb and Te approaches the solubility limit, coarsening becomes the predominant mechanism responsible for nanocrystal growth, and the time dependence of the radius is no longer given by ($R \propto t^2$) and becomes ($R \propto t^3$). The growth of the nanocrystals is governed by the classical atomic diffusion

Fig. 3.2 Square of average radius (R^2) as a function of time. (From Craievich et al. [3])

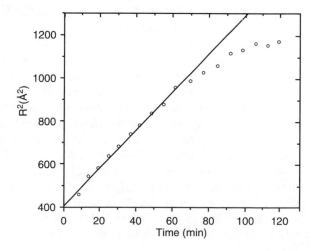

R²(Å²)

Time (min)

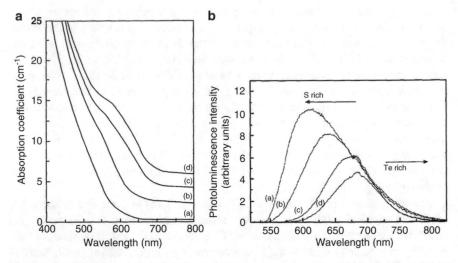

Fig. 3.3 (**a**) Absorption spectra of shell-core CdTeS quantum dots with 200 h annealing at 490 °C and annealing time at 540 °C of (a) 0 min, (b) 30 min, (c) 150 min, and (d) 300 min. (**b**) Photoluminescence spectra of shell-core CdTeS quantum dots with 200 h annealing at 490 °C and annealing time at 540 °C of (a) 0 min, (b) 30 min, (c) 150 min, and (d) 300 min. (From Liu et al. [4])

mechanism in the first stages and by coarsening in advanced stages of isothermal heat treatment.

The glass matrix contains 47.66% SiO_2, 16.55% B_2O_3, 30.57% Na_2O, and 5.22% ZnO (wt %) mixed with CdO, Te, and S in the $Cd_{0.6}Se_{0.4}S_{0.6}$ stoichiometry. Forty grams of fine powder were premixed and melted at t400 °C for 50 min. The molten glass was rapidly quenched by pouring onto a brass plate to avoid the immediate growth of semiconductor nanocrystallites. The photoluminescence spectra are shown in Fig. 3.3b). Comparing to the first group, the curve of single annealing shows no trap-related transitions. We also noticed a large change both in photoluminescence peaks and intensity as we increased the second annealing time. This is because of the increase in dot radii and the Te composition. The increase of Te composition was verified by the decrease of photoluminescence efficiency. The photoluminescence efficiency will not change with annealing time if the stoichiometry is not changed. In our case, the total amount of S source is constant. As S source was first used in the core growth, there would be less S than Te in the matrix during the second annealing. The longer the second annealing is, the less S (or the more le) source will be available. This is the cause of the gradual decrease of photoluminescence intensity. Figure 3.6 shows the absorption spectra of the second groups. The curves, except the lowest one, are shifted to aid clarity. Comparing with the first group, the single annealed sample (a) was much improved. Figure 3.7 shows the photoluminescence spectra of two samples, picked from Figs. 3.3 and 3.5, having the same second treatment but with different first treatment. Both samples were treated for 150 min during the second annealing. The difference between them is only the time of the first treatment. We can see that there is about 15 meV peak shift between the shell-core structure and the shell-seed structure. As both samples have the same second annealing, the amount of shell medium should be the same. Therefore, the one with the longer time of first annealing would be expected to have the larger size. If the shell and core have the same stoichiometry, the shell-core structure will shift to the right side (lower energy) of the shell-seed structure. Figure 3.7, however, shows a left side shift. This proves that the shell has more Te than the core. In conclusion, we have shown that there is the possibility to make structural quantum dots in glass-based materials by using a double annealing process. The idea may be extended to grow a three-dimensional quantum well structure inside a quantum dot in glass by, for example, a triple annealing process. The present problem is that one should choose better ternary sources which have the possibility to form distinctly different stoichiometry during different annealing processes.

3.4   Photon Conversion Processes of Rare-Earth-Doped Tellurite Glasses

The total solar power energy irradiated onto the Earth's surface is about 100,000 terawatts, which is 10,000 times more than that consumed globally. Nowadays, crystalline silicon (c-Si) photovoltaic (PV) cells are the most used among all types of solar cells on the market; however, even for single crystalline silicon (Si) PV cells with a rather small semiconductor bandgap (1.12 eV, corresponding to a wavelength of ~1100 nm), the transmission loss of sub-bandgap photons can still amount to about 20% of the sun's energy irradiated onto the Earth's surface. For photon conversion processesee as PV cells with a larger bandgap, such as amorphous Si (1.75 eV) solar cells, which are limited to absorb sunlight with wavelengths below 708 nm, they manifest even higher near-infrared transmission losses.

Yunfei Shang et al. [5] recently point that photon upconversion (UC) processes provides a means to circumvent transmission loss by converting two sub-bandgap photons into one above-bandgap photon, where the PV cell has high light responsivity. The inability to absorb infrared (IR) light (700–2500 nm), which constitutes 52% of the energy of the entire solar spectrum, forms the major energy loss mechanism of conventional solar cells. Figure 3.4 schematically illustrates the use of upconversion processes to convert the solar spectrum in the IR-Near IR (NIR)-short visible range into the peak (~500 nm) of sun radiation. There are three typical photon upconversion materials under investigation now outlined: rare-earth-doped microcrystals and nanocrystals (RED-UC) , which usually work with wavelengths above 800 nm; triplet-triplet annihilation upconversion (TTA-UC, response range $\lambda < 800$ nm) whereby the triplet states of two organic molecules interact with each other, exciting one molecule to its emitting state to produce fluorescence; and upconversion in quantum nanostructures (QN-UC, response range $\lambda < 800$ nm).

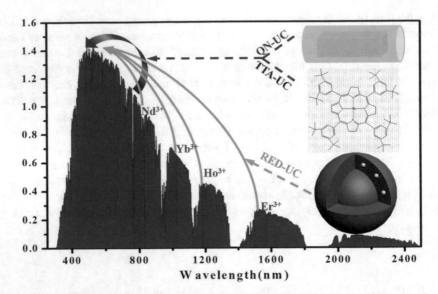

Fig. 3.4 The absorption and emission range of three types of upconvertion materials in reference to solar irradiance spectrum: QN-UC (purple), upconvertion in quantum nanostructures; TTA-UC (purple), triplet-triplet annihilation upconvertion; RED-UC (green), rare-earth-doped upconvertion materials. (From Yunfei Shang et al. [5])

Photon upconversion process (using rare-earth materials) may occur efficiently in materials with high concentrations of rare earths. For this purpose, tellurite glasses maybe be used to fabricate photovoltaic devices doped with high concentrations of rare earths, it is, because rare earths are very soluble in tellurite glasses. Figure 3.5 shows the UV-VIS-IR attenuation spectra of Er^{3+}-doped tellurite and Er^{3+}-Tm^{3+}-co-doped tellurite glasses with high concentrations of Er_2O_3 and Tm_2O_3. The glass matrix has the following composition: $70TeO_2$-$19WO_3$-$7Na_2O$-$4Nb_2O_5$ (%mol) [5]. In this figure, the following curves correspond to tellurite glass samples doped with: (a) 7500 ppm Er_2O_3, (b) 7500 ppm Er_2O_3 and 2500 ppm Tm_2O_3, (c) 7500 ppm Er_2O_3 and 5000 ppm Tm_2O_3, (d) 7500 ppm Er_2O_3 and 7500 ppm Tm_2O_3, (e) 7500 ppm Er_2O_3 and 10.000 ppm Tm_2O_3, and (f) 7500 ppm Er_2O_3 and 15.000 ppm Tm_2O_3. All curves show about 0.5 cm^{-1} minimum attenuation for wavelength above 1000 nm, but, with the exception of the first curve (a), they were vertically displaced arbitrarily to display more clearly the Tm^{3+} ions absorption bands.

Figure 3.6 shows the luminescence spectra tellurite glasses samples, doped with several concentrations of Er_2O_3 and Tm_2O_3 and excited with a 980 nm (120 mW) diode laser. Each curve was normalized to the peak intensity of the ($^4S_{3/2} \rightarrow {}^4I_{15/2}$) Er $^{3+}$ level transition. The ($^2H_{11/2} \rightarrow {}^4I_{15/2}$), ($^4S_{3/2} \rightarrow 4I_{15/2}$), ($^4F_{9/2} \rightarrow {}^4I_{15/2}$), and ($^4S_{3/2} \rightarrow {}^4I_{13/2}$) transitions of Er^{3+} levels, as well as the ($^3H_4 \rightarrow {}^3H_6$) transition of Tm3+ levels, are observed for different $T_3m_2O_3$ concentrations.

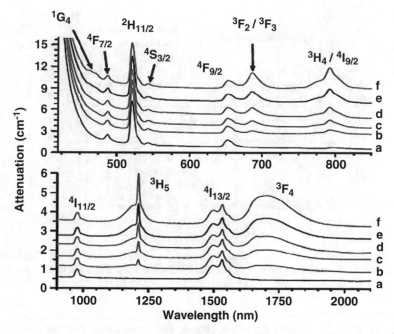

Fig. 3.5 Attenuation or extinction coefficient of Er^{3+}-doped and Er^{3+}-Tm^{3+}-co-doped tellurite glass. (From Chillcce et al. [6])

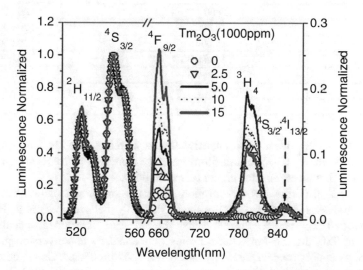

Fig. 3.6 Luminescence spectra in VIS-IR range for Er^{3+}-Tm^{3+}-co-doped tellurite fiber with 7500 ppm Er_2O_3 and diverse Tm_2O_3 concentrations and pumping with 980 nm (120 mW) diode laser. (From Chillcce et al. [6])

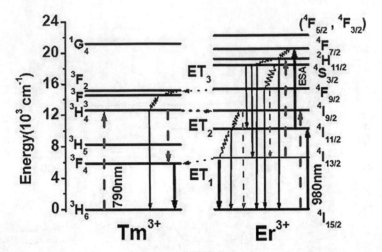

Fig. 3.7 Absorption, emission, and energy transfer processes between Er^{3+} and Tm3+ energy levels. Dark arrows represent processes when the Er^{3+}-Tm^{3+}-co-doped tellurite fiber is pumped with 980 nm laser and additionally dashed arrows when pumped with 790 nm laser only. (From Chillcce et al. [6])

Table 3.1 Glass (A–G): [69TeO$_2$-24WO$_3$-3Nb$_2$O$_5$-4Na$_2$O] (mol%) + 2000 ppm Al$_2$O$_3$ (wt%) +xEr$_2$O$_3$+yYb$_2$O$_3$ (ppm). Glass (H): [72.5TeO$_2$-20WO$_3$-1.5Nb$_2$O$_5$-6Na$_2$O] (mol%) + 2000 ppm Al$_2$O$_3$ (wt%)

Sample	Rare-earth composition (ppm)	
	x: Er$_2$O$_3$	y: Yb$_2$O$_3$
A	10.000	0
B	5000	20.000
C	10.000	20.000
D	15.000	20.000
E	10.000	40.000
F	0	100.000
G	0	0
H	0	0

The ($^4F_{9/2} \rightarrow {}^4I_{15/2}$) transition intensity always increase with the Tm$_2$O$_3$ concentration, while the ($^3H_4 \rightarrow {}^3H_6$) transition intensity increases up to 5000 ppm Tm$_2$O content and then decreases for higher concentrations.

Figure 3.7 also shows the absorption and emission processes and the energy transfer processes ET$_1$ (between $^4I_{13/2}$ and $^3F_4$), ET$_2$ (between $^4I_{9/2}$ and $^3H_4$), and ET$_3$ (between $^4F_{9/2}$ and $^3F_2$) levels. When the Er^{3+}-Tm^{3+}-co-doped optical fiber is pumped at 980 nm, the Tm^{3+} ion can only be excited by upconversion processes followed by energy transfer, specially the ET$_3$ and ET$_1$ processes. On the other hand, both ions can be excited directly from the ground state to the $^4I_{9/2}$ and $^3H_4$ levels, when the Er^{3+}-Tm^{3+}-co-doped optical fiber is pumped 790 nm.

Narro-Garcia et al. [7] have reported upconversion studies by using tellurite glass samples co-doped with Er^{3+} and Yb^{3+} ions. The glass compositions are indicated Table 3.1.

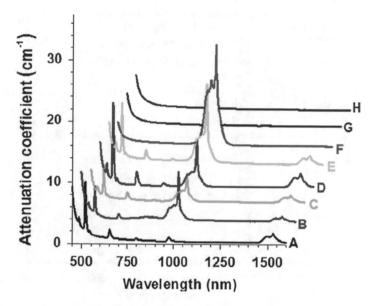

Fig. 3.8 Attenuation coefficient spectra of the tellurite glass samples from Table 3.1: (A) Er^{3+}-doped, (B, C, D, E) Er^{3+}-Yb^{3+}-co-doped, (F) Yb^{3+}-doped, and (G, H) un-doped tellurite glass samples. (From Narro-García et al. [7])

Figure 3.8 shows the attenuation spectra of the tellurite glass samples ranging from 450 to 1600 nm. The attenuation coefficient spectrum of the Er^{3+}-doped tellurite glass (curve A) clearly shows the characteristic transitions with its absorption peaks: $^4I_{15/2} \rightarrow {}^4F_{7/2}$(486 nm), $^4I_{15/2} \rightarrow {}^2H_{11/2}$ (520 nm), $^4I_{15/2} \rightarrow {}^4S_{3/2}$(542 nm), $^4I_{15/2} \rightarrow {}^4F_{9/2}$(651 nm), $^4I_{15/2} \rightarrow {}^4I_{9/2}$(796 nm), $^4I_{15/2} \rightarrow {}^4I_{11/2}$(977 nm), and $^4I_{15/2} \rightarrow {}^4I_{13/2}$(1530 nm). On the other hand, the attenuation coefficient spectrum of the Yb^{3+}-doped tellurite glass (curve F) shows only the broadband transition $^2F_{7/2} \rightarrow {}^2F_{5/2}$ (977 nm) that ranges from 900 nm to 1030 nm. Finally, the Er^{3+}-Yb^{3+}-co-doped tellurite glass spectra (curves B, C, D, and E) show both Er^{3+} and Yb^{3+} level transitions. Offsets in z axis were introduced to the spectra in order to observe the attenuation bands. The average attenuation coefficient at 1350 nm is 0.53 cm^{-1}. The background attenuation of the tellurite glasses in the NIR and IR regions may be described through the Rayleigh scattering and also the background attenuation in the UV–Blue region may be considered as caused by the Urbach tail (Fig. 3.8).

Fig. 3.9 shows the visible and near-infrared (NIR) emission spectra, in the range from 450 to 700 nm, of Er^{3+}-Yb^{3+}-co-doped glasses under 970 nm (diode laser) excitation. Three strong visible bands centered at 526, 548, and 660 nm, respectively, were observed, and the visible emission is due to the well-known upconversion process. The overall intensity of the upconverted signal increases monotonically with Yb^{3+} concentration but decreases with Er^{3+} being the maximum at 5000 ppm of Er_2O_3 (see Fig. 3.9a, b). Figure 3.9a shows no increment of the signal and is explained in terms of the optimization of Yb^{3+}, while Figure 3.9b shows a decrease of the signal which is presumable due to the quenching effect of Er^{3+}.

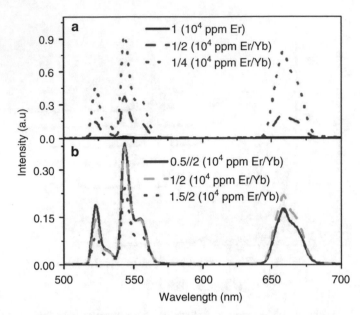

Fig. 3.9 Attenuation coefficient spectra of the tellurite glass samples from Table3: (A) Er^{3+} – doped, (B, C, D, E) Er^{3+}-Yb^{3+}-co-doped, (F) Yb^{3+}-doped, and (G, H) un-doped tellurite glass samples. (From Narro-García et al. [7])

The physical mechanism for both visible and NIR emissions can be described as follows: both Er^{3+} and Yb^{3+} ions are excited directly by the 980 nm pumping signal. However energy transfer (ET) is highly probably due to the larger absorption cross section of Yb^{3+} and the quasi-resonance between the $^2F_{5/2}$-$^2F_{7/2}$ and the $^4I_{15/2} \rightarrow {}^4I_{11/2}$ transition of Yb^{3+} and Er^{3+}, respectively. Part of the $^4I_{11/2}$ excited ions relax nonradioactive to the $^4I_{13/2}$ level and from here relax to the ground state producing the 1532 nm emission band. A second portion can be promoted to $^4F_{7/2}$ by the ET from the relaxation of another excited Yb^{3+} or Er^{3+} ion (excited state absorption (ESA) mechanism). The $^4F_{7/2}$ level ions decay nonradioactively to $^2H_{11/2}$ and $^4S_{3/2}$ due to phonon interactions. From here, the population decay to ground state producing the green emissions centered at 526 and 548 nm. For the red UC emission, population at $^4S_{3/2}$ state may decay nonradioactively to the $^4F_{9/2}$ state, and from there the $^4F_{9/2} \rightarrow {}^4I_{15/2}$ transition results on the red emission at 657 nm. This red band is also enhanced by increasing the concentration of both donor and acceptors. In this case, a part of the population in the $^4I_{13/2}$ level is promoted to $^4F_{9/2}$ by the energy transfer and ESA. These absorption, emission, and upconversion mechanisms, which correspond to Er^{3+}-Yb^{3+}-co-doped tellurite glasses, are well represented in Fig. 3.10.

From these results, that may be favorable to increase the efficiency of PV devices via upconversion process, and then we may use tellurite glass materials doped with several rare earths, such as Er, Tm, and Yb.

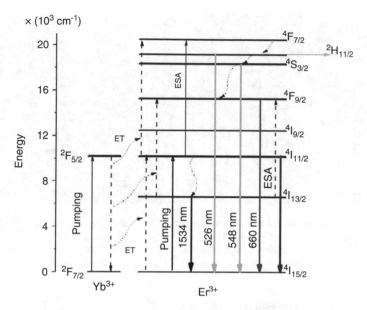

Fig. 3.10 Energy level diagram of Er^{3+} and Yb^{3+} and the possible mechanism for electronic transitions between the energy levels. (From Narro-García et al. [7])

3.5 PbTe Quantum-Dot Multilayers

Fabrication of PbTe quantum-dot multilayer is possible by using laser ablation technique. E. Rodriguez et al. [8] have reported the fabrication of structures containing layers of PbTe quantum dots (QDs) spaced by 15–20 nm thick SiO_2 layers. In this case, the QDs were grown by the laser ablation of a PbTe target using the second harmonic of Nd:YAG laser in an argon atmosphere. The SiO_2 layers were fabricated by plasma chemical vapor deposition using tetramethoxysilane as a precursor. Figure 3.11a shows the low magnification cross-section image of the 11 layer structure obtained, while Fig. 3.11b shows a high-resolution image of the Si substrate (right bottom corner) and the first QD layer. The Fourier transform shown at insert (i) agrees with the 0.31 nm spacing of the [200] planes of the face-centered cubic (fcc) PbTe structure. The presence of the rings in the electron diffraction pattern shown at inset (ii) indicates that there is no preferential orientation of the QDs inside the layers.

On the other hand, the influence of the ablation time on the size and size distribution of the QDs was studied by high-resolution transmission electron microscopy. Figure 3.12a shows one QD grown with 30 s ablation time, where the two-dimensional atom arrangement can be clearly seen. The Fourier transform of the nanoparticle (inset) shows the [200] and [220] directions of the fcc PbTe structure. Figure 3.12b shows the distribution with the average QDs diameter of 4.94 and 1.4 nm dispersion. Figure 3.13a shows the absorption spectra of the multilayer of

Fig. 3.11 (**a**) Low magnification cross-section image of a multilayer PbTe/SiO2 growth on a Si (100) substrate. (**b**) HRTEM image of the multilayer showing the nanocrystal embedded in the dielectric host. The inset (i) shows the Fourier transform of a PbTe nanoparticle. The inset (ii) is the electron diffraction pattern of a multilayer region. (From Rodríguez et al. [8])

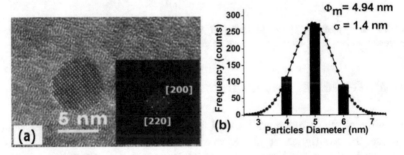

Fig. 3.12 (**a**) Nanoparticles of PbTe. The inset showing their Fourier transform of the nanoparticle. (**b**) QDs size distribution. (From Rodríguez et al. [8])

PbTe nanocrystals. The spectra were acquired at three lines P, Q, and R on the sample (Fig. 3.13b), each corresponding to regions where the quantum dots have different sizes.

Figure 3.13b shows schematic isothickness contours (dashed ellipses) corresponding to the in-plane distribution of the amount of PbTe deposited. The deposit has a central maximum and then decays with the distance from the target normal. On position P, the largest QD size is achieved and gradually decreases toward position R. The optical absorption measurements were done on the 60 bilayers 1.5 mm thick PbTe/SiO_2 structure. Even with the 60 bilayers, the absorption was too weak to be detected with a single pass through the film; therefore, we used a total internal reflection multipass arrangement for this measurement as shown in Fig. 3.13c. Absorption peaks clearly shift toward lower wavelength values by decreasing QD size. This behavior is related to quantum confinement effects. Using the experimental mean diameter obtained from the distribution in position P

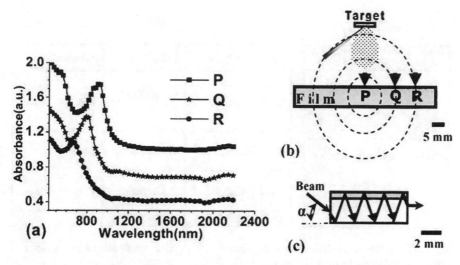

Fig. 3.13 (a) Absorption spectra of a SiO₂/PbTe QD multilayer grown alternately by the PbTe nanoparticles and SiO₂. (**b**) The measurement sample regions P, Q, and R, where the QDs have different sizes. Dashed rings are the isothickness contours corresponding to the in-plane distribution of the amount of PbTe deposited. (**c**) Schematic representation of the multiple pass total internal reflection geometry used for absorption spectra measurements. (From Rodríguez et al. [8])

(4.94 nm), it is possible to estimate the wavelength for the absorption peak as 894 nm, in excellent agreement with the experimental value shown in Fig. 3.13a.

3.6   CdS/CdTe Multijunctions for Solar Cell Applications

Solar cells based on cadmium telluride (CdTe) and cadmium selenide (CdSe) multijunction show great promise for high efficiency cells. The bandgap of CdTe multijunctions for solar cell applicatons is 1.44 eV, a value which is close to the optimal bandgap for single junction solar cell. CdTe is a direct bandgap material; consequently only a few micrometers of CdTe are required to absorb all the photons with an energy higher than the bandgap energy. On the other hand, the bandgap of CdSe is 1.74 eV. Based in this, recently, S. A. Amin et al. [8] have propose dual junction solar cell with high conversion efficiency of solar energy. Figure 3.14a shows the solar cell with the bottom layer made of CdTe semiconductor material and the top layer made of CdSe. These cells can be grown on large area Si substrates. Figure 3.14b shows a table that compares the calculated values for the single and multijunctions of solar cells under sun illumination. These results show that dual junction group CdSe/CdTe cell has better efficiency than other single and dual junction solar cells.

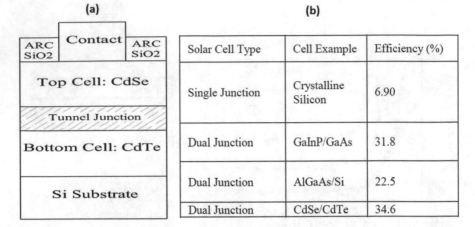

Fig. 3.14 (**a**) Design of CdSe/CdTe dual junction cell. (**b**) Calculated efficiencies for different single and multijunction cells. (From Amin et al. [9])

References

1. B.F. Tchanche, G. Lambrinos, A. Frangoudakis, G. Papadakis, Low-grade heat conversion into power using organic Rankine cycles—A review of various applications. Renew. Sust. Energ. Rev. **15**(8), 3963–3979 (2011). https://doi.org/10.1016/j.rser.2011.07.024
2. V.C.S. Reynoso, A.M. de Paula, R.F. Cuevas, J.A. Medeiros Neto, O.L. Alves, C.L. Cesar, L.C. Barbosa, PbTe quantum dot doped glasses with absorption edge in the 1.5pm wavelength region. El. Letters **31**(12) (1995)
3. A.F. Craievich, O.L. Alves, A.N.D.L.C. Barbosa, Formation and growth of semiconductor PbTe Nanocrystals in a borosilicate glass matrix. J. Appl. Crystallogr. **30**, 623–627 (1997)
4. Y. Liu, V.C.S. Reynoso, R.E.C. Rojas, C.H. Brito cruz, C.L. Cesar, H.L. Fragnito, L.C. Barbosa, O.L. Alves, Shell-core CdTeS quantum dots in glass. J. Mater. Sci. Letters **15**, 980–983 (1996)
5. S. Yunfei Shang, C.Y. Hao, G. Chen, Enhancing solar cell efficiency using photon upconversion materials. Nano **5**, 1782–1809 (2015)
6. E.F. Chillcce, E. Rodriguez, A.A.R. Neves, W.C. Moreira, C.L. Cesar, L.C. Barbosa, Er^{3+}-Tm^{3+} co-doped tellurite fibers for broadband optical fiber amplifier around 1550 nm band. Opt. Fiber Technol. **12**(2), 185–195 (2006)
7. R. Narro-García, H. Desirena, E.F. Chillcce, L.C. Barbosa, E. Rodriguez, E. De la Rosa, Optical and spectroscopic characterization of Er^{3+}- Yb^{3+} co-doped tellurite glasses and fibers. Opt. Commun. **317**, 93–101 (2014)
8. E. Rodríguez, E. Jimenez, L.A. Padilha, A.A.R. Neves, G.J. Jacob, C.L. César, L.C. Barbosa, SiO_2/PbTe quantum-dot multilayer production and characterization. App. Phys. Letters **86**, 113–117 (2005)
9. S.A. Amin, S.T. Salim, K.M.A. Salam, Cadmium Selenide and cadmium telluride based high efficiency multijunction photovoltaics for solar energy harvesting. Int. J. Elect. Energy **1** (1) (2013)

Chapter 4
Structural and Luminescence Properties of Tellurite Glasses for Laser Applications

P. Syam Prasad and P. Venkateswara Rao

Abstract In this chapter we will discuss in detail about structural and luminescence properties of heavy metal oxide-based TeO_2 glasses incorporated by rare-earth ions. The glasses were developed by conventional melt quenching method, and the structural analysis was done by XRD, FTIR, and Raman. The XRD patterns confirm the amorphous nature of the samples, and the FTIR characterization showed the formation of more non-bridging oxygen atoms in the glass network with the inclusion of rare-earth ions. Spectroscopic characterizations such as optical absorption, photo luminescence, and decay profile measurements were performed on the glasses. The Judd-Ofelt theory has been employed on optical absorption spectra to evaluate the Judd-Ofelt (*J-O*) intensity parameters Ω_λ ($\lambda = 2, 4, 6$). The measured J-O intensity parameters were used to determine the emission transition probability (A_R), stimulated emission cross section ($\sigma(\lambda_p)$), branching ratios (β_R), and radiative lifetimes (τ_R) for various emission transitions from the excited levels of rare-earth ions in the host glass network. The obtained results showed the use of the glasses for potential applications in the field of laser technology.

4.1 Introduction

A huge number of investigations have been carried out on tellurite-based glasses for the past decade, due to their potential applications in designing materials for optical communication systems, lasers, nonlinear optical, and optoelectronic devices. These applications of the tellurite glasses are attributed to their wide transparency window (0.4–6 μm), high linear and nonlinear refractive indices, low melting temperature, high thermal and chemical stability, high devitrification resistance, and low phonon energy [1–14]. Low phonon energy enables the tellurite host glasses to achieve high

P. Syam Prasad (✉)
Department of Physics, National Institute of Technology Warangal, Warangal, Telangana, India
e-mail: syamprasad@nitw.ac.in

P. Venkateswara Rao
Department of Physics, The University of the West Indies, Kingston, Jamaica

© Springer International Publishing AG, part of Springer Nature 2018
R. El-Mallawany (ed.), *Tellurite Glass Smart Materials*,
https://doi.org/10.1007/978-3-319-76568-6_4

45

quantum efficiency for the rare-earth ion doping. Further, tellurite glasses are not hygroscopic, which limits several applications of the phosphate glasses. Tellurite glasses have been exhibited extended transmission in the infrared region. Due to its easiness in the fiber drawing at low temperatures and its high solubility of rare-earth ions, tellurite glasses have been used in optical components for sensors, telecommunications, and medical applications [15]. It is well known that the extreme polarizability of the tellurium electron lone pair is responsible for the higher nonlinear optical susceptibility values of tellurite-based glasses [16, 17]. There is a chance of occurring of local redistribution of the electronic charge density, when the majority of glasses are exposed to high-intense laser pulse, and it affects the nanostructural modification in the glass matrix. Such sort of impact assumes a pivotal part during the process of nonlinear optical absorption, and this absorption eventually alters the refractive index of the glass material. Structural, thermal, and optical properties of glasses can be controlled by varying the composition of the glass.

Nowadays investigations on rare-earth ion-doped solid materials have been increasing due to their potential applications in various fields. In particular rare-earth-doped glasses have been considered as prime candidates for the essential optical applications such as color displays, optical amplifiers, sensors, optical data storage devices, lasers, and optoelectronic devices [18, 19]. Rare-earth (RE) ions exhibit 4f–4f or 5d–4f emission transitions and act as excellent activators in emitting sharp luminescence at different bands from UV to infrared region [20]. Among the trivalent lanthanide ions, dysprosium (Dy^{3+}) is one of the promising ions for commercial display applications and for laser devices as it exhibits several interesting optical properties with sharp emission bands in the visible and near-infrared regions. Dy^{3+} ions exhibit three emissions in the regions of blue (around 480 nm), yellow (575 nm), and red (660 nm) of visible light region which are due to their electronic transitions $^4F_{9/2} \rightarrow {}^6H_{15/2}$, $^4F_{9/2} \rightarrow {}^6H_{13/2}$, and $^4F_{9/2} \rightarrow {}^6H_{11/2}$, respectively [21]. In general blue and yellow emissions are more intense for the Dy^{3+} ion. The combination of suitable proportion of blue and yellow emissions can generate the white light. Further, in the CIE (Commission Intenationale de l'Eclairage) 1931 color chromaticity diagram, the line linking the blue and yellow wavelengths generally goes through the white light region [22]. Thus, the chromaticity coordinates for the Dy^{3+} ion-doped glasses can be tuned to the white light zone by the selection of a suitable intensity ratio of yellow to blue which can be useful for white light-emitting applications. Currently there has been a significant importance for the fabrication of Dy^{3+} ion-doped glasses for white LED's due to their high efficiency, low power consumption, long lifetime, low cost, and environmental-friendly nature [23]. A suitable composition of glass host for doping Dy^{3+} ions provides the possibility of the extraction of primary colors.

In the present chapter, we report on the spectroscopic properties of tellurite-based TeO_2-Sb_2O_3-WO_3 glass system doped with varied Dy^{3+} ion concentrations. So far, no spectroscopic properties were reported on Dy^{3+} ion-doped TeO_2-Sb_2O_3-WO_3 glass system. As our glass host consists of heavy metal oxides, it possesses low phonon energy, large emission cross sections, and less non-radiative losses, resulting in high quantum efficiencies for the electronic transitions between the Dy^{3+} ion energy levels.

This chapter discusses structural and spectroscopic properties of particular RE-doped tellurite glasses for laser applications in detail. Since it is highly impossible to cover all RE ion-doped tellurite glasses in a single chapter, we restricted for only Dy^{3+} ions.

4.2 Experimental

Dysprosium ion-doped glasses were prepared in the compositions of $(75-x)TeO_2$-$15Sb_2O_3$-$10WO_3$-xDy_2O_3 (where $x = 0.2, 0.5, 0.8, 1$, and 1.5 mol%) using conventional melt quenching and pressing method. The glasses are named as TSWD2, TSWD5, TSWD8, TSWD10, and TSWD15, respectively, and as a whole these samples are referred as TSWD glasses. The samples were prepared using the mixtures of high-purity chemicals TeO_2 (Sigma Aldrich, >99.9%), Sb_2O_3 (Sigma Aldrich, 99.99%), WO_3 (Himedia, 99.9%), and Dy_2O_3 (Sigma Aldrich, 99.99%). The appropriate molecular percentages of the components were weighed and mixed thoroughly by using an agate mortar and pestle to ensure the homogeneous mixture. Platinum crucible was used to melt the mixture in an electric furnace at 800 °C for 20 mins. The homogeneity of the mixture was maintained by stirring the melt for every 5 min. The stirring of the melt avoids the separation of the chemical components and also the formation of bubbles in the melt. The melt was quenched onto a preheated brass mold which had been kept at 300 °C to avoid the formation of internal strains, and it was immediately pressed with another brass plate. The obtained transparent samples were transferred to another furnace at a temperature of 300 °C and were annealed for 3 h in order to remove the retained internal strains and also to improve the mechanical strength of the samples. The samples were fine polished on both sides for further optical characterization yielding a thickness around 2.5 mm.

The amorphous nature of the glass samples was analyzed by the X-ray diffraction pattern recorded at room temperature by a PANalytical XPERT-PRO with Cu K_α ($\lambda = 1.5406$) radiation using the scanning rate of 4°/min and step size of 0.02 with the diffraction angle 2θ from 10° to 100°. The density values of the glasses were determined employing the Archimedes' principle using the xylene ($\rho = 0.86$ g/cc at room temperature) as the liquid for immersion. Refractive indices of the TSWD glasses were measured using an ellipsometer. The densities, refractive indices, and physical and other optical properties calculated as per the relations given in Ref. [24] are presented in Table 4.1. FTIR spectra of the samples were measured using a PerkinElmer 100S instrument in the wave numbers from 420 to 1100 cm^{-1} with a resolution of 1 cm^{-1} using the KBr pellet technique. Optical absorption spectra of each TSWD glass are obtained using a Jasco V-670 spectrophotometer at the wavelengths in the region 400–2000 nm with a resolution of 0.5 nm. The photoluminescence excitation and emission spectra of the TSWD glasses were

Table 4.1 Physical and optical parameters of TSWD glasses

S. no.	Parameter	TSWD2	TSWD5	TSWD8	TSWD10	TSWD15
1	Molar mass, M (g/mol)	187.039	187.679	188.319	188.746	189.813
2	Molar volume, V_m (cm^3/mol)	32.060	32.131	32.180	32.182	32.183
3	Path length, l (mm)	2.36	2.44	2.58	2.36	2.73
4	Density, ρ (g/cm^3)	5.834	5.841	5.852	5.865	5.898
5	Refractive index, n	2.122	2.153	2.141	2.166	2.192
6	Dy^{3+} ion concentration, C (10^{20} ions/cm^3)	0.376	0.937	1.497	1.871	2.807
7	Polaron radius, r_p (Å)	12.033	8.872	7.590	7.046	6.155
8	Inter-ionic distance, r_i (Å)	29.858	22.016	18.833	17.483	15.273
9	Field strength, F (10^{16}/cm^2)	0.456	0.838	1.146	1.330	1.742
11	Molar refractivity, R_m (cm^3)	17.270	17.604	17.517	17.754	17.995
10	Electronic polarizability, α_e (Å^3)	0.685	0.698	0.695	0.704	0.714
12	Reflection losses, R (%)	12.916	13.372	13.196	13.564	13.945
13	Dielectric constant (ε)	4.503	4.635	4.584	4.692	4.805
14	Metallization factor, M_t	0.461	0.452	0.456	0.448	0.441

measured by a Horiba Jobin Yvon Fluorolog-3-21 spectrofluorometer in the wavelength region 465–800 nm using the xenon arc lamp of 450 W as radiation source, and the decay profiles were measured using the xenon lamp of 60 W. All the experimental measurements were performed at room temperature.

4.3 Results and Discussion

4.3.1 Structural Properties: XRD and FTIR

X-ray diffraction patterns of the TSWD glasses presented in Fig. 4.1 shows a broad diffused hump at lower angles around 30° exhibiting the characteristic amorphous nature of the glasses.

Figure 4.2 shows the Fourier transform infrared (FTIR) spectra of the Dy^{3+} ion-doped TSWD glasses along with that of the pure base glass TSW15 [24]. FTIR characterization plays an important role in determining the structural properties of the glasses. The infrared spectra exhibit various intense absorption bands in the wave number region 420–1100 cm^{-1}. The intense absorption band around 652 cm^{-1} for all the TSWD glasses represents the stretching vibrations of Te-O-Te linkages in TeO$_4$ trigonal bi-pyramid units, and the band around 773 cm^{-1} is attributed to the asymmetric stretching vibrations of Te-O bonds in TeO$_{3+1}$ polyhedral or TeO$_3$ trigonal pyramid units [25, 26].

The shoulder appearing at around 848 cm^{-1} is assigned to the stretching vibrations of W-O-W linkages in WO$_4$ or WO$_6$ units [27]. The band positioned at around

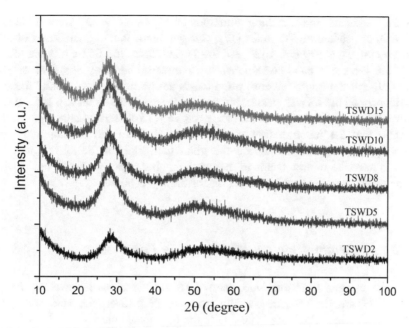

Fig. 4.1 X-ray diffraction patterns of TSWD glasses

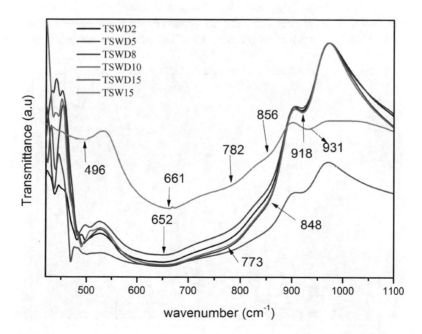

Fig. 4.2 FTIR spectra of base glass (TSW15) and dysprosium ion-doped TSWD glasses

918 cm^{-1} indicates the stretching vibrations of W=O and W-O$^-$ bonds in tetrahedral WO$_4$ or octahedral WO$_6$ units [17]. The symmetric bending vibrations of SbO$_3$ units, occurring at 579 cm^{-1} [28–30] are buried under the 652 cm^{-1} broadband, while the 496 cm^{-1} band corresponds to asymmetric bending vibrations of SbO$_3$ units [29]. All these infrared absorption bands are observed to be around the same positions for all the TSWD glasses, but the bands are shifted around 10 cm^{-1} toward lower wave number from that of the base glass TSW15 as shown in the Fig. 4.2. This indicates that the incorporation of Dy^{3+} ions leads to the formation of more non-bridging oxygen atoms inside the glass network. On the other hand, it is found to have the narrow bands in the wave number region 420–500 cm^{-1} which can be assigned to the bending vibrations of the Dy-O bonds.

4.3.2 Absorption Spectra and Bonding Parameters

Figure 4.3 presents the absorption spectra of TSWD glasses in the wavelength region of 400–2000 nm. Each spectrum exhibits various sharp absorption bands throughout the spectral region which originate due to the f-f-induced electric dipole transitions of Dy^{3+} ion from its ground state $^6H_{15/2}$ to various excited energy levels such as $^6H_{11/2}$, $^6F_{11/2}$, $^6F_{9/2}$, $^6F_{7/2}$, $^6F_{5/2}$, $^6F_{3/2}$, and $^4I_{15/2}$ located at 1689, 1280, 1098, 905, 804, 754, 473, and 453 nm, respectively. The absorption bands are assigned according to Carnall et al. [31]. The positions (in cm^{-1}) and assignment of the bands of all the

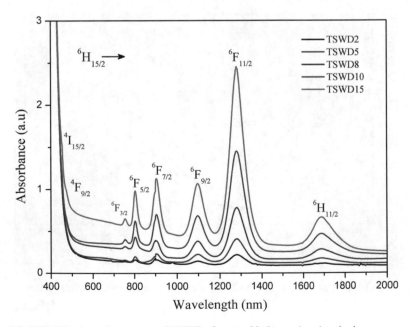

Fig. 4.3 VIS-NIR absorption spectra of TSWD glasses with dysprosium ion doping

Table 4.2 Peak positions (in cm^{-1}) of absorption transitions and bonding parameters ($\bar{\beta}$ and δ) of TSWD glasses

Transition from $^6H_{15/2} \rightarrow$	TSWD2	TSWD5	TSWD8	TSWD10	TSWD15	Aqua ion [32]
$^6H_{11/2}$	5888	5920	5918	5929	5920	5850
$^6F_{11/2}$	7800	7812	7809	7806	7809	7700
$^6F_{9/2}$	9099	9107	9111.6	9111.6	9107	9100
$^6F_{7/2}$	11,037	11,049	11,049	11,055	11,049	11,000
$^6F_{5/2}$	12,445	12,445	12,445	12,453	12,445	12,400
$^6F_{3/2}$	13,289	13,253	13,253	13,271	13,253	13,250
$^4F_{9/2}$	21,200	21,200	21,200	21,200	21,119	21,100
$^4I_{15/2}$	–	22,099	–	–	22,099	22,100
$\bar{\beta}$	1.0049	1.0050	1.0057	1.0063	1.0045	–
Δ	−0.4840	−0.5010	−0.5696	−0.6267	−0.4487	–

TSWD glasses are given in Table 4.2 along with that of the aqua ion [32]. The absorption bands in the wavelength region 400–500 nm are relatively weak as the host glass matrix exhibits a strong absorption near the absorption edge. The strong absorption band $^6H_{15/2} \rightarrow {}^6F_{11/2}$ located at around 1280 nm is found to be very sensitive to the rare-earth ion concentration, and as it follows the selection rule, $|\Delta S| = 0$, $|\Delta L| \leq 2$, and $|\Delta J| \leq 2$, this transition can be considered as hypersensitive transition [33]. In general the position and intensities of the absorption band corresponding to the hypersensitive transition are very sensitive to the rare-earth ion concentration and also to the host environment around the rare-earth ion.

The intensities of the absorption bands of the TSWD glasses are increasing with the Dy^{3+} ion concentration, while the peak positions of the absorption bands of all the TSWD glasses are located around the same positions. However a considerable shifting of the band toward lower wave number has been observed in comparison to that of the aqua ion. This shifting of the bands is called nephelauxctic effect which is due to the bonding nature of the Dy^{3+} ion with the surround ligand environment. The bonding nature of the Dy^{3+}-ligand bond was investigated by finding the nephelauxetic ratios and bonding parameters which are calculated from the absorption band positions of all the transitions. The nephelauxetic ratios of individual transitions were determined using the relation

$$\beta = \frac{\bar{\nu}_c}{\bar{\nu}_a} \tag{4.1}$$

where $\bar{\nu}_c$ and $\bar{\nu}_a$ are peak positions (in cm^{-1}) of the corresponding transitions of Dy^{3+} ion in glass matrix and aqua ion, respectively, which are given in Table 4.2. The bonding parameter δ of the each TSWD glass has calculated from the average values of β using the relation [34]

$$\delta = \frac{1-\overline{\beta}}{\overline{\beta}} \times 100 \qquad (4.2)$$

where $\overline{\beta}$ is the average value of the nephelauxetic ratios. The calculated values of nephelauxetic ratios and bonding parameters of the TSWD glasses are presented in Table 4.2. The δ value depends on the bonding nature of ligand-field environment, and it can be positive for covalent bonding and negative for ionic bonding [35]. The obtained negative δ values for all the TSWD glasses indicate the ionic bonding between the Dy^{3+} and the surrounding ligands in the glass matrix.

4.3.3 Judd-Ofelt Analysis

The intensity of the absorption bands of the TSWD glasses are represented by the experimental oscillator strengths (f_{exp}) which are determined using the integrated area under each absorption band following the relation [36]

$$f_{means} = \frac{2.303mc^2}{n\pi e^2} \int \varepsilon(v)dv = 4.32 \times 10^{-6} \int \varepsilon(v)dv \qquad (4.3)$$

where c is the velocity of light in vacuum; m and e are the rest mass and charge of an electron, respectively; N is the Avogadro's number; and $\varepsilon(\nu)$ is the molar absorptivity of the corresponding absorption band at wave number ν (in cm^{-1}) which are determined using Beer-Lambert's law. Based on the Judd-Ofelt theory [37, 38], the oscillator strengths (f_{cal}) for the induced electric dipole transitions from the ground state (ψJ) of Dy^{3+} ion to various excited states ($\psi'J'$) are calculated using the relation

$$f_{cal} = \frac{8\pi^2 mc\nu}{3he^2(2J+1)} \frac{n^2+2}{9n} \left(e^2 \sum_{\lambda=2,4,6} \Omega_\lambda \left(\psi J \parallel U^\lambda \parallel \psi'J' \right)^2 \right) \qquad (4.4)$$

where $e^2 \sum_{\lambda=2,4,6} \Omega_\lambda (\psi J \parallel U^\lambda \parallel \psi'J')^2$ is the line strength of electric dipole transitions, J and J' are the total angular momenta of the ground state and excited states, h is the Planck's constant, ν (cm^{-1}) is the wave number of the transition from initial state ψJ to the excited state $\psi'J'$, n is the refractive index of the sample, the factor $(n^2+2)/9n$ stands for the Lorenz electric field correction, Ω_λ ($\lambda = 2, 4, 6$) are the Judd-Ofelt intensity parameters, and $\parallel U^\lambda \parallel^2$ are the doubly reduced squared matrix elements of the unit tensor operator of the rank $\lambda = 2, 4,$ and 6 which are determined by using the method of intermediate coupling approximation for a transition $\psi J \rightarrow \psi'J'$ [36]. The measured experimental oscillator strengths (f_{exp}) and the calculated oscillator strengths (f_{cal}) for the present TSWD glasses are given in Table 4.3. In general, the magnitude of the reduced matrix elements ($\parallel U^\lambda \parallel^2$, $\lambda = 2, 4, 6$) of hypersensitive transitions for any rare-earth ion are very high. Consequently, the values of f_{exp} and f_{cal} are observed to be very high for the hypersensitive transition of all the TSWD glasses as shown in Table 4.3.

Table 4.3 Experimental (f_{exp}, 10^{-6}) and calculated (f_{cal}, 10^{-6}) oscillator strengths and r.m.s deviation (δ_{rms}, 10^{-6}) of Dy^{3+} ion-doped TSWD glasses

$^6H_{15/2} \rightarrow$	TSWD2		TSWD5		TSWD8		TSWD10		TSWD15	
	f_{exp}	f_{cal}	f_{exp}	f_{cal}	f_{exp}	f_{cal}	f_{exp}	f_{cal}	f_{exp}	f_{cal}
$^6H_{11/2}$	2.79	4.09	3.00	3.07	3.45	3.45	2.99	3.05	2.58	2.69
$^6F_{11/2}$	18.43	18.26	17.27	17.24	21.15	21.14	17.22	17.21	17.14	17.10
$^6F_{9/2}$	7.11	7.86	5.44	5.54	6.69	6.59	5.40	5.43	5.32	5.35
$^6F_{7/2}$	10.25	6.97	5.36	4.72	4.87	5.20	4.88	4.63	4.33	4.08
$^6F_{5/2}$	4.21	3.37	2.38	2.24	3.87	2.37	2.84	2.19	2.60	1.82
$^6F_{3/2}$	0.47	0.64	0.37	0.42	0.48	0.45	0.39	0.41	0.40	0.34
$^4F_{9/2}$	–	–	–	–	–	–	0.05	0.36	0.05	0.31
$^4I_{15/2}$	–	–	–	–	–	–	0.02	1.04	0.03	0.91
δ_{rms}	±0.163		±0.035		±0.066		±0.067		±0.044	

Judd-Ofelt intensity parameters are very much useful to study the local structure around the lanthanide ion in a glass matrix. The three phenomenological Judd-Ofelt (J-O) parameters (Ω_λ, $\lambda = 2, 4, 6$) were evaluated according to Judd-Ofelt theory by using the least square fitting for the equations of the experimental (f_{exp}) and calculated oscillator strengths (f_{cal}). The goodness of the fit has been found using

the δ_{rms} deviation ($\delta_{rms} = \sqrt{\dfrac{\sum_{i=1}^{N}\left(f_i^{exp} - f_i^{cal}\right)^2}{\sum_{i=1}^{N} f_{exp}^2}}$), and the obtained small δ_{rms} values

for the TSWD glasses indicate the quality of the fit between Eqs. (4.3) and (4.4) and also show the accuracy of the evaluated J-O parameters. The calculated J-O intensity parameters and their trends are given in Table 4.4 along with those of the Dy^{3+} ion-doped tellurite and other glass matrices [21–23, 41] reported earlier for comparison.

According to Jorgensen and Reisfeld [49], the value of Ω_2 intensity parameter depends on the covalence of ligand-metal bond as well as the asymmetry of local ion sites around the lanthanide ion, while the values of Ω_4 and Ω_6 are influenced by the bulk properties like rigidity and viscosity of the medium in which the ions are positioned. The factors covalency and asymmetry contribute the Ω_2 parameter, differently in different glass matrices [50]. The Ω_2 parameter of oxide glasses depends strongly on the asymmetry of ligand field surrounding the rare-earth ion, whereas the Ω_2 parameter for fluoride glass depends on the covalency [51, 52]. Further, the heavy metal oxides Sb_2O_3 and WO_3 containing in our TSWD glasses provide the higher asymmetry around the Dy^{3+} ion due to its high polarizability. Thus for the TSWD glasses, the higher values of Ω_2 parameter indicate the higher asymmetry of ion sites around the Dy^{3+} ion. Among the TSWD glasses, TSWD8 possesses higher magnitude of Ω_2 parameter (14.819×10^{-20} cm^2), and the Judd-Ofelt parameters of all the TSWD glasses are comparable to that of the reported Dy^{3+} ion-doped glasses as presented in Table 4.4 [21, 39–48].

Table 4.4 Judd-Ofelt intensity parameters Ω_2, Ω_4, and Ω_6 (10^{-20} cm^2) and their trend and spectroscopic quality factor (Ω_4/Ω_6) of TSWD glasses along with the reported Dy^{3+} ion-doped glasses

Glass	Ω_2	Ω_4	Ω_6	Ω_λ trend	Ω_4/Ω_6	References
TSWD2	12.404	3.458	5.224	$\Omega_2 > \Omega_6 > \Omega_4$	0.662	Present work
TSWD5	12.202	2.644	3.375	$\Omega_2 > \Omega_6 > \Omega_4$	0.783	Present work
TSWD8	14.819	3.726	3.607	$\Omega_2 > \Omega_4 > \Omega_6$	1.033	Present work
TSWD10	12.090	2.553	3.277	$\Omega_2 > \Omega_6 > \Omega_4$	0.779	Present work
TSWD15	11.359	3.087	2.663	$\Omega_2 > \Omega_4 > \Omega_6$	1.160	Present work
PbF$_2$-WO$_3$-TeO$_2$	5.190	1.930	1.070	$\Omega_2 > \Omega_4 > \Omega_6$	1.803	[39]
TWZDy10	6.910	0.990 .	1.010	$\Omega_2 > \Omega_6 > \Omega_4$	0.980	[40]
TeWK	14.310	9.427	1.682	$\Omega_2 > \Omega_4 > \Omega_6$	5.605	[41]
Dy:KLTB	9.86	3.31	2.41	$\Omega_2 > \Omega_4 > \Omega_6$	1.41	[42]
BTLN0.5D	6.329	1.715	1.141	$\Omega_2 > \Omega_4 > \Omega_6$	1.503	[43]
BLND	11.99	3.92	3.94	$\Omega_2 > \Omega_6 > \Omega_4$	0.995	[44]
CFBDy10	5.98	2.33	2.33	$\Omega_2 > \Omega_4 = \Omega_6$	1.00	[45]
G1 glass	14.83	4.01	2.87	$\Omega_2 > \Omega_4 > \Omega_6$	1.397	[46]
PbPKANDy10 glass	11.74	2.64	2.86	$\Omega_2 > \Omega_6 > \Omega_4$	0.923	[21]
TTWD10	3.37	0.30	1.07	$\Omega_2 > \Omega_6 > \Omega_4$	0.280	[47]
PKAZFDy	14.11	3.07	1.95	$\Omega_2 > \Omega_4 > \Omega_6$	1.574	[48]

4.3.4 Excitation-Emission Spectra

The intensity of photoluminescence emission of rare-earth ions depends on the excitation wavelength. Thus the excitation spectrum was recorded for the glass TSWD5 at wavelengths in the region 325–550 nm monitoring by a radiation of wavelength, 576 nm. Six distinguishable bands are observed with peak positions centered at 352, 365, 387, 426, 453, and 473 nm as presented in Fig. 4.4.

These excitation bands are due to the transitions from ground energy level $^6H_{15/2}$ to various excited energy levels $^5P_{7/2}$, $^5P_{9/2}$, $^4F_{7/2}$, $^4G_{11/2}$, $^4I_{15/2}$, and $^4F_{9/2}$, respectively. Among these excitation bands, the band at 453 nm corresponding to the transition $^6H_{15/2} \rightarrow {}^4I_{15/2}$ was found to be intense, and thus it was used as an excitation source to trace the emission spectra for all the TSWD glasses. The luminescence properties of the TSWD glass system were determined using the emission spectra recorded at room temperature in the wavelength region 465–800 nm as shown in Fig. 4.5. Each emission spectrum exhibits two intense bands, one in blue region at 484 nm ($^4F_{9/2} \rightarrow {}^6H_{15/2}$) and the other in yellow region at 574 nm ($^4F_{9/2} \rightarrow {}^6H_{13/2}$). Along with these bands, the emission spectra also contains the less intense bands at wavelengths 663 and 752 nm corresponds to the transitions from excited level $^4F_{9/2}$ of Dy^{3+} ion to its lower excited energy levels $^6H_{11/2}$ and $^6H_{9/2}$, respectively. The intensity of the emission bands is found to increase with increasing Dy^{3+} ion concentration up to 0.8 mol%, and then it decreases for higher concentrations 1 and 1.5 mol% which can be considered due to luminescence

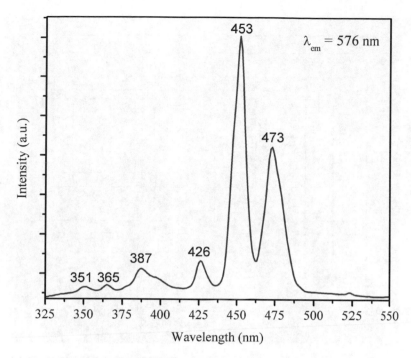

Fig. 4.4 Excitation spectra of TSWD5 glass monitoring at an emission wavelength $\lambda_{em} = 576$ nm

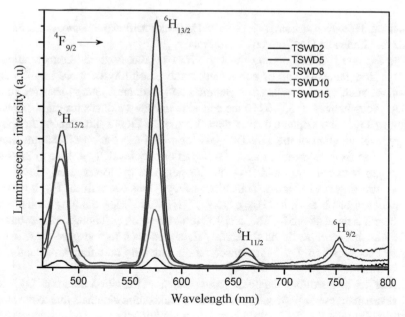

Fig. 4.5 Photoluminescence spectra of dysprosium ion-doped TSWD glasses measured by exciting at 453 nm

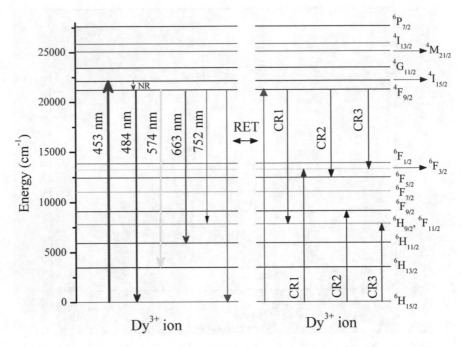

Fig. 4.6 Partial energy level diagram of TSWD8 glass showing emission transitions and also the different type of energy transitions quenching the luminescence at 453 nm excitation

quenching. However no change is observed in peak position or shape of the emission bands for change of Dy^{3+} ion concentration.

The average distance among the Dy^{3+} ions decreases as the concentration of the Dy^{3+} ion increase, and as consequently the possibility of more non-radiative transitions such as cross-relaxation channels and resonance energy transfer among the Dy^{3+} ions increases [53]. Thus the emission intensity reduces for higher concentration of Dy^{3+} ion content further than 0.8 mol%. Figure 4.6 depicts the partial energy level diagram of the TSWD8 glass for which a radiation of 453 nm excites the Dy^{3+} ion from its ground state to its higher energy level $^4I_{15/2}$. Due to the small energy gap between $^4I_{15/2}$ and $^4F_{9/2}$, the ion jumps to its lower energy level $^4F_{9/2}$ without emitting any radiation. Transitions of Dy^{3+} ions occur from $^4F_{9/2}$ level to its lower energy levels such as $^6H_{15/2}$, $^6H_{13/2}$, $^6H_{11/2}$, and $^6H_{9/2}$ emitting radiations of wavelength around 484, 574, 663, and 752 nm, respectively. Though, the probability of the transitions depends on different parameters such as the glass host matrix, chemical composition, Dy^{3+} ion concentration, pumping radiation wavelength, and heat treatment.

Figure 4.6 also exhibits the possible non-radiation transition modes of Dy^{3+} ions such as resonance energy transfer and the cross-relaxation channels due to the strong interaction among the Dy^{3+} ions when the separation between the ions decreases.

From Fig. 4.5, it is observed that the intensity of the emission band at 574 nm related to the transition $^4F_{9/2} \rightarrow {}^6H_{13/2}$ is found to be very sensitive to the rare-earth ion environment, and it obeys the selection rules $\Delta L = \pm 2$ and $\Delta J = \pm 2$; thus, it is called as hypersensitive transition [54]. In general the intensity of the hypersensitive transition is influenced by the ligand environment surrounding the Dy^{3+} ions and dominates the intensity of the other emission transitions. Yellow emission $^4F_{9/2} \rightarrow {}^6H_{13/2}$ being an electric dipole in nature dominated the blue emission $^4F_{9/2} \rightarrow {}^6H_{15/2}$ (magnetic dipole transition) for all the prepared TSWD glasses indicating the higher asymmetry of Dy^{3+} ion site [55, 56].

4.3.5   Radiative Properties

The luminescence properties of the Dy^{3+} ion-doped TSWD glasses were evaluated by determining the parameters such as spontaneous emission radiative transition probabilities (A_R) radiative lifetimes (τ_R), stimulated peak emission cross section ($\sigma(\lambda_P)$), and branching ratios (β_R). The J-O intensity parameters and the refractive index values of the TSWD glasses were used to calculate the radiative properties. Radiative transition probabilities of the TSWD glasses from initial level ψJ ($^4F_{9/2}$) to various energy levels $\psi'J'$ ($^6H_{J'}$, $J' = 15/2, 13/2, 11/2$, and $9/2$) were determined from the following equation [21]

$$A_R(\psi J, \psi' J') = \frac{64\pi^4 \nu^3}{3h(2J+1)} \left[\frac{n(n^2+2)^2}{9} S_{ed} + n^3 S_{md} \right] \qquad (4.5)$$

where S_{ed} is the electric dipole line strength, S_{md} is the magnetic dipole line strength, ν is the peak position of the emission, the factor $n(n^2 + 2)^2/9$ is for the local-field correction for the electric dipole transitions, and n^3 for magnetic dipole transitions. The electric (S_{ed}) and magnetic (S_{md}) dipole line strengths are obtained using the relations

$$S_{ed} = e^2 \sum_{\lambda=2,4,6} \Omega_\lambda \left(\psi J \| U^\lambda \| \psi' J' \right)^2 \qquad (4.6)$$

$$S_{md} = \frac{e^2 h^2}{16\pi^2 m^2 c^2} \left(\psi J \| L + 2S \| \psi' J' \right)^2 \qquad (4.7)$$

The total radiative transition probabilities of the TSWD glasses were obtained by adding up the transition probabilities of all the transitions from the excited energy level $^4F_{9/2}$ to various lower energy levels $\psi'J'$ as in the following equation

$$A_T(\psi J) = \sum_{\psi' J'} A_R(\psi J, \psi' J') \qquad (4.8)$$

The radiative lifetime (τ_R) of the excited level $^4F_{9/2}$ was determined by the inverse of the total radiative transition probability $A_T(\psi J)$,

$$\tau_R = \frac{1}{A_T(\psi J)} \tag{4.9}$$

The luminescence branching ratio (β_R) corresponding to the emission from the excited level ψJ ($^4F_{9/2}$) to a lower level $\psi' J'$ can be calculated from the ratio of the transition probability of the particular transition to the total transition probability of the ψJ ($^4F_{9/2}$) by using the expression

$$\beta_R = \frac{A_R(\psi J, \psi' J')}{A_T(\psi J)} \tag{4.10}$$

The experimental branching ratios of the TSWD glasses were measured by the ratio of integral intensity of the corresponding emission peak to the total integral intensity of the emission. An important laser parameter-stimulated emission cross section is a measure of potential laser performance, and it is used to estimate the laser gain of the prepared glasses. The stimulated emission cross section, $\sigma(\lambda_p)(\psi J \rightarrow \psi' J')$, of a transition having a transition probability of $A_R(\psi J, \psi' J')$ is

$$\sigma(\lambda_p) = \left(\frac{\lambda_p^4}{8\pi c n^2 \Delta\lambda_{\text{eff}}} \right) A_R(\psi J, \psi' J') \tag{4.11}$$

where λ_p is the wavelength of the emission peak corresponding to the particular transition and $\Delta\lambda_{\text{eff}}$ is the effective bandwidth of the emission band obtained by the relation, $\Delta\lambda_{\text{eff}} = \int \frac{I(\lambda)d\lambda}{I_{\text{max}}}$. The optical gain bandwidth ($\sigma(\lambda_p) \times \Delta\lambda_{\text{eff}}$) of the TSWD glasses were calculated using values of the stimulated emission cross section. The calculated values of radiative parameters (A_R, $\sigma(\lambda_p)$, β_R) for all the obtained emission transitions of the TSWD glasses are presented in Table 4.5 along with the values of effective bandwidths, experimental branching ratios (β_{exp}) and the optical gain bandwidths ($\sigma(\lambda_p) \times \Delta\lambda_{\text{eff}}$).

Calculated radiative lifetimes of the glasses are given in Table 4.7. The difference in the values of β_{exp} and β_R is due to the non-radiative transitions among the Dy^{3+} ions. The value of branching ratio of a transition is a measure of attaining stimulated emission, and the transition possessing $\beta_{\text{exp}} > 0.5$ can be useful for potential laser action [57]. The branching ratio of the emission transition $^4F_{9/2} \rightarrow ^6H_{13/2}$ of all the TSWD glasses is higher than that of the other transitions and also satisfies the condition $\beta_{\text{exp}} > 0.5$. Further, as presented in Table 4.5, the values of stimulated emission cross section and the optical gain bandwidth ($\sigma(\lambda_p) \times \Delta\lambda_{\text{eff}}$) of the transition $^4F_{9/2} \rightarrow ^6H_{13/2}$ of all the glasses are higher than the reported tellurite and other glass matrices doped with Dy^{3+} ion [47, 48, 58, 59]. Thus the glasses can be used for potential yellow laser applications and for continuous wave laser action [48].

Table 4.5 Peak wavelength of emission (λ_p, nm), spontaneous emission transition probability (A_R, s^{-1}), effective bandwidth ($\Delta\lambda_{eff}$, nm), stimulated emission cross section ($\sigma(\lambda_p)$, 10^{-22} cm^2), experimental (β_{exp}) and measured (β_R) branching ratios, and optical gain bandwidth ($\sigma(\lambda_p) \times \Delta\lambda_{eff}$, 10^{-28} cm^3) for the emission transitions of Dy^{3+} ion-doped TSWD glasses

Glass	$^4F_{9/2}\rightarrow$	λ (nm)	A_R	$\Delta\lambda_{eff}$	$\sigma(\lambda_p)$	β_{exp}	β_R	$\sigma(\lambda_p) \times \Delta\lambda_{eff}$
TSWD2	$^6H_{15/2}$	484	669.78	15.50	6.99	0.378	0.189	10.83
	$^6H_{13/2}$	574	2586.05	14.09	58.74	0.572	0.730	82.74
	$^6H_{11/2}$	663	286.33	17.89	9.12	0.050	0.081	16.31
TSWD5	$^6H_{15/2}$	481	473.87	15.95	4.55	0.365	0.147	7.26
	$^6H_{13/2}$	574	2375.03	14.94	49.42	0.557	0.739	73.82
	$^6H_{11/2}$	662	283.18	16.41	9.49	0.039	0.088	15.57
	$^6H_{9/2}$	753	81.90	11.74	6.42	0.039	0.025	7.54
TSWD8	$^6H_{15/2}$	484	503.67	15.82	5.06	0.320	0.136	8.00
	$^6H_{13/2}$	574	2772.22	15.78	55.21	0.596	0.748	87.13
	$^6H_{11/2}$	663	333.70	16.92	11.03	0.046	0.090	18.67
	$^6H_{9/2}$	752	97.00	14.10	6.37	0.038	0.026	8.98
TSWD10	$^6H_{15/2}$	483	464.54	16.07	4.45	0.393	0.144	7.15
	$^6H_{13/2}$	574	2393.56	14.92	49.26	0.584	0.742	73.50
	$^6H_{11/2}$	662	286.30	15.53	10.02	0.019	0.089	15.56
	$^6H_{9/2}$	751	82.97	11.37	6.56	0.004	0.026	7.47
TSWD15	$^6H_{15/2}$	484	413.59	14.55	4.31	0.371	0.133	6.27
	$^6H_{13/2}$	573	2330.68	14.76	47.01	0.605	0.750	69.40
	$^6H_{11/2}$	661	282.48	15.43	9.66	0.021	0.091	14.90
	$^6H_{9/2}$	751	82.04	10.84	6.65	0.003	0.026	7.21

4.3.6   Yellow to Blue Intensity Ratio and White Light Generation

Dy^{3+} ion has been given importance in solid-state lighting applications due to its emission of the primary colors which are required for the generation of white light. In general white light can be produced by the proper combination of blue and yellow light. The ratio between integrated intensity of yellow emission band to that of blue emission band for Dy^{3+} ion-doped glasses are called intensity ratio, and it is used to analyze the generation of white light. The intensity ratio mainly depends on the excitation source, Dy^{3+} ion concentration, and the host glass matrix. The calculated Y/B intensity ratios of the TSWD glasses are given in Table 4.6.

The *Y/B* intensity ratios of the glasses are found to have values 1.51, 1.53, 1.86, 1.49, and 1.63 for TSWD2, TSWD5, TSWD8, TSWD10, and TSWD15 glasses, respectively, and are near to 1 that represents the generation of light in white region. However the values of the intensity ratios are found to increase with Dy^{3+} ion concentration up to 0.8 mol% which indicates the increase of asymmetry of Dy^{3+} ion local environment [59]. The larger values of the Y/B intensity ratios of the TSWD glasses indicate the higher asymmetric nature of the ligand field around the Dy^{3+} ions [21, 61] which is also confirmed from the higher values of Ω_2 intensity parameters.

Table 4.6 Yellow to blue intensity ratio (*Y/B*), CIE color chromaticity coordinates (*x*, *y*), and correlated color temperature (CCT, *K*) of the light emitted by the TSWD glasses under the excitation of 452 nm light

Glass	*Y/B* ratio	*x*	*y*	CCT	References
TSWD2	1.51	0.35	0.40	4971	Present work
TSWD5	1.53	0.36	0.40	4687	Present work
TSWD8	1.86	0.38	0.42	4265	Present work
TSWD10	1.49	0.33	0.38	5591	Present work
TSWD15	1.63	0.32	0.35	6039	Present work
TTWD10	2.02	0.38	0.41	–	[47]
LSBP0.5Dy	1.81	0.39	0.42	3943	[59]
TeWK	1.29	0.34	0.34	5178	[41]
0.5DZTFB	1.12	0.36	0.41	4769	[60]

4.3.7 CIE Chromaticity Coordinates and CCT Values

The emitting color of the TSWD glasses excited by 453 nm has been investigated using the Commission Internationale de l'Eclairage (CIE) 1931 chromaticity diagram. The color chromaticity coordinates of the glasses were determined using the tristimulus values X, Y, and Z which are calculated from the luminescence spectra of the TSWD glasses [50]. The obtained CIE color chromaticity coordinates are (0.35, 0.40), (0.36, 0.40), (0.38, 0.42), (0.33, 0.38), and (0.32, 0.35) for TSWD2, TSWD5, TSWD8, TSWD10, and TSWD15 glasses, respectively, as presented in the CIE 1931 chromaticity diagram shown in Fig. 4.7.

Interestingly all these chromaticity coordinates are found to be in the white light region as shown in Fig. 4.7. The coordinates of the glasses TSWD10 and TSWD15 are located very near to the equal energy point (*x* = 0.33, *y* = 0.33) in the CIE diagram, which ensures the suitability of the glasses for white light-emitting applications under the excitation of 453 nm. The quality of the emitted light was evaluated by determining the correlated color temperature (CCT) values using the McCamy's approximate relation [57]

$$CCT = -449n^3 + 3525n^2 - 6823n + 5520.33 \qquad (4.12)$$

where $n = (x - x_e)/(y - y_e)$ is the inverse slope line and ($x_e = 0.332$ and $y_e = 0.186$) is the epicenter. The calculated CCT values are presented in Table 4.6 along with that of the reported Dy^{3+} ion-doped glasses. The CCT values are observed to decrease with the increase of Dy^{3+} ion content up to 0.8 mol% and then increases for higher concentrations. Further, the CCT value of the TSWD10 glass is very close to the reported sunlight (5500 K) showing the white bright lightening of the glass under the excitation of 453 nm. The obtained Y/B intensity ratios, chromaticity coordinates and the CCT values of the TSWD glasses along with that of the reported glass systems [41, 47, 59, 60] are presented in Table 4.6, which reveal that the studied TSWD glasses are potential candidates for the white LED and laser applications.

Fig. 4.7 CIE chromaticity diagram of the TSWD glasses

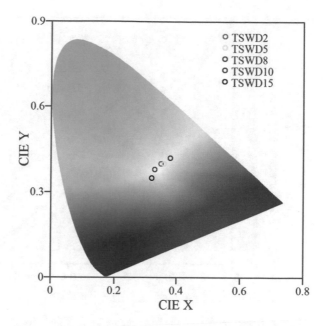

4.3.8 Decay Curve Analysis

The luminescence decay profiles of the TSWD glasses were measured to find out the lifetime of the excited energy level $^4F_{9/2}$ by monitoring an excitation at 453 nm and an emission at 574 nm. Figure 4.8 shows the logarithmic plots of normalized decay profiles of the TSWD glasses.

All the measured decay profiles are observed to be single exponential in nature. The experimental lifetimes (τ_{exp}) of the studied glasses are determined by fitting the decay profiles to the following equation [62]

$$I_t = I_0 e^{\frac{-t}{\tau_{exp}}} \tag{4.13}$$

where I_0 and I_t are the emission intensities of the of the decay curves at time $t = 0$ and $t = t$ s, respectively. The measured experimental lifetime values are presented in Table 4.7 along with the calculated radiative lifetime (τ_R) values.

The obtained τ_{exp} values are 195, 176, 153, 120, and 102 μs corresponding to the TSWD2, TSWD5, TSWD8, TSWD10, and TSWD15 glasses, respectively. The tabulated radiative lifetimes are found to be differing more from the experimental lifetimes which are due to the non-radiative transitions in the glasses. Even though our glass host possesses lower phonon energy reducing the multiphonon relaxation rates, the resonance energy transfer modes and the cross-relaxation channels are strongly influenced to quash the lifetimes of the TSWD glasses. The τ_{exp} values are observed to decrease with increasing Dy^{3+} ion concentration. This is due to the more non-radiative transitions due to the strong interaction among the Dy^{3+} ions.

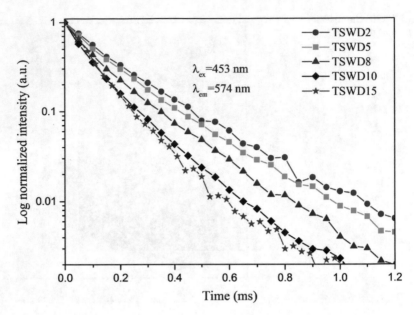

Fig. 4.8 Decay profiles of the $^4F_{9/2}$ energy level of Dy^{3+} ions in TSWD glasses as a function of Dy^{3+} ion concentration

Table 4.7 The experimental (τ_{exp}, μs) and calculated (τ_R, μs) lifetimes, quantum efficiency (η, %), and non-radiative transition rate (W_{NR}, s^{-1}) of the TSWD glasses as a function of Dy^{3+} ion content

Glass	τ_{exp}	τ_R	η	W_{NR}
TSWD2	195	282	69	1582
TSWD5	176	311	57	2466
TSWD8	153	266	58	2777
TSWD10	120	310	39	5108
TSWD15	102	321	32	6689

The quantum efficiency (η) is defined as the ratio of the number of photons emitted to that of the photons observed. For the glasses doped with rare-earth ions, η can be expressed as [63]

$$\eta = \frac{\tau_{exp}}{\tau_R} \qquad (4.14)$$

The calculated η values are presented in Table 4.7, and the η values are found to decrease with the increase of Dy^{3+} ion concentration due to the increase of non-radiative transition rates. The non-radiative transition rate (W_{NR}) can be calculated by the difference of inverses of the experimental and radiative lifetime values ($W_{NR} = 1/\tau_{exp} - 1/\tau_R$). The obtained W_{NR} values are decreasing with the increase of Dy^{3+} ion content as given in Table 4.7.

4.4   Conclusions

Heavy metal oxide-based, dysprosium ion-doped TSWD glasses were prepared by conventional melt quenching method and are investigated through FTIR, optical absorption, photoluminescence, and decay profile measurements. FTIR spectra of the glasses showed the existence of Dy-O bonds and also formation of more non-bridging oxygen atoms in the glass network with the incorporation of Dy^{3+} ions in the glass. The obtained negative values of the bonding parameters of the TSWD glasses show the domination of ionic nature of the ligand field around the Dy^{3+} ion. The higher magnitudes of Ω_2 parameters and yellow to blue intensity ratios (Y/B) of the glasses are the representation of higher asymmetric nature of the ligand environment around the Dy^{3+} ion. The luminescence spectra measured under the excitation of 453 nm exhibits the emission of primary colors around 484 (blue), 574 (yellow), and 663 nm (red) along with light at 752 nm. The intensity of emission increases with Dy^{3+} ion concentration up to 0.8 mol% and then decreases for higher concentrations due to non-radiative transitions such as resonance energy transfer and cross-relaxation modes among the Dy^{3+} ions. Spontaneous transition probability (A_R), stimulated emission cross section ($\sigma(\lambda_p)$), branching ratios (β_R), and radiative lifetimes (τ_R) were determined using the Judd-Ofelt intensity parameters. The emission transition $^4F_{9/2} \rightarrow {}^6H_{13/2}$ is found to possess higher values of branching ratios and stimulated emission cross sections recommending the use of the glasses for potential laser applications. The calculated CIE color chromaticity coordinates are found to be near to the equal energy point (0.33, 0.33), recommending the glasses for white light generation and white LED applications. Lifetime values and quantum efficiencies of the glasses were determined by measuring the decay profiles of the $^4F_{9/2}$ excited energy level under the excitation of 453 nm. The lifetime of the TSWD glasses decreases with increasing Dy^{3+} ion concentration and are in the range of 102–195 nm.

References

1. N.S. Hussain, G. Hungerford, R. El-Mallawany, M.J.M. Gomes, M.A. Lopes, N. Ali, J.D. Santos, S. Buddhudu, Absorption and emission analysis of RE^{3+} (Sm^{3+} and Dy^{3+}): Lithium Boro Tellurite glasses. J. Nanosci. Nanotechnol. **9**(6), 3672–3677 (2009)
2. I.Z.Hager, R.El-Mallawany, A.Bulou, Luminescence spectra and optical properties of TeO_2–WO_3–Li_2O glasses doped with Nd, Sm and Er rare earth ions. Phys. B Condens. Matter **406**(4), 972–980 (2011)
3. I.Z. Hager, R. El-Mallawany, Preparation and structural studies in the (70– x) TeO_2–$20WO_3$–$10Li_2O$–xLn_2O_3 glasses. J. Mater. Sci. **45**(4), 897 (2010)
4. M.M. El-Zaidia, A.A. Ammar, R.A. El-Mallwany, Infra-red spectra, electron spin resonance spectra, and density of $(TeO_2)100$– x–$(WO_3)x$ and $(TeO_2)100$– x–$(ZnCl_2)x$ glasses. Phys. Status Solidi A **91**(2), 637–642 (1985)
5. A. Abdel-Kader, R. El-Mallawany, M.M. Elkholy, Network structure of tellurite phosphate glasses: Optical absorption and infrared spectra. J. Appl. Phys. **73**(1), 71–74 (1993)

6. R. El-Mallawany, M. Sidkey, A. Khafagy, H. Afifi, Ultrasonic attenuation of tellurite glasses. Mater. Chem. Phys. **37**(2), 197–200 (1994)

7. M.M. Elkholy, R.A. El-Mallawany, Ac conductivity of tellurite glasses. Mater. Chem. Phys. **40**(3), 163–167 (1995)

8. A. El-Adawy, R. El-Mallawany, Elastic modulus of tellurite glasses. J. Mater. Sci. Lett. **15**(23), 2065–2067 (1996)

9. H.M.M. Moawad, H. Jain, R. El-Mallawany, T. Ramadan, M. El-Sharbiny, Electrical conductivity of silver vanadium tellurite glasses. J. Am. Ceram. Soc. **85**(11), 2655–2659 (2002)

10. R. El-Mallawany, Specific heat capacity of semiconducting glasses: Binary vanadium tellurite. Physica Status Solidi(a) **177**(2), 439–444 (2000)

11. R. El-Mallawany, A. Abd El-Moneim, Comparison between the elastic moduli of tellurite and phosphate glasses. Physica status solidi (a) **166**(2), 829–834 (1998)

12. R.S. Kundu, S. Dhankhar, R. Punia, K. Nanda, N. Kishore, Bismuth modified physical, structural and optical properties of mid-IR transparent zinc boro-tellurite glasses. J. Alloys Compd. **587**, 66–73 (2014)

13. Y. Wang, S. Dai, F. Chen, T. Xu, Q. Nie, Physical properties and optical band gap of new tellurite glasses within the TeO_2–Nb_2O_5–Bi_2O_3 system. Mater. Chem. Phys. **113**, 407–411 (2009)

14. D. Linda, J.R. Duclère, T. Hayakawa, M. Dutreilh-Colas, T. Cardinal, A. Mirgorodsky, et al., Optical properties of tellurite glasses elaborated within the TeO_2–Tl_2O–Ag_2O and TeO_2–ZnO–Ag_2O ternary systems. J. Alloys Compd. **561**, 151–160 (2013)

15. J.S. Wang, E.M. Vogel, E. Snitzer, Tellurite glasses: A new candidate for fiber devices. Opt. Mater. (Amst.) **3**, 187–203 (1994)

16. O. Noguera, S. Suehara, High nonlinear optical properties in TeO_2-based materials: Localized hyperpolarisability approach. Ferroelectrics **347**, 162–167 (2007)

17. S. Suehara, P. Thomas, a. Mirgorodsky, T. Merle-Méjean, J.C. Champarnaud-Mesjard, T. Aizawa, et al., Non-linear optical properties of TeO_2-based glasses: Ab initio static finite-field and time-dependent calculations. J. Non-Cryst. Solids **345-346**, 730–733 (2004)

18. C. Madhukar Reddy, B. Deva Prasad Raju, N. John Sushma, N.S. Dhoble, S.J. Dhoble, A review on optical and photoluminescence studies of RE^{3+} (RE=Sm, Dy, Eu, Tb and Nd) ions doped LCZSFB glasses. Renew. Sust. Energ. Rev. **51**, 566–584 (2015)

19. G. Blasse, B.C. Grabmaier, *Luminescent Materials* (Springer-Verlag, Berlin, 1994)

20. S. Tanabe, Optical transitions of rare earth ions for amplifiers: How the local structure works in glass. J. Non-Cryst. Solids **259**, 1–9 (1999)

21. K. Linganna, C. Srinivasa Rao, C.K. Jayasankar, Optical properties and generation of white light in Dy^{3+} doped lead phosphate glasses. J. Quant. Spectrosc. Radiat. Transf. **118**, 40–48 (2013)

22. S. Babu, V. Reddy Prasad, D. Rajesh, Y. Ratnakaram, Luminescence properties of Dy^{3+} doped different fluoro-phosphate glasses for solid state lighting applications. J. Mol. Struct. **1080**, 153–161 (2015)

23. S. Arunkumar, G. Venkataiah, K. Marimuthu, Spectroscopic and energy transfer behavior of Dy $^{3+}$ ion in B_2O_3-TeO_2-PbO-PbF-Bi_2O_3-CdO glasses for laser and WLED applications, Spectrochim. Acta Part A Mol. Biomol. Spectroscopy **136**, 1684–1697 (2015)

24. V.H. Rao, P.S. Prasad, P.V. Rao, L.F. Santos, N. Veeraiah, Influence of Sb_2O_3 on tellurite based glasses for photonic applications. J. Alloys Compd. **687**, 898–905 (2016)

25. A.E. Ersundu, M. Çelikbilek, S. Aydin, Characterization of B_2O_3 and/or WO_3 containing tellurite glasses. J. Non-Cryst. Solids **358**, 641–647 (2012)

26. Y. Zhou, Y. Yang, F. Huang, J. Ren, S. Yuan, G. Chen, Characterization of new tellurite glasses and crystalline phases in the TeO_2-PbO-Bi_2O_3-B_2O_3 system. J. Non-Cryst. Solids **386**, 90–94 (2014)

27. G. Upender, C. Sameera Devi, V. Kamalaker, V. Chandra Mouli, The structural and spectroscopic investigations of ternary tellurite glasses, doped with copper. J. Alloys Compd. **509**, 5887–5892 (2011)

28. M. Nalin, Y. Messaddeq, S.J.L. Ribeiro, M. Poulain, V. Briois, G. Brunklaus, C. Rosenhahn, B.D. Mosel, H. Eckert, R. Cedex, Structural organization and thermal properties of the Sb_2O_3 – $SbPO_4$ glass system. J. Mater. Chem. **14**, 3398–3405 (2004)
29. T. Kentaro, T. Hashimoto, T. Uchino, A.-H. Kim, T. Yoko, Structure and nonlinear optical properties of Sb_2O_3-B_2O_3 binary glasses. J. Ceram. Soc. Japan **104**, 1008–1014 (1996)
30. S.Y. Marzouk, F.H. Elbatal, Infrared and UV–visible spectroscopic studies of gamma-irradiated Sb_2O_3–B_2O_3 glasses. J. Mol. Struct. **1063**, 328–335 (2014)
31. W.T. Carnall, G.L. Goodman, K. Rajnak, R.S. Rana, A systematic analysis of the spectra of the lanthanides doped into single crystal LaF_3. J. Chem. Phys. **90**, 3443–3457 (1989)
32. W.T. Carnall, P.R. Fields, K. Rajnak, Electronic energy levels in the trivalent lanthanide aquo ions. I. Pr^{3+}, Nd^{3+}, Pm^{3+}, Sm^{3+}, Dy^{3+}, Ho^{3+}, Er^{3+}, and Tm^{3+}. J. Chem. Phys. **49**, 4424–4442 (1968)
33. C.K. Jorgensen, B.R. Judd, Hypersensitive pseudoquadrupole transitions in lanthanides. Mol. Phys. **8**, 281–290 (1964)
34. S.P. Sinha, *Complexes of the Rare Earths* (Pergamon, Oxford, London, 1966)
35. R.T. Karunakaran, K. Marimuthu, S. Surendra Babu, S. Arumugam, Structural, optical and thermal investigations on Dy^{3+} doped NaF-Li_2O-B_2O_3 glasses. Phys. B Condens. Matter **404**, 3995–4000 (2009)
36. C.K. Jayasankar, E. Rukmini, Optical properties of Sm^{3+} ions in zinc and alkali zinc borosulphate glasses. Opt. Mater. **8**, 193–205 (1997)
37. B.R. Judd, Optical absorption intensities of rare-earth ions. Phys. Rev. **127**, 750–761 (1962)
38. G.S. Ofelt, Intensities of crystal spectra of rare-earth ions. J. Chem. Phys. **37**, 511–520 (1962)
39. A. Mohan Babu, B.C. Jamalaiah, J. Suresh Kumar, T. Sasikala, L. Rama Moorthy, Spectroscopic and photoluminescence properties of Dy^{3+} –doped lead tungsten tellurite glasses for laser materials. J. Alloys Compd. **509**, 457–462 (2011)
40. G. Venkataiah, C.K. Jayasankar, Dy^{3+} –doped tellurite based tungsten-zirconium glasses: Spectroscopic study. J. Mol. Struct. **1084**, 182–189 (2015)
41. C.B. Annapurna Devi, S. Mahamuda, M. Venkateswarlu, K. Swapna, A. Srinivasa Rao, G. Vijaya Prakash, Dy^{3+} ions doped single and mixed alkali fluoro tungsten tellurite glasses for LASER and white LED applications. Opt. Mater. **62**, 569–577 (2016)
42. S.A. Saleem, B.C. Jamalaiah, M. Jayasimhadri, A. Srinivasa Rao, K. Jang, L. Rama Moorthy, Luminescent studies of Dy^{3+} ion in alkali lead tellurofluoroborate glasses. J. Quant. Spectrosc. Radiat. Transf. **112**, 78–84 (2011)
43. V. Uma, K. Maheshvaran, K. Marimuthu, G. Muralidharan, Structural and optical investigations on Dy^{3+} doped lithium tellurofluoroborate glasses for white light applications. J. Lumin. **176**, 15–24 (2016)
44. R.T. Karunakaran, K. Marimuthu, S. Surendra Babu, S. Arumugam, Dysprosium doped alkali fluoroborate glasses — Thermal, structural and optical investigations. J. Lumin. **130**, 1067–1072 (2010)
45. J. Suresh Kumar, K. Pavani, A. Mohan Babu, N.K. Giri, S.B. Rai, L. Rama Moorthy, Fluorescence characteristics of Dy^{3+} ions in calcium fluoroborate glasses. J. Lumin. **130**, 1916–1923 (2010)
46. N. Luewarasirikul, H.J. Kim, P. Meejitpaisan, J. Kaewkhao, White light emission of dysprosium doped lanthanum calcium phosphate oxide and oxyfluoride glasses. Opt. Mater. **66**, 559–566 (2017)
47. L. Jyothi, G. Upender, R. Kuladeep, D. Narayana Rao, Structural, thermal, optical properties and simulation of white light of titanium-tungstate-tellurite glasses doped with dysprosium. Mater. Res. Bull. **50**, 424–431 (2014)
48. V.B. Sreedhar, D. Ramachari, C.K. Jayasankar, Optical properties of zincfluorophosphate glasses doped with Dy^{3+} ions. Phys. B Condens. Matter **408**, 158–163 (2013)
49. C.K. Jorgensen, R. Reisfeld, Judd-Ofelt parameters and chemical bonding. J. Less-Common Met. **93**, 107–112 (1983)

50. M. Vijayakumar, K. Marimuthu, Structural and luminescence properties of Dy^{3+} doped oxyfluoro-borophosphate glasses for lasing materials and white LEDs. J. Alloys Compd. **629**, 230–241 (2015)
51. H. Ebendorff-heidepriem, D. Ehrt, M. Bettinelli, A. Speghini, Effect of glass composition on Judd-Ofelt parameters and radiative decay rates of Er^{3+} in fluoride phosphate and phosphate glasses. J. Non-Cryst. Solids **240**, 66–78 (1998)
52. H. Ebendorff-Heidepriem, D. Ehrt, Tb^{3+} f-d absorption as indicator of the effect of covalency on the Judd-Ofelt Ω_2 parameter in glasses. J. Non-Cryst. Solids **248**, 247–252 (1999)
53. V.K. Rai, S.B. Rai, D.K. Rai, Optical studies of Dy^{3+} doped tellurite glass: Observation of yellow-green upconversion. Opt. Commun. **257**, 112–119 (2006)
54. B.C. Jamalaiah, J.S. Kumar, T. Suhasini, K. Jang, H.S. Lee, H. Choi, L. Rama Moorthy, Visible luminescence characteristics of Dy^{3+}-doped LBTAF glasses. J. Alloys Compd. **474**, 382–387 (2009)
55. J. Kaewkhao, N. Wantana, S. Kaewjaeng, S. Kothan, H.J. Kim, Luminescence characteristics of Dy^{3+} doped Gd_2O_3-CaO-SiO_2-B_2O_3 scintillating glasses. J. Rare Earths **34**, 583–589 (2016)
56. B. Shanmugavelu, V.V. Ravi Kanth Kumar, Luminescence studies of Dy^{3+} doped bismuth zinc borate glasses. J. Lumin. **146**, 358–363 (2014)
57. C.S. McCamy, Correlated color temperature as an explicit function of chromaticity coordinates. Color. Res. Appl. **17**, 142–144 (1992)
58. O. Ravi, C. Madhukar Reddy, B. Sudhakar Reddy, B. Deva Prasad Raju, Judd – Ofelt analysis and spectral properties of Dy^{3+} ions doped niobium containing tellurium calcium zinc borate glasses. Opt. Commun. **312**, 263–268 (2014)
59. R. Vijayakumar, G. Venkataiah, K. Marimuthu, Structural and luminescence studies on Dy^{3+} doped boro-phosphate glasses for white LED's and laser applications. J. Alloys Compd. **652**, 234–243 (2015)
60. P. Suthanthirakumar, K. Marimuthu, Investigations on spectroscopic properties of Dy^{3+} doped zinc telluro-fluoroborate glasses for laser and white LED applications. J. Mol. Struct. **1125**, 443–452 (2016)
61. G. Lakshminarayana, J. Qiu, Photoluminescence of Pr^{3+}, Sm^{3+} and Dy^{3+}-doped SiO_2-Al_2O_3-BaF_2-GdF_3 glasses. J. Alloys Compd. **476**, 470–476 (2009)
62. N. Vijaya, K. Upendra Kumar, C.K. Jayasankar, Spectrochimica acta part a : Molecular and biomolecular spectroscopy Dy^{3+} –doped zinc fluorophosphate glasses for white luminescence applications. Spectrochim. Acta Part A Mol. Biomol. Spectrosc. **113**, 145–153 (2013)
63. C. Basavapoornima, C.K. Jayasankar, P.P. Chandrachoodan, Luminescence and laser transition studies of Dy^{3+}:K-Mg-Al fluorophosphate glasses. Phys. B Condens. Matter **404**, 235–242 (2009)

Chapter 5
Optothermal Properties of Vanadate-Tellurite Oxide Glasses and Some Suggested Applications

Dariush Souri

Abstract Transition metal oxide-containing glasses (TMOGs) have special and unique optothermal properties. One can verify how the optical, thermal, and thermoelectric properties vary with composition. This work tries to give more insight into the subject of the thermal stability and its effect on the optothermal properties of some vanadate-tellurite oxide glasses. In other words, optical and thermal characterization of such glassy systems can be investigated versus the composition with the aim of finding the more potential candidates in optical applications. Moreover, besides thermal stability, different parameters such as elastic moduli, optical bandgap, molar volume (V_m), oxygen molar volume (V_O^*), oxygen packing density (OPD), molar refraction (R_m), metallization criterion (M), and the concentration of non-bridging oxygen ions (NBOs) can be evaluated and discussed as the most important factors on the properties and applications of a material. In brief, optical applications (such as active material in optical fibers) of oxide glasses need the high thermal stable glasses with narrower bandgap. It should be noted that any suggestion for optical applications needs precise determination of optical properties such as energy bandgap, which affect the evaluation of other related optical parameters and so in optical device manufacturing and applications; in the case of bandgap determination, derivative absorption spectrum fitting method (abbreviated as DASF) has been recently proposed; this method is briefly introduced in this work. Also, such thermal stable glasses with good thermoelectric properties are promising materials which can be used in solar cells and photovoltaic (PV) panels as heat pumps to elevate the PV efficiency. An attempt has been made to discuss these subjects, giving more light in the field of optothermal aspects.

D. Souri (✉)
Department of Physics, Faculty of Science, Malayer University, Malayer, Iran
e-mail: d.souri@malayeru.ac.ir

© Springer International Publishing AG, part of Springer Nature 2018
R. El-Mallawany (ed.), *Tellurite Glass Smart Materials*,
https://doi.org/10.1007/978-3-319-76568-6_5

5.1   Introduction

Transition metal oxide-containing glasses (TMOGs) have special and unique optothermal properties. One can verify how the optical, thermal, and thermoelectric properties vary with composition. Several works tried to give more insight into the subject of the thermal stability and its effect on the optothermal properties of some vanadate-tellurite oxide glasses. In other words, optical and thermal characterization of such glassy systems can be investigated versus the composition with the aim of finding the more potential candidates in optical applications. Moreover, besides thermal stability, different parameters such as elastic moduli, optical bandgap, molar volume (V_m), oxygen molar volume (V_O^*), oxygen packing density (OPD), molar refraction (R_m), metallization criterion (M), and the concentration of non-bridging oxygen ions (NBOs) can be evaluated and discussed as the most important factors on the properties and applications of a material.

Multicomponent Te-based and tellurite-vanadate glasses [1–28] (as series of transition metal oxide-containing glasses (TMOGs)) have attracted considerable attention among the oxide glasses like phosphate [29–35] and borate glasses [36–38], avoiding especially the ambient limitation of hygroscopic nature of phosphates. Te-based glasses are still attracting interest from the basic point of view of fundamental researches as well as due to the suggested applications in antibacterial materials [2], solar cells and waveguides [39], heat pumps in photovoltaic panels [40], acousto-optical materials [40 nocmech], laser [41], photochromic glasses [42, 43], and so forth. In the case of optical and optothermal applications, it is noticeable that any suggestion for optical applications needs precise determination of optical properties such as energy bandgap, which affect the evaluation of other related optical parameters and so in optical device manufacturing and applications; in the case of bandgap determination, derivative absorption spectrum fitting method (abbreviated as DASF) has been recently proposed [44] and employed in different glassy systems [11, 43–45] and also in nanomaterials [47, 48].

Research works on TeO_2-based glasses have been made by many groups, because they show several advantages when compared with other glasses. As mentioned, these glasses are peculiarly important due to their unique optical, thermal, mechanical, and chemical properties which encourage and allow the glass processing to be characterized and suggested in optothermal applications. Some of the characteristics/advantages of TeO_2-based glasses are as follows [27, 39, 48]:

  (i) Their low melting temperatures (generally $<\sim 1000\ ^\circ C$)
 (ii) Their relatively high ambient and chemical stability (durability) and their long-time resistance to ambient moisture
(iii) Their high vitreous stability, high glass-forming ability, and high solubility of different transition metal oxides
(iv) Their large optical transmittance window and other optical aspects such as high nonlinear optical properties and high refractive index

(v) Their low glass transition temperature and large width between glass transition and crystallization temperatures, in which made them as high thermal stable glasses

In the present chapter, some of our previous findings on binary and ternary tellurite-vanadate glasses are reviewed. Section 5.2 deals with the experimental details of methods for the fabrication of expected glassy samples and also their characterization. In Sect. 5.3, different approaches of bandgap determination were reviewed as well as the recently proposed new method (known as DASF approach). Also some optical and structural parameters are introduced. In Sect. 5.4, different tellurite-vanadate glasses are reviewed for their optical, calorimetric, thermal, and thermoelectric properties, aiming their evaluation for probable application in optical devices. Finally, in Sect. 5.5, a conclusion (as a summary of the results) and further comments on the potential of TMOGs in optical and optothermal applications are given.

5.2 Experimental Procedures for TeO$_2$-Based Glasses: Sample's Fabrication and Characterization Methods

The glassy systems, which are reviewed and used here, have the composition of $40TeO_2$-$(60-x)V_2O_5$-xA_yO_z (or A) (in which A_yO_z and A are another oxide and another element as the third component, respectively); the third components were MoO_3, NiO, Sb_2O_3, Ag_2O, Bi_2O_3, and Sb. Furthermore, the compositional range of x (in mol%) was different in each series depending to the glass-forming capability. All of the samples were prepared/fabricated by using the well-known rapid melt quenching method. All high-purity chemical materials of reagent grades (as starting raw materials) were bought from Merck & Co., without any further purification. Generally, after dehydration process which prevents any volatilization, during the melting and processing of each sample in porcelain/alumina crucible, the liquid was stirred every about 5 min to obtain the homogeneous melts and so to prevent the separation of the initial glass constituents; generally, melting temperatures of glass series were within the range of 640–980 °C, depending on the composition and the molar ratio of the components. Samples (corresponding to the needed characterization) were provided in bulk or film states. Film samples were obtained from the blowing of the viscous melt by a fine tube such as bore-silica. To obtain the bulk samples, the obtained melts were poured on to a steel or brace mold and immediately pressed by another polished mold (to obtain disks of uniform, parallel, and flat surfaces), where both blocks were kept at room temperature. The fabricated bulk glasses were immediately annealed at a suitable fixed temperature (corresponding to the composition and in better word by considering the glass transition temperature T_g) for a period of about 2 h to achieve more homogeneous structure and also to remove mechanical stresses resulting from the quenching process [50]; as mentioned before, annealing temperature must be chosen well below glass transition region and also well below glass transition temperature (~ at least 50 °C lower than T_g) to ensure the amorphous nature avoiding any crystallization occurrence; annealing process can assure the reliability and validity of the results of experiments on the structural parameters of glass such as those of thermal, mechanical, and elastic experiments.

Initial structural information of the under-reviewed samples were taken from XRD diffraction patterns of the fine powdered samples at room temperature, which confirmed their vitreous nature. Sometimes, to more analysis and to certify the XRD results, SEM or FESEM images were provided to obtain more light on their microstructure. In optical absorption experiments, the outputs of UV-visible spectra (~190–1100 nm) were recorded at room temperature from the amorphous layer samples; from these spectra, some important optical data such as optical bandgap, Urbach energy, and the nature of optical charge carrier transition were obtained to give more and valuable awareness on the energy band structure. Besides our glasses, the compositions of $60Bi_2O_3$-$(40-x)TeO_2$-xV_2O_5 ($x = 0$, 5, 10, and 15) abbreviated as BTVx [51] have been reviewed comparatively.

Some other worthy parameters in study and in application of a glassy material are characteristic temperatures such as glass transition temperature T_g, crystallization temperature T_{Cr}, and the onset temperature of crystallization process T_x, which were taken from differential scanning calorimetry (DSC) experiments at different heating rates; these temperatures help the determination of thermal and more structural information of the glass and so the thermal stability and glass-forming tendency. Furthermore, the existence of flat baseline before the glass transition region confirmed the amorphous nature of the studied glasses; more details of DSC experiments are given in Sect. 5.4.2.1.

Thermoelectric effect and so Seebeck coefficient (S) for some of the mentioned glass series was measured within the relatively wide temperature range. The Seebeck coefficient of the glasses containing MoO_3, Sb_2O_3, and Sb was measured in the range of 250–470 K in a calibrated system using a cryogenic system of Janis-CCS 450(USA) under suitable vacuum. The thermoelectric potential difference was measured by a digital multimeter (Tti-Thurlby, 1906 GP, UK, or Keithley 196 system DMM microvoltmeters), with a design in which two spring loaded copper-constantan thermocouples were placed on two opposite highly polished parallel faces of the glass (as hot and cold faces with a temperature gradient of about $\Delta T = 4$–6 K). To produce the temperature gradient, small heat sources (heaters having a DC power of about 7 W) were employed. Upon another research work by H. Mori and H. Sakata [52], in the case of glasses having the compositions of $(96-x)$ V_2O_5-$4Sb_2O_3$-$xTeO_2$ ($x = 33,40$, 52 and 71) and $(94-x)V_2O_5$-$6Bi_2O_3$-$xTeO_2$ ($x = 25$, 34, 64, and 74), high-temperature Seebeck coefficient has been estimated to be within the temperature range of 373–473 K, using two thermocouples of Cu-Co between two opposite surfaces of the sample which suffer a temperature gradient. To produce a gradient temperature, an electric furnace has been employed. In the mentioned work, temperature difference has been ΔT~5–10 K, in which the hot face of the samples has been placed near the coil of the electric furnace in the ambient pressure, and when $\Delta T > 10$ K, argon gas has been flowed on the opposite face. For a more clear study of Seebeck effect and some structural parameters such as oxygen molar volume, the quantity of reduced ratio of transition metal ions should be determined; its determination method (as titration) will be discussed in Sect. 5.4.3.1.

5.3 Different Approaches on the Determination of Optical Bandgap with the Emphasis on DASF Method; Introduction to Some Optical and Structural Parameters

As known, many theories in different scientific fields have been improved once or step by step during the time to remove the whole or some parts of the inadequacies or simplifications of the older versions. In the field of optical absorption and determination of optical bandgap, we encounter such revolutions, or in other words improvements have been occurred. What is noticeable is the use of an easy, rapid, reliable, and more precise approach for evaluation of the optical bandgap (E_g) and also the exact nature of charge carrier optical transitions, which is related to the direct outputs of experimental equipment, not on different side parameters. Crucially, the more accurate the evaluations are, the less unwanted errors there will be, resulting in great design and excellent device optimization. In continuation, based on literature and reviewing the older methods and some difficulties with them, the newest revised method (reported by D. Souri & Z. E. Tahan 2015 [44]) is introduced.

5.3.1 Tauc's Method

The energy band structure of semiconducting crystalline, amorphous, and nano-structured materials is fundamentally different; the existence of non-sharp band edges and so the appearance of band tails in the gap in the vicinity of band edges or deep in the gap mean the somewhat difficult bandgap determination of amorphous ones. Such problem arises from the complexity in the definition of the true optical bandgap, in which one can define the pseudo-gap.

Upon Tauc's formalism [53], the following equation has been introduced for absorption coefficient of semiconductors as:

$$\alpha \hbar \omega = B \left(\hbar \omega - E_{gap} \right)^m \tag{5.1}$$

where $\hbar \omega$, E_g, and m are correspondingly incident photon energy, optical bandgap, and the index which can possess different 1/2, 3/2, 2, and 3 values corresponding to the type of carrier transition; B is a constant called the band tailing parameter [54]. In Eq. 5.1, the absorption coefficient, $\alpha(\omega)$, has been defined as [54]:

$$\alpha(\omega) = \left(\frac{1}{d}\right) \ln \left(\frac{I_o}{I_t}\right) = \frac{2.303 A}{d} \tag{5.2}$$

where d, A, I_o, and I_t are thickness of the sample, optical absorbance, intensity of incident, and transmitted beam, respectively.

By using (choosing) the optimum m value, Tauc's plot as $(\alpha \hbar \omega)^{1/m}$ vs. $\hbar \omega$ results in determination of E_{gap} from the extrapolation of the linear part of plot to intercept $\hbar \omega$ at $(\alpha \hbar \omega)^{1/m} = 0$.

5.3.2   Absorption Spectrum Fitting Method (ASF)

As is clear from Tauc's formalism, one of the error sources in bandgap evaluation is the needed thickness of the sample/concentration of the solution. Such parameter cannot be measured precisely because of the limited accuracy of equipment and also due to the probable nonuniform thickness or concentration; so, absorption coefficient is not precise and affects the estimated bandgap values. Revised Tauc's formalism called as ASF method was suggested by D. Souri and K. Shomalian (2009) [55] for semiconductors; in the mentioned formalism, employing the Beer-Lambert law $(\alpha(\lambda) = (2.303/d)A)$ in Eq. 5.1, one can reach the following equation [55]:

$$A(\lambda) = C\lambda \left(\lambda^{-1} - \lambda_g^{-1}\right)^m \tag{5.3}$$

where $C = [B(hc)^{m-1} d/2.303]$, in which λ_g, h, and c are wavelength corresponding to the optical gap ($E_{gap}^{ASF} = hc/\lambda_g = 1239.83/\lambda_g$), Planck's constant, and the velocity of the light, accordingly. In this method, there is only the need for direct absorption data rather than the absorption coefficient.

Similar to Tauc's approach, the first one has to examine different values of m and obtain its optimum value and then achieve the optical bandgap by plotting the curve of $(A\lambda^{-1})^{1/m}$ vs. λ^{-1}. The advantage of ASF with respect to Tauc's is that of using the direct absorption spectra without any need for film thickness or the concentration of solution, which removes one of the error sources in bandgap determination.

5.3.3   Derivation of Absorption Spectrum Fitting Method (DASF)

Although Tauc's and ASF have been extensively used [11, 44, 53, 54], as stated in Sects. 5.3.1 and 5.3.2, one should examine different m values for the index of optical charge carrier transition, choose its optimum value by checking the linear correlation coefficient of the linear part of the plots, and then estimate the bandgap using the selected value of m. In other words, presumption of transition index is necessary, which is assisted with using of least square methods; such manner accompanies with some errors. In an advanced approach, we must avoid any need to sample thickness and also avoid any need to presumption of transition index, which surely affects bandgap values and so device optimizations. Upon these comments, revised ASF method abbreviated as DASF was proposed (as new method on determination of bandgap and exact type of carrier transition) by D. Souri and Z. E. Tahan 2015 [44]; this method only needs the direct output of a spectrometer as absorbance spectrum, preventing any need for any further factors. This approach has been recently used in different systems [11, 43–47].

To give more insight about DASF, formula extraction is reviewed in brief; Eq. 5.3 can be rewritten as follows:

$$\ln\left[A(\lambda)\lambda^{-1}\right] = \ln(C) + m \ln\left(\lambda^{-1} - \lambda_g^{-1}\right) \tag{5.4}$$

where $C = [B(hc)^{m-1} d/2.303]$, and then

$$\frac{d\left\{\ln\left[A(\lambda)\lambda^{-1}\right]\right\}}{d\left(\lambda^{-1}\right)} = \frac{m}{\left(\lambda^{-1} - \lambda_g^{-1}\right)} \tag{5.5}$$

As is clear from Eq. 5.5, a discontinuity is observed in the $\frac{d\left\{\ln\left[A(\lambda)\lambda^{-1}\right]\right\}}{d\left(\lambda^{-1}\right)}$ vs. (λ^{-1}) plot at $\lambda^{-1} = \lambda_g^{-1}$; employing the value of λ_g^{-1} at the discontinuity of such plot, bandgap value is directly obtained from $E_{gap}^{DASF} = 1239.83 \times \lambda_g^{-1}$ (in eV) without any presumption about the value of m (as the nature of transition).

As can be deduced from the abovementioned characteristics of DASF, this approach has no deficiency of linear extrapolation in determination of bandgap; as known, it is because any extrapolation involve the large sensitivity to the range over which the extrapolation is taken, while DASF avoids such inadequacy since the whole of spectrum is fitted. Now, upon Eq. 5.4, by employing the precise λ_g^{-1} values, the index of optical transition (m) can be estimated (without any presumption) from the slope of the plot of $\ln[A(\lambda)\lambda^{-1}]$ versus $\ln\left(\lambda^{-1} - \lambda_g^{-1}\right)$.

5.3.4 Some Optical and Structural Parameters: Optical Basicity, Electronic Polarizability, Molar Refractivity, Metallization Factor, and Refractive Index

- *Molar volume, optical basicity, and electronic polarizability*

The molar volume of a multicomponent glass is obtained using the following equation:

$$V_{\mathrm{m}} = \sum_i {}^{M_i}/_\rho \tag{5.6}$$

where $M_i = x_i\, m_i$ ($x_i\, C_i$ and m_i as molar concentration and the molecular weight of the ith component, respectively) represents the molar mass of glass.

Optical basicity of a glass can be obtained from [55]:

$$\Lambda_{th} == \sum_i x_i \Lambda_i \tag{5.7}$$

where $\Lambda_{TeO_2}, \Lambda_{V_2O_5}$ denotes optical basicity of ith component which can be taken from Ref. [1].

Polarizability might be considered as some microscopic and macroscopic chemical and physical properties such as optical absorption and ionic refraction (i.e., refraction for each ion which is proportional to the radius of ion). The oxide ion polarizability (α_0^{2-}) can be introduced as [55, 56]:

$$\Lambda_{th} = 1.67 \left(1 - \frac{1}{\alpha_0^{2-}}\right) \tag{5.8}$$

- *Molar refractivity and metallization factor*

The molar refractivity (R_{m}) (in proportionality with the polarizabilities of the glass constituent ions) is defined as follows [37, 55]:

$$R_{\mathrm{m}} = V_{\mathrm{m}} \left(\frac{n^2 - 1}{n^2 + 2}\right) \tag{5.9}$$

where V_{m} is the molar volume of glass and n is refractive index. Besides the different approaches (such as temperature dependence of electrical conduction, optical spectra, and thermoelectric experiments), one of the conditions/ways to predict the conductivity nature of a sample can be defined based on metallization factor; this parameter is given as [38]:

$$M = 1 - R_{\mathrm{m}}/V_{\mathrm{m}} \tag{5.10}$$

Then,

(i) If $R_{\mathrm{m}}/V_{\mathrm{m}} > 1$, solid has metallic behavior.
(ii) If $R_{\mathrm{m}}/V_{\mathrm{m}} < 1$, solid has semiconducting nature.

- *Oxygen packing density and oxygen molar volume*

Another noticeable structural parameter which can be used in explanation of optical absorption is oxygen packing density (OPD). OPD and molar volume depict the compactness/rigidity in glass structure; so these two parameters can give trends of the concentration of bridging (BOs) and non-bridging oxygen atoms (NBOs). OPD is given as [55]:

$$\text{OPD} = \rho N_O / M = N_O / V_m \tag{5.11}$$

where ρ, N_O, M, and V_m represent the density, the number of oxygen atoms per formula unit, molecular mass, and molar volume, accordingly.

In addition, another worthy structural quantity is the oxygen molar volume V_O^*, in which increase of V_O^* means the increase of NBOs and so increase in fragility [58]. V_O^* is defined as follows for TeO_2-V_2O_5- (A_yO_z or A) systems:

$$V_O^* = \frac{\left\{ (M_{V_2O_5} - 16 C_V) x_{V_2O_5} + M_{TeO_2} x_{TeO_2} + M_{A_yO_z \text{ or A}} x_{A_yO_z \text{ or A}} \right\}}{\left\{ \rho(5 - C_V) x_{V_2O_5} + 2 x_{TeO_2} + z x_{A_yO_z \text{ or A}} \right\}} \tag{5.12}$$

where M_i is the molar weight and x_i represents the mole fraction of each of the glass components. Also, $C_V = [V^{4+}]/V_{tot}$ is the ratio of reduced vanadium ions as $[V^{4+}]/V_{tot}$, which will be introduced in Sect. 5.4.3.1.

- *Refractive* index *and description of optical susceptibility* $\chi^{(3)}$

Optical bandgap (E_g) and refractive index (n) are in specific relation as proposed by Dimitrov and Sakka [59]:

$$\frac{n^2 - 1}{n^2 + 2} = 1 - \left(E_g/20 \right)^{1/2} \tag{5.13}$$

So, employing the data of E_g, n is estimated. Furthermore, another parameter related to bandgap and refractive index is known as nonlinear optical susceptibility $\chi^{(3)}$. This parameter is in inverse relation with bandgap [55, 58] and is a vital factor in suggesting a material for optical applications, as will be discussed in Sect. 5.4.

5.4 Optothermal Properties of Multicomponent Tellurite-Vanadate Oxide Glasses

Nowadays, to suggest a material in special application, some factors play important roles; in other words, different materials compete by their environmental and speed of operation and also in the point of view of economics. Additionally, high reliability and high durability in ambient atmosphere, simplicity, and cost in material preparation are vital. Upon these circumstances and with those properties mentioned in Sect. 5.1, multicomponent TeO_2-based glasses are beneficial materials, because they have some exceptional properties which other engineering materials lacked. These glassy systems, along with the search for new glasses with specific characteristics, are still evolving and comprehensive. Generally, with the view of special requests, different fields of verification are important and necessary; according to the optothermal field, optical, thermal, mechanical, and thermoelectric investigations are equally imperative to study the material features. Optothermal characterization of different tellurite-vanadate glasses is reviewed in this section.

5.4.1 Optical Bandgap and Some Related Optical and Structural Properties of TeO_2-V_2O_5-A_yO_z (or A) Glasses

Measuring the bandgap is important in the amorphous, crystalline, and nanostructured semiconductors. The analysis of optical absorption, especially the shape and shift of absorption edge, has been always an influential way in realizing the nature of optical charge carrier transitions and the energy band structure [54, 55, 60–64]; therefore, UV-visible spectroscopy is the most dominant instrument to achieve the band structure of semiconductors [65–67]. Because of the discrepancies between the reported data of optical bandgap for different materials, the great importance of material quality, and enough optical knowledge in optical device manufacturing [68], one must use a reliable and precise method. All of the methods for determination of optical bandgap and optical carrier transition feature use the crop of UV-visible spectroscopy; but, some methods need direct absorption spectrum, while others need additional factors (such as film thickness/the concentration of solution) besides the absorption. As stated in Sect. 5.3, Tauc's, ASF, and DASF were compared due to their deficiencies and advantages. DASF technique can be used, as a more accurate method, to escape non-precise reports on optical gap.

In this section, optical and some structural properties of our glasses with the composition of $40TeO_2$-$(60\text{-}x)V_2O_5$-xA_yO_z (or A) are reviewed. Upon the chemical formula of the third component (MoO_3, NiO, Sb_2O_3, Ag_2O, and Sb), glasses having these components were abbreviated as TVMx, TVNx, TVSx, TVAgx, and TVSbx, accordingly. All glasses have been prepared by rapid melt quenching method. x as the mole fraction of the third component was varied between 0 mol% and a maximum relating to the composition and in better word relating to the glass-forming region of each system. Table 5.1 denotes the variety in composition for different glass systems. Densities were measured by Archimedes' principle, and melting points (T_m) are those estimated during the melting in an oven in air (see Table 5.1).

From Table 5.1, it is observed that the mole fraction of the third component is more limited for systems containing Sb and Sb_2O_3. The more extended glass-forming region was devoted to compositions containing MoO_3 and Ag_2O, because of the good glass-forming ability of molybdenum and silver oxides. Related to the subject of optical bandgap and optical properties, it should be declared that the bandgap and the nature of optical carrier transition of TVMx, TVNx, and BTVx glasses have been obtained by using Tauc's method [53]; for those of TVSx and TVSbx, ASF approach [55] has been employed; but in system of TVAgx, DASF [44] was employed as the newest, accurate, and reliable approach and then compared with the outputs of ASF. Upon the discussions in Sect. 5.3, reported data on optical (stating the method used) and structural parameters of the studied glasses are presented in Table 5.1. Typically, besides the absorption spectra, ASF and DASF plots of TVAgx glass system are presented in Figs. 5.1a, b, and c. As is clear from Fig. 5.1c, in DASF manner, λ_g^{-1} corresponds to E_g^{DASF} which can be obtained

Table 5.1 Some tellurite-vanadate glasses, collected data of density (ρ), molar volume (V_m), melting temperature (T_m), optical bandgap (E_g) and the related determination approach, and also m values as the index of the nature of the optical transition of charge carriers

Glass abbreviation	Detailed composition	ρ (gr cm^{-3})	V_m (cm^3 mol^{-1})	Melting temperature T_m (°C)	E_g (eV)	Method of bandgap determination	m-index
TVMx	TVM0: 40TeO$_2$-60V$_2$O$_5$	3.857 [68]	45.722 [69]	600	1.70 [72]	Tauc	2
					–	ASF	
	TVM10: 40TeO$_2$-50V$_2$O$_5$-10MoO$_3$	3.880 [68]	44.325 [69]	630	1.9 [72]	Tauc	
					–	ASF	
	TVM20: 40TeO$_2$-40V$_2$O$_5$-20MoO$_3$	4.053 [68]	40.805 [59]	650	2.00 [72]	Tauc	
					2.03 [73]	ASF	
	TVM30: 40TeO$_2$-30V$_2$O$_5$-30MoO$_3$	4.290 [68]	37.667 [69]	680	2.17 [72]	Tauc	
					2.20 [73]	ASF	
	TVM40: 40TeO$_2$-20V$_2$O$_5$-40MoO$_3$	4.394 [68]	35.908 [69]	720	2.25 [72]	Tauc	
					2.26 [73]	ASF	
	TVM50: 40TeO$_2$-10V$_2$O$_5$-50MoO$_3$	4.596 [68]	33.506 [69]	750	2.55 [72]	Tauc	
					2.57 [73]	ASF	
	TVM60: 40TeO$_2$-60MoO$_3$	4.895 [68]	30.686 [69]	780	2.85 [72]	Tauc	
					2.86 [73]	ASF	

(continued)

Table 5.1 (continued)

Glass abbreviation	Detailed composition	ρ (gr cm^{-3})	V_m (cm^3 mol^{-1})	Melting temperature T_m (°C)	E_g (eV)	Method of bandgap determination	m-index
TVNx	TVN0: 40TeO$_2$ -60V$_2$O$_5$	3.856 [53]	44.858 [70]	720	2.02 [53]	Tauc	2
	TVN5: 40TeO$_2$ -55V$_2$O$_5$-5NiO	3.872 [53]	43.289 [70]	750	1.99 [53]		
	TVN10: 40TeO$_2$ -50V$_2$O$_5$-10NiO	3.900 [53]	41.604 [70]	820	1.83 [53]		
	TVN20: 40TeO$_2$ -40V$_2$O$_5$-20NiO	4.019 [53]	37.703 [70]	900	1.57 [53]		
	TVN30: 40TeO$_2$ -30V$_2$O$_5$-30NiO	4.243 [53]	33.189 [70]	980	1.64 [53]		
TVSbx	TVSb0: 40TeO$_2$ -60V$_2$O$_5$	3.710 [49]	46.623 [49]	680 [26, 49]	1.73 [49]	ASF	2
	TVSb5: 40TeO$_2$ -55V$_2$O$_5$-5Sb	3.759 [49]	46.240 [49]	700 [26, 49]	2.14 [49]		
	TVSb8: 40TeO$_2$ -52V$_2$O$_5$-8Sb	3.882 [49]	44.260 [49]	705 [26, 49]	1.57 [49]		
	TVSb10: 40TeO$_2$ -50V$_2$O$_5$-10Sb	3.888 [49]	43.836 [49]	710 [26, 49]	1.85 [49]		
	TVSb12: 40TeO$_2$ -48V$_2$O$_5$-12Sb	4.031 [49]	41.353 [49]	720 [26, 49]	–		
	TVSb15: 40TeO$_2$ -45V$_2$O$_5$-15Sb	4.144 [49]	40.320 [49]	730 [26, 49]	–		

	Composition						
TVSx	TVS0: 40TeO$_2$-60V$_2$O$_5$	3.710 [54]	46.623 [71]	680 [13, 71]	1.74 [54]	ASF	2
	TVS5: 40TeO$_2$-55V$_2$O$_5$-5Sb$_2$O$_3$	3.778 [54]	47.235 [71]	720 [13, 71]	1.98 [54]		
	TVS8: 40TeO$_2$-52V$_2$O$_5$-8Sb$_2$O$_3$	3.848 [54]	47.230 [71]	740 [13, 71]	1.72 [54]		
	TVS10: 40TeO$_2$-50V$_2$O$_5$-10Sb$_2$O$_3$	3.922 [54]	46.898 [71]	750 [13, 71]	1.37 [54]		
TVAgx	TVAg0: 40TeO$_2$-60V$_2$O$_5$	3.587 [43]	48.214 [43]	660 [55]	2.41 [55]	ASF	0.5
					2.35 [55]	DASF	0.623
	TVAg10:40TeO$_2$-50V$_2$O$_5$-10Ag$_2$O	4.038 [43]	44.066 [43]	680 [55]	2.29 [55]	ASF	0.5
					2.30 [55]	DASF	0.510
	TVAg20: 40TeO$_2$-40V$_2$O$_5$-20Ag$_2$O	4.756 [43]	38.457 [43]	700 [55]	2.16 [55]	ASF	0.5
					2.19 [55]	DASF	0.483
	TVAg30: 40TeO$_2$-30V$_2$O$_5$-30Ag$_2$O	5.126 [43]	36.657 [43]	710 [55]	2.41 [55]	ASF	0.5
					2.30 [55]	DASF	0.869
	TVAg40: 40TeO$_2$-20V$_2$O$_5$-40Ag$_2$O	5.717 [43]	33.736 [43]	720 [55]	2.72 [55]	ASF	0.5
					2.72 [55]	DASF	0.551
	TVAg50: 40TeO$_2$-10V$_2$O$_5$-50Ag$_2$O	6.398 [43]	30.925 [43]	730 [55]	-	-	-
					-	-	-

(continued)

Table 5.1 (continued)

Glass abbreviation	Detailed composition	ρ (gr cm^{-3})	V_m (cm^3 mol^{-1})	Melting temperature T_m (°C)	E_g (eV)	Method of bandgap determination	m-index
BTVx	BTV0: 60Bi$_2$O$_3$-40TeO$_2$	5.950 [50]	57.700 [50]	–	2.95 [50]	Tauc	0.5
	BTV5: 60Bi$_2$O$_3$-35TeO$_2$ -5V$_2$O$_5$	5.411 [50]	63.654 [50]	–	2.86 [50]		
	BTV10: 60Bi$_2$O$_3$-30TeO$_2$ -10V$_2$O$_5$	4.670 [50]	76.618 [50]	–	2.83 [50]		
	BTV15: 60Bi$_2$O$_3$-25TeO$_2$ -15V$_2$O$_5$	4.448 [50]	82.203 [50]	–	2.81 [50]		

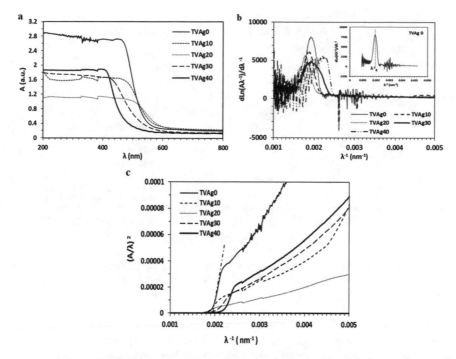

Fig. 5.1 $40TeO_2$-$(60-x)V_2O_5$-xAg_2O (TVAgx) glasses: (**a**) UV-Vis absorbance, (**b**) plots of $(A/\lambda)^2$ vs. λ^{-1} in ASF approach, which has been drawn using the optimum value of $m = 1/2$. As depicted by dashed-dotted line, band edge was obtained in the linear absorption region from the intercept of the linear part at $(A/\lambda)^2 = 0$; (**c**) depiction of DASF plots (upon Eq. 5.5), in which the inset clearly shows the discontinuity in DASF plot as indicated by λ_g^{-1} [44]

directly and accurately from the discontinuity observed in $\dfrac{d\{\ln[A(\lambda)\lambda^{-1}]\}}{d(\lambda^{-1})}$ vs. (λ^{-1}) curves.

From the data listed in Table 5.1 and Fig. 5.2, in TVMx systems a regularly increasing trend is observed for E_g [69], ranging from 1.7 to 2.85 eV. For that of TVSbx, there is a minimum of 1.57 eV for TVSb8, mentioning the value of 1.73 eV for TVSb0; in the case of TVSx glasses, E_g ranged from 1.74 to 1.37 eV, (corresponding to the start point of $x = 0$ and end point as $x = 10$ mol% of vanadium pentoxide), with a maximum of 1.98 eV for TVS5. BTVx glasses have been pointed out a decreasing trend with increase of V_2O_5 content, in which the data have been 2.83 eV for BTV0, 2.79 eV for BTV5, 2.67 eV for BTV10, and 2.53 for BTV15. Figure 5.2 shows the comparative plot of the variation trends of optical bandgap for different reviewed glasses, besides mentioning the theory and reference number which were used. As known, in optical applications such as optical fibers, we need glassy samples with narrower bandgap and hence with higher optical suscep-tibility χ^3; thus, reviewing Table 5.1, samples of the lowest optical bandgap can be initially/conditionally selected between all other here-studied samples as those have higher capability in optical devices such as optical fibers; but further

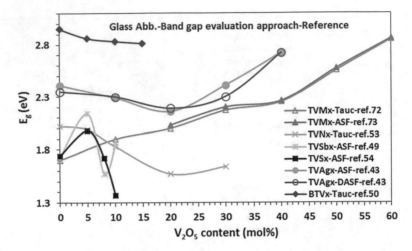

Fig. 5.2 Comparative plot of the variation trends of optical bandgap (E_g) against V_2O_5 content for different tellurite-vanadate glasses; in the legend of each series, bandgap approach method and its related reference were written

considerations/conditions on thermal and mechanical aspects must be included; what is necessary to be noted is that, in such applications, the only requirement is not the mentioned criterion; one should consider other vital conditions; in better word, device operation under different ambient situations means the necessity of good thermal durability and stability of the sample used. Such variants can guarantee the better signal transfer. Further discussion on the calorimetric aspects, thermal stability and its different criteria and so glass-forming tendency, will be presented in Sect. 5.4.2. From the compositional details of the reviewed glasses, one can sometimes observe different values of E_g for the same glasses; such results can be somewhat due to the preparation conditions; but, essentially it seems to be related to the estimation errors of older methods.

In continuation, an attempt is done to review some tellurite glasses; but for more clarity and to more easy tracking of the discussions, TVAgx system has been discussed in more detail (other systems have been briefly data reported in Tables 5.1 and 5.2); thus as mentioned before, the outputs of ASF and DASF performance on TVAgx glasses were shown in Figs. 5.1a, b, and c; summarized results of such comparison are listed in Table 5.1 and presented in Fig. 5.2, confirming the validity of DASF technique, which has the minimum error in evaluations; Fig. 5.2 implies on two points of view for TVAgx samples, namely, high coincidence of ASF and DASF results for TVAgx (with a few percent discrepancy between them) and, also, the more accurate DASF gap determination without any presumption on the optical charge carrier transition and no any use of extrapolation method. In the mentioned work, ASF plots were depicted for m = 1/2, which is the optimum m value that was obtained by testing different m values.

DASF approach has been employed in several glassy [11, 43–45] and nanostructured [47, 48] systems. For instance, because of the great importance of accurate

Table 5.2 Here-reviewed ternary tellurite-vanadate oxide glasses: oxygen packing density (OPD), oxygen molar volume (V^*_O), metallization parameter (M), optical basicity (Λ_{th}), oxide ion polarizability (α_o^{2-}), refractive index (n), and different glass thermal stability criteria of S_1, S_2, and S_3 at the definite heating rate

General glass name	Composition	OPD (g-atom L⁻¹)	V^*_O (cm³ mol⁻¹)	M	Λ_{th}	α_o^{2-}	n	Thermal stability criteria		
								$S_1 = T_{cr}-T_g$	$S_2 = T_x-T_g$	$S_3 = K_{gl} = (T_{cr}-T_g)/(T_m-T_{cr})$
TVSbx	TVSb0	81.50 [74]	12.281 [57]	0.431	–	–	2.238	39.46 [49]	21.56 [57]	0.13 [49]
	TVSb5	76.77 [74]	12.686 [57]	0.433			2.218	78.25 [49]	37.60 [57]	0.21 [49]
	TVSb8	76.82 [74]	12.655 [57]	0.440			2.193	–	–	–
	TVSb10	75.28 [74]	12.861 [57]	0.441			2.191	119.43 [49]	67.42 [57]	0.29 [49]
	TVSb12	77.38 [74]	12.621 [57]	0.448			2.163	191.29 [49]	81.56 [57]	0.65 [49]
	TVSb15	75.64 [74]	12.615 [57]	0.455			2.143	146.45 [49]	78.81 [57]	0.43 [49]
TVAgx	TVAg0	78.82 [27, 55]	12.70 [27]	0.343 [55]	1.026 [43]	2.593 [55]	2.598 [55]	48.22 [55]	29.17 [55]	0.129 [55]
	TVAg10	77.16 [27, 55]	12.99 [27]	0.339 [55]	1.030 [43]	2.609 [55]	2.617 [55]	51.75 [55]	33.20 [55]	0.135 [55]
	TVAg20	78.01 [27, 55]	12.36 [27]	0.330 [55]	1.036 [<3]	2.634 [55]	2.658 [55]	88.33 [55]	63.75 [55]	0.226 [55]
	TVAg30	68.20 [27, 55]	14.21 [27]	0.339 [55]	1.044 [43]	2.667 [55]	2.617 [55]	–	–	–
	TVAg40	65.21 [27, 55]	15.44 [27]	0.369 [55]	1.054 [43]	2.711 [55]	2.477 [55]	–	–	–
	TVAg50	58.21 [27, 55]	17.29 [27]	–	1.069 [43]	2.779 [55]	–	46.17 [55]	29.12 [55]	0.085 [55]

(continued)

Table 5.2 (continued)

General glass name	Composition	OPD (g-atom L^{-1})	V_O^* (cm^3 mol^{-1})	M	Λ_{th}	a_o^{2-}	n	Thermal stability criteria		
								$S_1 = T_{cr} - T_g$	$S_2 = T_{X} - T_g$	$S_3 = K_{gl} = (T_{cr}-T_g)/(T_m-T_{cr})$
BTVx	BTV0	45.06 [50]	–	0.376 [50]	1.303 [50]	4.552 [50]	2.441 [50]	–		
	BTV5	43.202 [50]	–	0.374 [50]	1.328 [50]	4.889 [50]	2.453 [50]			
	BTV10	39.192 [50]	–	0.360 [50]	1.373 [50]	5.628 [50]	2.514 [50]			
	BTV15	39.134 [50]	–	0.355 [50]	1.381 [50]	5.797 [50]	2.536 [50]			

scientific optical reports, as reported in the work on ZnSe nanocrystals (NCs) done by M. Molaei et al. [63], reported trend of bandgap (which has been obtained qualitatively) was revised and corrected by using DASF as a separate work by D. Souri et al. [47], which shows different trend in respect with the first qualitative reports of M. Molaei et al. [63].

As is obvious from Fig. 5.2, DASF outputs on TVAgx system showed two regions for bandgap such as (i) decreasing trend region from 2.35 eV to 2.19 eV for $0 \leq x$ (mol%) ≤ 20 and (ii) increasing trend region from 2.19 to 2.72 eV for $20 \leq$ (mol%) ≤ 40.

Generally, in oxide glasses, the behavior of bandgap is attributed to the change in the content of non-bridging oxygen (NBO) atoms. It is generally believed that the increase of NBOs changes the glass structure and then shifts the band edge toward red. Therefore, in TVAgx glasses or in any oxide glasses, any decrease in energy bandgap value might be due to the eruption of the bands and creation of dangling bands such as NBOs [11, 43, 49, 54]. Such explanations can be validate using the exploring of Urbach energy as the tail of the localized stated in the adjacent of the band edges; in other words, if the addition of the concentration of a component in a supposed oxide glass causes the decreasing of bandgap, it could be explained upon the increase of the lattice disorder and so the more extension of the localized states within the gap according to Mott and Davies [62]. Another worthy mentionable subject is the introducing of the m-index by using DASF bandgap; based on Eq. 5.4, employing the precise λ_g^{-1} values, the index of optical transition (m) can be estimated (without any presumption) from the slope of the plot of $\ln[A(\lambda)\lambda^{-1}]$ versus $\ln \left(\lambda^{-1} - \lambda_g^{-1} \right)$; such plots are presented in Fig. 5.3 for TVAgx system.

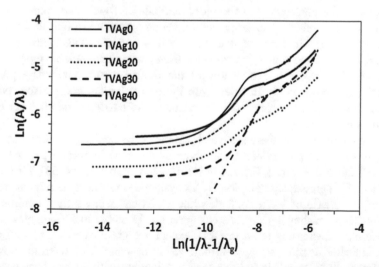

Fig. 5.3 m-index estimation of the compositions TVAgx in DASF approach, which is obtained by plotting $\ln(A\lambda^{-1})$ vs. $\ln(\lambda^{-1} - \lambda_g^{-1})$ using the correspond values of λ_g^{-1} values of each sample; upon Eq. 5.4, the slope of such plots indicate m (middle part of a curve corresponds to absorption region, as indicated by dashed-dotted line for $x = 30$ mol%) [44]

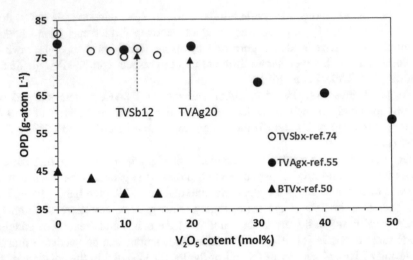

Fig. 5.4 Comparative plot of OPD versus V_2O_5 content for different tellurite-vanadate glasses; in the legend of each series, the related reference was mentioned

As is observed in Table 5.1, different glasses have been reviewed by several properties such as bandgap; the interface of such systems is the investigation of tellurite-vanadate glasses without the third component, which have nearly same compositions in details. As is clear from the data of Table 5.1, the scattered bandgap values have been reported for TeO_2-V_2O_5 glass in separate works, although the samples have been prepared by the same method as melt quenching. These scattered values show the inadequacy of the older methods of Tauc's and somewhat ASF, because of their defects (such as presumptions on optical transitions and especially the extrapolation) which are explained in detail in Sect. 5.3. This is not reasonable that in relatively same conditions, there be discrepant and not reliable reports. These points should move researchers toward the more accurate methods with more simplicity in performance. In brief, with the above limited comparison, one can validate the DASF as starting point in evaluation of bandgap and all other optical properties belong to E_g.

As mentioned before, there is several parameters recognizing the trend of optical and also structural properties, which some of the most important parameters are listed in Table 5.2. Figure 5.4 shows the OPD behavior of TVAgx [55], TVSbx [75], and BTVx [51] glasses as examples. BTVx system shows a fairly regular decrease in OPD with increase of V_2O_5 content, while TVSx cases present the nonlinear and complex behavior showing a drastic change at $x = 12$ mol%. In TVAgx glasses there is a relatively decreasing trend in OPD, except a slight increase at $x = 20$ mol%, which indicates its more strong and more stable structure. Moreover, the nonlinear optical susceptibility $\chi^{(3)}$ is in inverse proportionality with energy bandgap (or in direct proportionality with the refractive index), and so it elevates if the energy bandgap decreases [18, 27, 59]; so, higher $\chi^{(3)}$ and then lower E_g means the good optical capabilities in industry; such properties of this glass composition ($x = 20$ in

TVAgx) can be a reason to be introduced as promising material in optical device optimizations such as optical fibers; it should be mentioned that any suggestion of an optical material in optical fibers needs additional condition such as high thermal stability, which will be discussed in the next Sect. (Sect. 5.4.2.2); it means that simultaneous optimum capabilities should exist.

Several valuable parameters help us to decide about the advantages of glass, i.e., density, molar volume, optical basicity, molar refraction, polarizability, OPD, characteristic calorimetric temperatures, and thermopower.

Usually, density and the molar volume exhibit an inverse behavior. Such relation could be seen for different glass systems as listed in Tables 5.1 and 5.2.

Another helpful parameter is the optical basicity and can be explained in terms of change in NBOs and coordination number. There is an expected increase of bandgap with decrease in optical basicity, if a drastic and critical structural change does not occur. In the case of TVAgx, an increase in Ag_2O content (as a network modifier) causes two distinct bandgap regions; in the $0 \leq x$ (mol%) ≤ 20 region, the expected trends for E_g^{DASF} and optical basicity is observed, denoting the increase of NBOs; but, for $20 \leq x$ (mol%) ≤ 40, nevertheless of increasing in optical basicity, optical bandgap shows an increasing trend, which might be attributed to the structural modifications. Such structural changes could be observed in XRD patterns of these glasses as reported and explained in more detail in ref. [44] and also SEM images reported in Ref. [28]. The data of some other tellurite glasses are tabulated in Table 5.2.

Also, ion polarizability (α_0^{2-}) of oxide glasses can be estimated by Eq. 5.8, using the data of Λ_{th}. Generally, polarizability is defined in which to be related to some microscopic and macroscopic features like optical absorbance and ionic refraction (i.e., refraction for each ion in direct proportionality with the radius of ion). The data of polarizability are available for some glass systems in Table 5.2. For the tellurite-vanadate glasses containing silver oxide, the increasing trend of polarizability suggests the increase in ionic nature of glass with addition of Ag_2O, which is expected output and is in good accordance with the DC electrical conductivity data [76] on the same glasses (i.e., electrical conduction mechanism has shown a clear change from polaronic to mixed polaronic-ionic for $x \geq 20$, meaning the critical x value as also seen in optical bandgap). Using Eq. 5.9 for molar refractivity (R_m), one can identify the conductivity behavior of a glass by employing the metallization parameter $(M = 1 - R_m/V_m)$; conditions of $R_m/V_m > 1$ $R_m/V_m < 1$ are correspond to metallic and semiconducting behavior, respectively. For example, in TVAgx, the result of $R_m/V_m < 1$ has been obtained confirming the semiconducting behavior, which is in good matching with the absorption data [44] and also the temperature dependency of their electrical conductivity [76]. Also, BTVx and TVSbx systems show the semiconducting nature. Moreover, as is clear from Table 5.2, a critical molar ratio of silver oxide (x) could be observed from the trend of M, which in similar with the trend of E_g, two same compositional regions are denoted. The minimum value of M is devoted to $x = 20$ mol%, which might be interpreted as widening the width of both valence and conduction bands in the compositional range of $0 \leq x \leq 20$ and especially at $x = 20$; such interpretation is in accordance with the minimum bandgap of this glass with twenty percentage of silver oxide. A relevant optical parameter is the refractive index (n) which should have the inverse trend with the optical gap [38] as will be presented in continuation; employing Eq. 5.13, refractive index (n) at the absorption edge can be evaluated and discussed. Collected data of refractive index for different glasses are tabulated in Table 5.2, generally indicating the mentioned/expected inverse relation with energy bandgap. In the focused glass system of TVAgx, again and again one can observe the critical x value at 20 mol%, in reasonable matching with the previous optical, mechanical, elastic moduli, thermal, and structural aspects [27, 43, 55]; such explanations on the results for refractive index can be understood based on the data of OPD and V_O^*; collected data of these factors are presented in Table 5.2 in different oxide tellurite glasses. Having the data of ρ, N_O, M, and V_m (described in Sect. 5.3.4.), OPD can be estimated. Commonly, OPD and V_m are demonstrations of the compactness/rigidity of a glass structure and are reasonably in inverse relation. Trend of OPD or V_m signifies that glass of lower OPD has lower rigid/ compact structure (weakening of the structure), implying on the increment in free volume. Plots of OPD are depicted in Fig. 5.4 and the data of V_m are presented in Table 5.1. In Fig. 5.4 and Table 5.1, some Te-based glasses have been compared. It is clear the inverse proportionality of OPD and V_m. Also, as mentioned before, for TVAgx glasses, there is a behavior change at $x = 20$ as a slight increase, which is in good agreement with the previously explained data and will be justified with the results of oxygen molar volume. Alternatively, glass containing 20 mol% of silver oxide is introduced as a good candidate for optoelectronic application because of its strength and stable structure besides the optimum optical properties. On the other hand, glasses of higher OPD/lower molar volume/lower oxygen molar volume should have higher ultrasonic velocities and then higher mechanical/elastic moduli. In the case of TVAgx, such mechanical outputs, confirming the all before explained results, have been reported in Ref. [28]. Also, structural and elastic and microhardness properties of TVSbx [75] have been reported in relating to their optical properties. All of these findings reveal the general expected relations between the different quantities, as presented during the previous discussions. As TVAg20 is the optimum optical glass in TVAgx glasses, it has been found that TVSb12, TVS8, and BTV0 are optimum materials in the systems with the compositions of $40TeO_2\text{-}(60\text{-}x)V_2O_5\text{-}xSb_2O_3$ [55], $40TeO_2\text{-}(60\text{-}x)V_2O_5\text{-}xSb$ [50], and $(40\text{-}x)TeO_2\text{-}xV_2O_5\text{-}60Bi_2O_3$ [51], respectively.

Moreover, another valuable and informative parameter, which can guide the readers about the structural and related optical properties of a glass, is the oxygen molar volume (V_O^*) as expressed by Eq. 5.12. To evaluate V_O^*, it is necessary to measure $C_V = [V^{4+}]/V_{tot}$) as the ratio of reduced vanadium ions, which the method for its determination will be introduced in Sect. 5.4.3.1. Table 5.2 represents the comparative data of this parameter for different here-reviewed glasses. It is known that decreasing of V_O^* means the increase in glass stability or decrease in structural fragility. The valley value of V_O^* is devoted to TVAg20 in TVAgx glasses (~ 12.860 cm^3 mol^{-1}) and TVSb12 in TVSbx (~ 12.621 cm^3 mol^{-1} (. Thus glasses possessing lower V_O^* are higher thermal stable glasses, which confirms the results of

OPD (see Fig. 5.4). As stated before, additional calorimetric analysis is needed for more precise judgment about the promising materials in applications.

The refractive indices of glasses of TVSx, TVSbx, TVAgx, and BTVx have been reported at the absorption edge (see Table 5.2); but as reported in Ref. [54] for TeO_2-V_2O_5-NiO, one can obtain the dispersion of refractive index at the whole of wavelengths above the absorption edge by employing the Swanepoel method [77] which has been suggested based on Manifacier approach [78]. In this method one can benefit the probable interference pattern with a sharp fall of transmittance at the band edge; so, transmittance spectra in UV-Vis spectrum is used by plotting an upper and lower envelopes on the transmission spectrum in the weak absorption region beyond the absorption edge. From the envelopes, parameters such as T_{min} and T_{max} can be obtained as the values of the envelopes at the wavelengths in which the upper and lower envelopes at the experimental transmittance spectrum are tangent, correspondingly. Therefore, retaining the mentioned single oscillator model of Swanepoel, the refractive index $n(\lambda)$ can be calculated by:

$$n(\lambda) = \sqrt{P + \sqrt{P^2 - N_S^2}} \qquad (5.14)$$

where

$$P = 2N_S \left(\frac{T_{max} - T_{min}}{T_{max} \times T_{min}} \right) + \frac{N_S^2 + 1}{2} \qquad (5.15)$$

where N_S is the refractive index of the substrate of the understudied film.

5.4.2 Calorimetric Aspects and Thermal Stability

5.4.2.1 Calorimetric Analysis and Characteristic Temperatures

We can complete all aforementioned aspects and findings by introducing the calorimetric properties. Nowadays, glass compositions are extensively investigated, in which the increasing of the number of these glasses leads to the extension of their compositional range and also their usage; thus, efforts to get as much as possible information about their various properties such as thermal aspects seem to be mandatory, desirable, and satisfactory. On the other hand, nevertheless that crystalline materials are properly comprehended for their physical properties, glassy materials are not sufficient; moreover, in the case of amorphous materials, the extension of experimental investigations can somewhat remove the difficulties raised from theoretical complications and fairly poor experimental information.

Development of oxide glasses, especially Te-based ones, needs the suitable understanding of their thermal properties; thus thermal and calorimetric analysis (such as DSC, DTA, etc.) has a noteworthy role in this subject. DTA and DSC can directly give the thermodynamic characteristic temperatures, which play a dominant role in reviewing the potential benefits of glasses and also in understanding of their physical properties. Calorimetric experiments are assisted with the heating of glass under a definite heating rate; in a typical DSC output/curve, a flat baseline and at higher temperatures an endothermic region (attributed to the glass transition temperature (T_g)) are clearly observed before the exothermal peak/peaks of crystallization process, in which show amorphous nature and a change in vitreous structure, respectively. So, T_g is the lowest temperature in which with increase of temperature above T_g, a vitreous system is able to move toward steadiness in its atomic configuration and so crystallization. Characteristic temperatures such as T_g are heating rate (φ) dependent, and this dependency of T_g leads the determination of glass transition activation energy [6–8, 12–13,18,23–26,32,35,44,55,57,69–71]. From calorimetric analysis, one can deduce the strong/fragile character of a glass, relating to the degree of structural reorganization with temperatures near the glass transition region.

In the glass systems reviewed in this chapter, differential scanning calorimetry (DSC) has been used under dynamic N_2 gas atmosphere with a fixed flux to guarantee the constant pressure during the experiment; to obtain the glasses with the same initial thermal conditions, an isothermal process has been executed; in such isothermal process, the glass was first heated at a definite heating rate up to ~30 K higher than its T_g keeping it at this temperature for about 5 min to remove the previous thermal history of the glass relating to the melt quenching. Then glasses should be cooled at the same heating rate to a temperature about ~100 °C below their T_g. Now, samples are ready to be experimented via DSC at different desired heating rates to obtain their calorimetric characteristics. As is clear from Fig. 5.5, TVAgx glasses were investigated by DSC at rates (φ) of 2.5, 5, 7.5, and 10 K/min; in the case of TVNx [79], TVSbx [8, 26], and TVSx [24, 25], DSC was performed at the rates of 3, 6, 9, 10, and 12 K/min within the temperature range of 20–500 °C. Finally, DSC thermograms were obtained at several heating rates. In such thermograms, clear glass transition endothermic region and crystallization peak/peaks appeared enabling the determination of glass transition and crystallization kinetics, respectively. The sample weight needs in each DSC experiment is about 5–8 mg.

From a DSC thermogram, characteristic temperatures as glass transition temperature (T_g: as the midpoint of endothermic region of DSC curve), the first and successive crystallization temperatures (T_{cr} and T'_{cr}) and also the onset temperature of crystallization process (T_x) can be obtained; such characteristic temperatures are elucidated typically in the inset of Fig. 5.5. These characteristic temperatures show the increasing trend with increase of heating rate, in which such dynamic features give more light about the fragile/strong of a glass by obtaining the glass transition activation energy and crystallization activation energy [35, 58, 80–82]. One of the main usages of characteristic temperatures is in obtaining the thermal stability and glass-forming tendency of a glass, as will be discussed in Sect. 5.4.2.2. Further information can be obtained from the DSC thermograms. Some of these quantities are glass transition activation energy, crystallization activation energy, Avrami index relating to the dimension of crystalline growth in the amorphous matrix, kinetics of

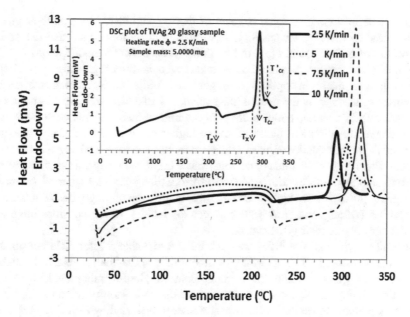

Fig. 5.5 DSC thermograms of a selected $40TeO_2$-$40V_2O_5$-$20Sb_2O_3$ (TVAg20) glass composition at the rates (φ) of 2.5–10 K/min; the picked inset plot is devoted to the rate of 2.5 K/min, to indicate more clear the characteristic temperatures for the reader [55]

glass transition and crystallization processes, fragility, viscosity, etc., which are out of the scope of this chapter.

5.4.2.2 Thermal Stability

Thermal stability and glass-forming tendency in oxide glasses can be achieved by using the calorimetric outputs. The glass stability and glass-forming ability parameters are directly related to the fragile/strong nature of a glass, and one can estimate them from the values of characteristic temperatures T_g, T_x, T_{cr}, and T_m which are glass transition temperature, the onset temperature of crystallization process, crystallization temperature, and melting point, respectively [18, 27, 55, 58, 71, 83, 84]. Evaluation of glass-forming ability and stability can guide us well to the capability of glass advantage in optoelectronic devices and signal transport. Different definitions exist for stability, which can be summarized as follows:

(a) $S_1 = T_{cr} - T_g$ [27, 55, 58].
(b) $S_2 = T_x - T_g$ [58].
(c) Hruby parameter $S_3 = K_{gl} = (T_{cr} - T_g)/(T_m - T_{cr})$ interpreted as glass-forming tendency [85].
(d) Modified Hruby parameter $S_4 = (T_x - T_g)/T_g$ [86].
(e) $S_5 = (T_x - T_g)/T_m$ [87].

These definitions give same concept of thermal stability, in which their greater values lead to their higher stability. Also, it should be mentioned that the optical properties of a glass are in close link with the glass network modifications [28, 50, 75] and in better word with the thermal and mechanical findings; so, aggregation of such aspects is useful in application suggestions. Additionally, thermal stability is a significant parameter in the field of glass science. Parameters of S_1 and S_2 define the thermal stability as the temperature range within which the glass does not tend toward crystallization and then lead to the higher value of glass-forming tendency (S_3) and the higher quality of glass. Glasses may be used in optical devices such as in optical fibers and waveguides and must have higher glass-forming tendency and thermal stability besides the good optical properties such as higher χ^3 (narrower bandgap); glasses with high thermal stability do not undergo the formation of crystallites (which can act as scattering centers leading to the considerable signal weakening) inside their glass matrix.

Results of thermal stabilities are summarized in Table 5.2 for different glasses; i.e., as is obvious, in TVAgx system glass with $x = 20$ has the relatively highest capability in glass forming (S_3 is equivalent to glass-forming tendency which sometimes called as K_{gl}) and has the highest thermal stability (S_1 and S_2), which are in excellent agreement with the aforementioned findings on E_g, OPD, and oxygen molar volume reviewed in Sect. 5.4.1. As seen in Table 5.2, tellurite-vanadate glasses containing Sb, Sb_2O_3, and Ag_2O have their higher stability at $x = 12$, 8, and 20, respectively (see Fig. 5.6). These results are in excellent accordance with the results of OPD and V_O^* presented before. In brief, in TVAgx system, glass with $x = 20$ has relatively high thermal stability, high glass-forming tendency, and higher OPD and V_O^* and so lower non-bridging oxygen ions (NBOs), which show that it is a suitable option and promising material for optical and

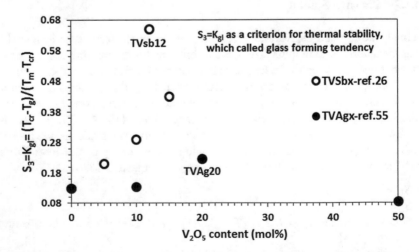

Fig. 5.6 Comparative plot of glass-forming tendency (thermal stability) versus V_2O_5 content for TVSbx and TVAgx glasses; in the legend of each series, the related reference was mentioned

optoelectronic devices such as base semiconductor in solar cells and optical fibers because of its resistance against thermal attacks and also its narrower bandgap.

Different reports on the applicability of thermal stable tellurite and vanadate glasses in silicon solar cell [39], in heat pump in solar cells to improve the photovoltaic system efficiency [40], and in optical fibers [59] confirm their potential in optical and optoelectronic applications.

5.4.3 Thermopower: Seebeck Effect

5.4.3.1 Titration Procedure

In vanadate glasses such as what this work deals with, one of the key parameters in determining oxygen molar volume (Sects. 5.3.4 and 5.4.1) and also to study the details of Seebeck effect is the measurement of $(C_V = [V^{4+}]/V_{tot})$ as the ratio of reduced vanadium ions; titration process is the common process; in order to do this, there are different ways such as cyclic voltammetry, titration, X-ray photoelectron spectroscopy, etc. In multicomponent vanadate glasses, we commonly use a chemical reaction of V_2O_5 and $FeSO_4(NH_4)_2SO_4$ solution for the titration, considering that V_2O_5 (V^{5+}) reduces to VO_2 (V^{4+}) during melting in melt quenching method. Titration process can be done by an algorithm involving three steps, which is expressed here in more detail as guide for the readers.

- *The first step of titration: preparation of required solutions*

At this stage, we should provide two types of sample solutions (as sample solutions A and B); one consists of only V^{5+} and another of a normal solution owing different valence states of vanadium ions.

Sample solution (A) is provided by adding an oxidizing agent in order to $V^{4+} \rightarrow V^{5+}$ (vanadate glass of about 0.2 g is dissolved in this solution). Diagram of the preparation of this solution is seen in Fig. 5.7.

Sample solution B another sample solution is prepared in which any agent is not used and V^{5+} and V^{4+} ions are present (the same glass of equal to 0.2 g is dissolved in this solution); see diagram of this preparation step in Fig. 5.8.

- *The second step of titration: titration by $FeSO_4(NH_4)_2SO_4$*

The solutions A and B are titrated separately by $FeSO_4(NH_4)_2SO_4$ solution, and then V ions are analyzed, which reach to (result A) and (result B), accordingly. Indicator used in this process is diphenylamine with the chemical formula of $(C_6H_5)_2NH$; see Fig. 5.9 as a diagram of titration process.

It should be noted that, by testing the factor of $KMnO_4$ and the $FeSO_4(NH_4)_2SO_4$ solution, one can obtain that 1 ml $FeSO_4(NH_4)_2SO_4$ solution reacts with 0.009226 g V_2O_5 in sample solutions.

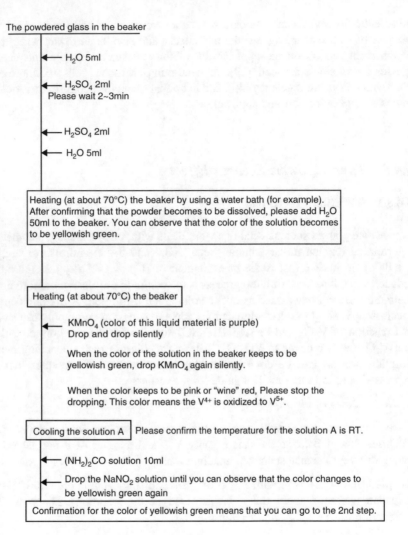

Fig. 5.7 Step-by-step diagram of the preparation of sample solution A at the first stage

- *The third step of titration: determination of V^{4+} ions*

The concentration of V^{4+} ions in the composition can be determined from the difference between "results of A and B." Moreover, we know V_2O_5 contains 56% of V. The final calculations to obtain V^{4+}, V_{tot}, and C_V can be followed as illustrated in Table 5.3.

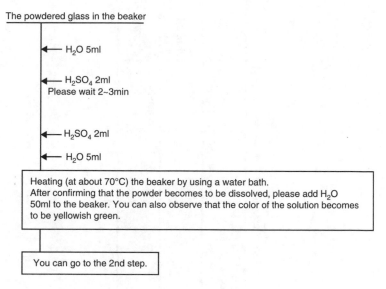

Fig. 5.8 Step-by-step diagram of the preparation of sample solution B at the first stage

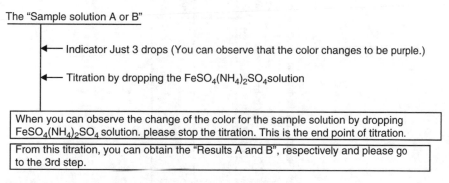

Fig. 5.9 Diagram of titration process for sample solutions A and B; results A and B are denoted hereafter by f_A and f_B, respectively

5.4.3.2   Thermopower and Applicability of Thermal Stable Glasses in Photovoltaic Pannels

The thermoelectric effect (as a characteristic of thermoelectric materials) means the Seebeck effect if temperature gradient generates an electric potential or Peltier effect if an electric potential produces a temperature gradient; moreover, thermoelectric materials are good options in applications such as power generation and refrigeration. An environmentally friendly material to be used in special thermoelectric application should have high figure of merit (as $F = S^2 \sigma / k_t$, where S, σ, and k_t show the thermoelectric power, electrical conduction, and thermal conductivity, respectively); in figure of merit, thermal and electrical conductivity compete. Glasses

Table 5.3 The representative table for recording the obtained data in titration process of a vanadate glass to determine V^{4+} ion content and so C_V

Glass name	Mass of the powdered glass (α) (g)	Result of the titration ($\beta\beta$) (ml)	V_2O_5 V_2O_5 (g) in the solution ($\beta \times 0.009226\beta \times 0.009226$) ($\gamma$)	$V_2O_5 V_2O_5$ in 1 gr of glass sample (g) ($\mu = \gamma l$) (α)	$f = \mu \times 0.56$	$V^{4+} = f_A - f_B$ ($= V_{tot} - V^{5+}$)	$C_V = V^{4+}/V_{tot}$
Sample solution A	α_A	β_A	γ_A	μ_A	f_A	Is obtained	Is obtained
Sample solution B	α_B	β_B	γ_B	μ_B	f_B	Is obtained	Is obtained

may be good materials in this field, because they have low thermal conductivity [88]; in addition, glassy semiconductors are more promising candidates for thermoelectric applications, because of their electrical conductivity nature. Thermoelectric oxide glasses have been highly interesting due to their thermoelectric effect, driven by the need to more efficient materials for electric or power generation. Such materials become more important in the regime of reduction of the dependency to fossil fuels. Searching for good thermoelectric materials leads to numerous works. In this section, we review some glassy systems. As regards tellurite-vanadate glass to be investigated in respect to Seebeck coefficient, we should have knowledge about the reduced fraction of transition metal ions (C_V), which was introduced by experimental titration method in the previous section. In Seebeck effect (as described experimentally in Sect. 5.2), a temperature gradient (ΔT) was supplied between two opposite highly polished faces of a glass; such temperature gradient generates a voltage difference (ΔV) in microvolt; dividing the ΔV to ΔT gives the Seebeck coefficient (in microvolt/K) which can be obtained within a desired temperature range. Seebeck coefficient of oxide glasses obeys the Heikes formula [89, 90] expressed by:

$$S = (K_B/e)\left[\ln\left(\frac{C_V}{1 - C_V}\right) + \alpha'\right] \tag{5.16}$$

where K_B, e, C_V, and α' are the Boltzmann's constant, electronic charge, the reduced ratio of transition metal ions (i.e., V ions in vanadate glasses), and a proportionality constant between the heat transfer and the kinetic energy of an electron, correspondingly. It should be noted that negative/positive sign of S refers to electron/hole as majority charge carriers.

Employing Heikes equation and estimation of α', one can determine/confirm the electrical conduction mechanism in samples having low charge carrier mobility [40, 91, 92]; so, condition of $\alpha' \langle\langle 1$ implies to small polaron hopping conduction mechanism [40, 71], while $\alpha' \rangle\rangle 2$ implies on large polarons. Heikes formula has been applied to the experimental outputs of Seebeck experiments in different glasses to certify (i) the applicability (validity) of Heikes formula and (ii) to obtain α'; in order to certify the strength of Heikes equation, S vs. $\ln(C_V/1 - C_V)$ plots should be drawn, in which the slope of such graphs should be nearly equal to the theoretical value of $(K_B/e)_{th} = 86.18 \ \mu VK^{-1}$. Figure 5.10 shows such plot for TVSx glasses [91]; in this system experimentally obtained slope of $(K_B/e)_{exp}$ and α' were 87.79, 82.56, and 89.81 μVK^{-1} at 310, 370, and 435 K, respectively, and -5.76, -5.81, and -5.69 at 310, 370, and 435 K, respectively, confirming the SPH mechanism of DC electrical conductivity and the validity of Heikes formula, respectively. Such values of α' have been reported previously for TeO_2-V_2O_5-Bi_2O_3 [52], TeO_2-V_2O_5-Sb [52, 91], TeO_2-V_2O_5-Sb_2O_3 [40], and TeO_2-V_2O_5-MoO_3 [92]; the reported thermoelectric studies justify electrical findings of DC electrical conductivity for TVSbx [93], TVSx [94], and TVMx [69].

What is important in this section is the figure of merit. Data reported in Refs [40, 91, 92]. show the plots and values of power factor ($S^2 \sigma$ in the numerator of the formula of figure of merit) at different temperatures for different oxide glasses; as is

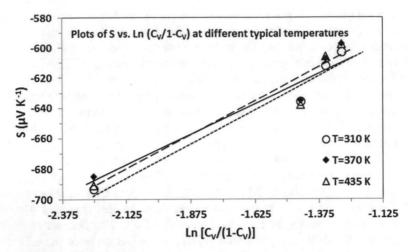

Fig. 5.10 Variation of the Seebeck coefficient (S) against ln ($C_V/1-C_V$) for TVSx glasses at the selected temperatures of 310, 370, and 435 K. experimental slopes (K_B/e)$_{exp}$ were 87.79, 82.56, and 89.81 μVK^{-1} at 310, 370, and 435 K, respectively, indicating the near fit to the theoretical value of (K_B/e) = 86.18 μVK, [91]

clear, this parameter has its highest value at $x = 10$ mol% for the system of TVMx, at $x = 0$ mol% in the cases of TVSbx and TVSx, which suggest that these compositions are more promising and powerful materials in designing.

In the field of applications of thermal stable oxide Te-based glasses, there are several reports in silicon solar cells [39], in heat pump in solar cells to improve the photovoltaic system efficiency [40], and in optical fibers [41]. Benefits of such glasses in optical fibers were explained in the previous sections. In order to give more light in the field of solar cells, the heat pumping feature of thermoelectric materials should be examined more closely. Today, solar energy is one of the most useful renewable energy sources, which can be harnessed through solar cells/photovoltaic panels. Expending on renewable energy resources instead on fossil fuels can reduce environmental pollution and risk of global warming. Power generation from photovoltaic (PV) technology is simple, free, available, reliable, and clean [95–99]. Relatively low efficiency of photovoltaic (PV) systems is a problem, which needs more systematic attempts to be improved. Moreover, different factors such as solar irradiation, temperature, and array voltage affect the PV panel output power. It should be noted that solar irradiation raises the panel temperature, in which the increasing of temperature reduces the PV efficiency. It has been suggested the pumping out of the heat from PV panel can be done by using thermal stable thermoelectric materials as thermoelectric coolers (TECs) [40]. As mentioned, overheating is the major reason for insufficient efficiency of PV panels, in which there is reduction between 0.25% (for amorphous cells) and 0.5% (most crystalline cells) for each degree of temperature [99]. Researchers are trying to improve the efficiency solar cells. The record at the moment stands at an efficiency of around 40%, using multijunction cells, each layer trap different frequencies of light. This type of solar cell (SC) will be expensive to produce and attempts are focused on the employing of single junction SCs. Researchers try to increase the amount of light entering the solar cells. Solar cells are supported with non-reflective layers to ensure that as much light as possible enters the solar cell, because silicon is a shiny substance. Solar cells used in photovoltaic panels are usually of single junction type with an efficiency of somewhere around 15% [101]. Panel temperatures in the summer and in warm climates can easily reach 50 °C resulting in a 12% reduction in output compared to the nominal output at 25 °C. Self-cooling PV panels are a solution to this problem [101]. Therefore, employing thermal stable oxide glasses of higher figure of merit (which are simple in preparation and of low cost and act as TECs) is a reliable way to remove the overheating [40]. From Fig. 10, the typical samples with higher thermoelectric cooling features can be selected.

5.5   Conclusions

In this chapter, transition metal oxide-containing glasses (TMOGs) were reviewed due to their special and unique optothermal properties. TeO_2-V_2O_5-A_yO_z (or A) oxide glasses were reviewed to elucidate that the optical, thermal, and thermoelectric properties vary with composition. This work tried to give more insight into the subject of the thermal stability and its effect on the optothermal properties of some vanadate-tellurite oxide glasses, searching the more potential candidates in optical applications. In brief, optical applications (such as active material in optical fibers) of oxide glasses need the narrower bandgap and higher χ^3 glasses owing to high thermal stability to prevent any crystallization in the amorphous matrix and to achieve strong signal transport. It should be noted that any suggestion for optical applications needs precise determination of optical properties such as energy bandgap, which affect the evaluation of other related optical parameters in optical device manufacturing and applications; in the case of bandgap determination, derivative absorption spectrum fitting method (abbreviated as DASF) has been used. It is concluded that such thermal stable oxide glasses with good thermoelectric properties are promising materials which can also be used in solar cells and photo-voltaic (PV) panels as heat pumps to elevate the PV efficiency. It was concluded that tellurite-vanadate glasses owing to high figure of merit are good options in thermo-electric applications.

In conclusion, the works reviewed here demonstrated the large potential of tellurite-vanadate and also Te-based glasses for optical and optoelectronic devices.

Acknowledgment   The author gratefully acknowledges Dr. Hidetsugu Mori for his support on titration.

References

1. R. El-Mallawany, Introduction to Tellurite glasses. Springer Series Mater. Sci. **254**, 1–13 (2017)
2. S.H. Elazoumi, H.A.A. Sidek, Y.S. Rammah, R. El-Mallawany, M.K. Halimah, K.A. Matori, M.H.M. Zaid, Effect of PbO on optical properties of tellurite glass. Res. Phys. **8**, 16–25 (2018)
3. H.M.M. Moawad, H. Jain, R. El-Mallawany, T. Ramadan, M. El-Sharbiny, Electrical conductivity of silver vanadium tellurite glasses. J. Am. Ceram. Soc. **320**(11), 2655 (2002)
4. R. El-Mallawany, N. El-Khoshkhany, H. Afifi, Ultrasonic studies of $(TeO_2)_{50}$–$(V_2O_5)_{50-x}(TiO_2)_x$ glasses. Mater. Chem. Phys. **95**, 321 (2006)
5. R. El-Mallawany, A. Abousehly, E. Yousef, Elastic moduli of tricomponent tellurite glasses TeO_2-V_2O_5-Ag_2O. J. Mater. Sci. Lett. **19**, 409 (2000)
6. R. El-Mallawany, Specific heat capacity of semiconducting glasses: binary vanadium tellurite. Phys. Status Solidi A **177**, 439 (2000)
7. D. Souri, Z. Torkashvand, Thermomechanical properties of Sb_2O_3-TeO_2-V_2O_5 glassy systems: Thermal stability, glass-forming tendency and Vickers hardness. J. Electron. Mater. **4**(2017), 46 (2158)
8. D. Souri, The study of glass transition temperature in Sb–V_2O_5–TeO_2 glasses at different heating rates. Indian J. Phys. **12**(2015), 89 (1277)
9. R. El-Mallawany, M. Sidkey, A. Khafagy, H. Afifi, Ultrasonic attenuation of tellurite glasses. Mater. Chem. Phys. **37**(2), 197 (1994)
10. I.Z. Hager, R. El-Mallawany, A. Bulou, Luminescence spectra and optical properties of TeO_2–WO_3–Li_2O glasses doped with Nd, Sm and Er rare earth ions. Phys. B Condens. Matter **406**(4), 972 (2011)
11. A.A. Ali, Y.S. Rammah, R. El-Mallawany, D. Souri, FTIR and UV spectra of pentaternary borate glasses. Measurement **105**, 72 (2017)
12. D. Souri, The study of crystallization kinetics and determination of Avrami index in TeO_2-V_2O_5-NiO amorphous samples by calorimetric analysis. Iranian J. Cer. Sci. Eng. **5**(3), 73 (2016)
13. D. Souri, Y. Shahmoradi, Calorimetric analysis of non-crystalline TeO_2- V_2O_5-Sb_2O_3: Determination of crystallization activation energy, Avrami index and stability parameter. J. Therm. Anal. Calorim. **129**, 601 (2017)
14. A. El-Adawy, R. El-Mallawany, Elastic modulus of tellurite glasses. J. Mater. Sci. Lett. **15**, 2065 (1996)
15. I.Z. Hager, R. El-Mallawany, Preparation and structural studies in the $(70- x) TeO_2$–$20WO_3$–$10Li_2O$–xLn_2O_3 glasses. J. Mater. Sci. **45**(4), 897 (2010)
16. M.M. El-Zaidia, A.A. Ammar, R.A. El-Mallwany, Infra-red spectra, electron spin resonance spectra, and density of $(TeO_2)_{100- x}$–$(WO_3)_x$ and $(TeO_2)_{100- x}$–$(ZnCl_2)_x$ glasses. Phys. Status Solidi A **91**(2), 637 (1985)
17. R.A. Montani, M.A. Frechero, The conductive behavior of silver vanadium molybdenum tellurite glasses: Part II. Solid State Ionics **158**, 327 (2003)
18. N.S. Hussain, G. Hungerford, R. El-Mallawany, M.J.M. Gomes, M.A. Lopes, N. Ali, J.D. Santos, S. Buddhudu, Absorption and emission analysis of RE^{3+} (Sm^{3+} and Dy^{3+}): Lithium Boro Tellurite glasses. J. Nanosci. Nanotechnol. **9**(6), 3672 (2009)
19. R. El-Mallawany, A. Abd El-Moneim, Comparison between the elastic moduli of tellurite and phosphate glasses. Phys. Status Solidi A **166**(2), 829 (1998)
20. M.A. Sidkey, R. El-Mallawany, A. Abousehly, Y.B. Saddeek, Elastic properties of tellurite glasses. Glass Sci. Technol.: Glastechnische Berichte **75**, 87 (2002)
21. M.M. Elkholy, R.A. El-Mallawany, Ac conductivity of tellurite glasses. Mater. Chem. Phys. **40**(3), 163 (1995)
22. D. Souri, R. Ghasemi, M. Shiravand, The study of high-dc electric field effect on the conduction of V_2O_5–Sb–TeO_2 glasses and the applicability of an electrothermal model. J. Mater. Sci. **50**(6), 2554 (2015)

23. D. Souri, Glass transition and fragility of telluro-vanadate glasses containing antimony oxide. J. Mater. Sci. **47**, 625 (2012)
24. D. Souri, Study of the heating rate effect on the glass transition properties of $(60 -x)V_2O_5$- x Sb_2O_3-$40TeO_2$ oxide glasses using differential scanning calorimetry (DSC). Measurement **44**, 2049 (2011)
25. S.A. Salehizadeh, D. Souri, The glassy state of the amorphous V_2O_5-NiO-TeO_2 samples. J. Phys. Chem. Solids **72**, 1381 (2011)
26. D. Souri, H. Zaliani, E. Mirdawoodi, M. Zendehzaban, Thermal stability of Sb-V_2O_5-TeO_2 semiconducting oxide glasses using thermal analysis. Measurement **82**, 19 (2016)
27. D. Souri, F. Honarvar, Z.E. Tahan, Characterization of semiconducting mixed electronic-ionic TeO_2-V_2O_5-Ag_2O glasses by employing ultrasonic measurements and Vicker's microhardness. J. Alloys Compd. **699**, 601 (2017)
28. P.Y. Shih, S.W. Yung, C.Y. Chen, H.S. Liu, T.S. Chiu, The effect of SnO and $PbCl_2$ on properties of Stanous Chlorophosphate glasses. Mater. Chem. Phys. **50**, 63 (1997)
29. K. Pradeesh, J.C. Oton, V.K. Agotiya, M. Raghavendra, G.V. Prakash, Optical properties of Er $^{3+}$ doped alkali chlorophosphate glasses for optical amplifiers. Opt. Mater. **31**, 155 (2008)
30. R.K. Brow, Review: The structure of simple phosphate glasses. J. Non-Cryst. Solids **263/264**, 1 (2000)
31. S.S. Das, B.P. Baranwal, C.P. Gupta, P. Singh, Characteristics of solid-state batteries with zinc/ cadmium halide-doped silver phosphate glasses as electrolytes. J. Power Sources **114**, 346 (2003)
32. M. Shapaan, Effect of heat treatment on the hyperfine structure and the dielectric properties of $40P_2O_5$-$40V_2O_5$-$20Fe_2O_3$ oxide glass. J. Non-Cryst. Solids **356**, 314 (2010)
33. M. Altaf, M.A. Chaudhry, Physical properties of lithium containing cadmium phosphate glasses. J. Mod. Phys. **1**, 201 (2010)
34. A. Abdel-Kader, R. El-Mallawany, M.M. Elkholy, Network structure of tellurite phosphate glasses: Optical absorption and infrared spectra. J. Appl. Phys. **73**(1), 71 (1993)
35. M.S. Dahiya, S. Khasa, A. Agarwal, Thermal characterization of novel magnesium oxyhalide bismo-borate glass doped with VO^{2+} ions. J. Therm. Anal. Calorim. **123**(1), 457 (2016)
36. M.S. Dahiya, S. Khasa, A. Agarwal, Optical absorption and heating rate dependent glass transition in vanadyl doped calcium oxy-chloride borate glasses. J. Mol. Struct. **1086**, 172 (2015)
37. Y.B. Saddeek, A. Aly, S.A. Bashier, Optical study of lead borosilicate glasses. Phys. B Condens. Matter **405**, 2407 (2010)
38. X.X. Pi, X.-H. Cao, Z.-X. Fu, L. Zhang, P.D. Han, L.X. Wang, Q.T. Zhang, Application of Te-based glass in silicon solar cells. Acta Metall. Sin. (Engl. Lett.) **28**(2), 223 (2015)
39. D. Souri, Suggestion for using the thermal stable thermoelectric glasses as a strategy for improvement of photovoltaic system efficiency: Seebeck coefficients of tellurite-vanadate glasses containing antimony oxide. Sol. Energy **139**, 19 (2016)
40. J. Koen, M. Res, R. Heckroodt, V. Hasson, Investigation of the photochromic effect in erbium-doped tellurite glasses. J. Phys. D. Appl. Phys. **9**, 13 (1976)
41. R. Braunstein, Photochromic and electrochromic properties of tungstate glasses. J Solid State Commun **28**, 839 (1978)
42. I. Morozova, A. Yakhind, Sov. J. Glas. Phys. Chem. **6**, 83 (1980)
43. D. Souri, Z.E. Tahan, A new method for the determination of optical band gap and the nature of optical transitions in semiconductors. Appl. Phys. B Lasers Opt. **119**(2), 273 (2015)
44. R. El-Mallawany, Y.S. Rammah, A. El Adawy, Z. Wassel, Optical and thermal properties of some tellurite glasses. Am. J. Optics Photon. **5**(2), 11 (2017)
45. M.H. Ehsani, R. Zarei Moghadam, H.R. Gholipour Dizaji, P. Kameli, Surface modification of ZnS films by applying an external magnetic field in vacuum chamber. Mater Res Expr **4**(9), 096408 (2017)

46. D. Souri, A.R. Khezripour, M. Molaei, M. Karimipour, ZnSe and copper-doped ZnSe nanocrystals (NCs): Optical transmittance and precise determination of energy band gap beside their exact optical transition type and Urbach energy. Curr. Appl. Phys. **17**, 41 (2017)
47. A. Kirsch, M.M. Murshed, M. Schowalter, A. Rosenauer, T.M. Gesing, Nanoparticle precursor into polycrystalline $Bi_2Fe_4O_9$: An evolutionary investigation of structural, morphological, optical, and vibrational properties. J. Phy. Chem. C **120**(33), 18831 (2016)
48. T. Katsuhisa, Y. Toshinobu, Y. Hiroyoki, K. Kanichi, Structure and ionic conductivity of LiCl-Li_2O-TeO_2 glasses. J. Non-Cryst. Solids **103**, 250 (1988)
49. D. Souri, M. Mohammadi, H. Zaliani, Effect of antimony on the optical and physical properties of Sb-V_2O_5-TeO_2 glasses. Electron. Mater. Lett. **10**(6), 1103 (2014)
50. H.S. Farhan, Study of some physical and optical properties of Bi_2O_3-TeO_2-V_2O_5 glasses. Aust. J. Basic Appl. Sci. **11**(9), 171 (2017)
51. H. Mori, H. Sakata, Seebeck coefficient of V_2O_5-R_2O_3-TeO_2 (R=Sb or Bi) glasses. J. Mater. Sci. **31**, 1621 (1996)
52. J. Tauc, A. Menth, States in the gap. J. Non-Cryst. Solids **8**, 569 (1972)
53. D. Souri, S.A. Salehizadeh, Effect of NiO content on the optical band gap, refractive index and density of TeO_2-V_2O_5-NiO glasses. J. Mater. Sci. **44**, 5800 (2009)
54. D. Souri, K. Shomalian, Band gap determination by absorption spectrum fitting method (ASF) and structural properties of different compositions of (60-x) V_2O_5–$40TeO_2$–xSb_2O_3 glasses. J. Non-Cryst. Solids **355**, 1597 (2009)
55. D. Souri, Physical and thermal characterization and glass stability criteria of amorphous silver-vanadate-tellurate system at different heating rates: Inducing critical Ag_2O/V_2O_5 ratio. J. Non-Cryst. Solids **475**, 136 (2017)
56. J.A. Duffy, M.D. Ingram, Optical basicity—IV: Influence of electronegativity on the Lewis basicity and solvent properties of molten oxyanion salts and glasses. J. Inorg. Nucl. Chem. **37**, 1203 (1975)
57. D. Souri, Crystallization kinetic of Sb–V_2O_5–TeO_2 glasses investigated by DSC and their elastic moduli and Poisson's ratio. Phys. B Condens. Matter **456**, 185 (2015)
58. V. Dimitrov, S. Sakka, Electronic oxide polarizability and optical basicity of simple oxides.1. J. Appl. Phys. **79**, 1736 (1996)
59. H. Fritzsche, Optical and electrical energy gap in amorphous semiconductors. J. Non-Cryst. Solids **6**, 49 (1971)
60. J.T. Edmond, Measurement of electrical conductivity and optical absorption in chalcogenide glasses. J. Non-Cryst. Solids **1**, 39 (1968)
61. N.F. Mott, E.A. Davis, *Electronic Processes in Non-crystalline Materials*, 2nd edn. (Clarendon Press, Oxford, 1979)
62. M. Molaei, A.R. Khezripour, M. Karimipour, Synthesis of ZnSe nanocrystals (NCs) using a rapid microwave irradiation method and investigation of the effect of copper (Cu) doping on the optical properties. Appl. Surf. Sci. **317**, 236 (2014)
63. L.E. Alarcon, A. Arrieta, E. Camps, S. Muhl, S. Rudil, E. V. Santiago; an alternative procedure for the determination of the optical band gap and thickness of amorphous carbon nitride thin films. Appl. Surf. Sci. **254**, 412–415 (2007)
64. S.D. Hart, G.R. Maskaly, B. Temelkuran, External reflection from omnidirectional dielectric mirror fibers. Science **296**, 510 (2002)
65. J.S. Lou, J.M. Olson, Y. Zhang, A. Mascarenhas, Near-band-gap reflectance anisotropy in ordered $Ga_{0.5}In_{0.5}P$. Phys. Rev. B **55**, 16385 (1997)
66. C. Kittel, *Introduction to Solid State Physics*, 7th edn. (Singapore, Wiley (ASIA) Pte. Ltd., 1996)
67. L. Changshi, L. Feng, Natural path for more precise determination of band gap by optical spectra. Opt. Commun. **285**, 2868 (2012)
68. D. Souri, M. Elahi, The DC electrical conductivity of TeO_2-V_2O_5-MoO_3 amorphous bulk samples. Phys. Scr. **75**(2), 219 (2007)

69. D. Souri, Fragility, DSC and elastic moduli studies on tellurite-vanadate glasses containing molubdenum. Measurement **44**, 1904 (2011)
70. D. Souri, S.A. Salehizadeh, Glass transition, fragility, and structural features of amorphous nickel–tellurate–vanadate samples. J. Therm. Anal. Calorim. **112**(2), 689 (2013)
71. D. Souri, DSC and elastic moduli studies on tellurite-vanadate glasses containing antimony oxide. Eur. Phys. J. B **84**, 47 (2011)
72. M. Elahi, D. Souri, Study of optical absorption and optical band gap determination of thin amorphous TeO_2-V_2O_5-MoO_3 blown films. Indian J. Pure Appl. Phys. **44**, 468 (2006)
73. D. Souri, Effect of molybdenum tri-oxide molar ratio on the optical and some physical properties of tellurite-vanadate-molybdate glasses. Measurement **44**, 717 (2011)
74. D. Souri, Ultrasonic velocities, elastic modulus and hardness of ternary Sb-V_2O_5-TeO_2 glasses. J. Non-Cryst. Solids **470**, 112 (2017)
75. D. Souri, Z.E. Tahan, S.A. Salehizadeh, DC electrical conductivity of Ag_2O -TeO_2-V_2O_5 glassy systems. Indian J. Phys. **90**(4), 407 (2016)
76. R. Swanepoel, Determination of the thickness and optical constants of amorphous silicon. J. Phys. E: Sci. Instr. **16**, 1214 (1983)
77. J.C. Manifacier, J. Gasiot, J.P. Fillard, A simple method for the determination of the optical constants n, k and the thickness of a weakly absorbing thin film. J. Phys. E: Sci. Instr. **9**, 1002 (1976)
78. D. Souri, Investigation of glass transition temperature in (60-x)V_2O_5-40TeO_2-xNiO glasses at different heating rates. J. Mater. Sci. **46**, 6998 (2011)
79. K. Aida, T. Komatsu, V. Dimitrov, Thermal stability, electronic polarizability and optical basicity of ternary tellurite glasses. Phys. Chem. Solids **42**(2), 103 (2001)
80. C.T. Moynihan, A.J. Easteal, J. Wilder, J. Tucker, Dependence of the glass transition temperature on heating and cooling rate. J. Phys. Chem. **78**, 2673 (1974)
81. A.A. Abu-Sehly, M. Abu El-Oyoun, A.A. Elabbar, Study of the glass transition in amorphous se by differential scanning calorimetry. Thermochemica Acta **472**, 25 (2008)
82. S. Weyer, H. Huth, C. Schick, Application of an extended tool-Narayanaswamy-Moynihan model. part 2. Frequency and cooling rate dependence of glass transition from temperature modulated DSC. Ploymer **46**, 12240 (2005)
83. S. Grujic, N. Blagojevic, M. Tosic, V. Zivanovic, J. Nikolic, Crystallization kinetics of $K_2O·TiO_2·3GeO_2$ glass studied by DTA. Sci. Sinter. **40**, 333 (2008)
84. A. Hruby, Evaluation of glass-forming tendency by means of DTA. Czechoslovak J. Phys. B **22**, 1187 (1972)
85. M. Saad, M. Poulain, Glass forming ability criterion. Mater. Sci. Forum **19**, 11 (1987)
86. P. Subbalakshmi, N. Veeaiah, Study of CaO-WO_3-P_2O_5 glass system by dielectric properties, IR spectra and differential thermal analysis. J. Non-Cryst. Solids **298**, 89 (2002)
87. D.M. Rowe, *Thermoelectrics Handbook* (CRC Press, Boca Raton, 2005), p. 60
88. R.R. Heikes, A.A. Maradudine, R.C. Miller, Une etude des properietes de transport des semiconducteures de valence mixte. Ann. Phys. NY **8**, 733 (1963)
89. R.R. Heikes, in *Thermoelectricity*, ed. by R. R. Heikes, R. W. Ure (Eds), (Interscience, New York, 1961), p. 2502
90. D. Souri, Z. Siahkali, M. Moradi, Thermoelectric power measurements of xSb-(60-x)V_2O_5-40TeO_2 glasses. J. Electron. Mater. **45**(1), 307 (2016)
91. D. Souri, Seebeck coefficient of Tellurite- vanadate glasses containing molybdenum. J. Phys. D: Appl. Phys **41**, 105102 (2008.) (3pp)
92. D. Souri, P. Azizpour, H. Zaliani, Electrical conductivity of V_2O_5–TeO_2–Sb glasses at low temperatures. J. Electron. Mater. **43**(9), 3672 (2014)
93. D. Souri, Small polaron hopping conduction in tellurium based glasses containing vanadium and antimony. J. Non-Cryst. Solids **356**, 2181 (2010)
94. A. Keyhani, M.N. Marwali, M. Dai, *Integration of Green and Renewable Energy in Electric Power System* (Wiley, Hoboken, 2009)

95. S. Leva, D. Zaninelli, Technical and financial analysis for hybrid photovoltaic power generation systems. WSEAS Transact. Power Syst. **5**(1), 831 (2006)
96. S. Leva, D. Zaninelli, R. Contino, Integrated renewable sources for supplying remote power systems. WSEAS Transact. Power Syst. **2**(2), 41 (2007)
97. G.K. Singh, Solar power generation by PV (photovoltaic) technology. Renew. Sust. Energ. Rev. **53**, 1013 (2013)
98. B. Parida, S. Iniyan, R. Goic, A review of solar photovoltaic technologies. Renew. Sust. Energ. Rev. **15**, 1625–1636 (2011)
99. Photovoltaic Efficiency – Inherent and System, solar facts, http://www.solar-facts.com/panels/panel-efficiency.php (Accessed 2015-6-5)

Chapter 6
Optical Properties of Tellurite Glasses Embedded with Gold Nanoparticles

Saman Q. Mawlud

Abstract Three series of samarium-doped sodium tellurite glass embedded with gold nanoparticles (Au NPs) in the compositions $(80-x)$ TeO_2-$20Na_2O$-xSm_2O_3 ($x = 0, 0.3, 0.6, 1, 1.2, 1.5$ mol%), $(79-x)TeO_2$-$20Na_2O$-$1Sm_2O_3$-$xAuCl_3$ ($x = 0, 0.2, 0.4, 0.6, 0.8, 1$ mol%), and $(80-x)TeO_2$-$20Na_2O$-$xAuCl_3$ ($x = 0.2, 0.4, 0.6, 0.8, 1$) were prepared using conventional melt quenching technique. The homogeneous distribution and growth of spherical and non-spherical Au NPs (average size $\sim3.36 \pm 0.076$ to $\sim10.62 \pm 0.303$ nm) in the glassy matrix was evidenced from the transmission electron microscopy (TEM) analysis. The UV-Vis-NIR absorption spectra showed nine bands corresponding to transition bands from ground state $^6H_{5/2}$ to excited states $^6P_{3/2}$, $^4I_{11/2}$, $^6F_{11/2}$, $^6F_{9/2}$, $^6F_{7/2}$, $^6F_{5/2}$, $^6F_{3/2}$, $^6H_{15/2}$, and $^6F_{1/2}$ in which the most intense bands were $^6F_{9/2}$, $^6F_{7/2}$, $^6F_{5/2}$, and $^6F_{3/2}$. The absorption spectrum of Sm^{3+} ions free glass sample containing Au NPs displayed two prominent surface plasmon resonance (SPR) band located at ~550 and ~590 nm. The infrared to visible frequency downconversion emission under 404 nm excitation showed four emission bands centered at 577 nm, 614 nm, 658 nm, and 718 related to the transitions $^4G_{5/2} \rightarrow {}^6H_{5/2}$, $^4G_{5/2} \rightarrow {}^6H_{7/2}$, $^4G_{5/2} \rightarrow {}^6H_{9/2}$, and $^4G_{5/2} \rightarrow {}^6H_{11/2}$, respectively, corresponding to Sm^{3+} transitions. An enhancement in downconversion emission intensity of both green and red bands was observed in the presence of gold NPs either by increasing annealing time or by NPs concentration. For glass series II, the enhancement in photoluminescence (PL) intensity of glass containing 0.4 mol% $AuCl_3$ showed the maximum enhancement by a factor of 1.90:1.82:1.97:2.25 times for all transitions bands. The enhancement was mainly ascribed to the highly localized electric field of Au NPs positioned in the vicinity of Sm^{3+} ion. The enhancement of downconversion emission was understood in terms of the intensified local field effect due to gold NPs.

S. Q. Mawlud (✉)
University of Salahaddin/College of Education/Physics Department, Iraq-Kurdistan, Erbil, Iraq
e-mail: saman.mawlud@su.edu.krd

© Springer International Publishing AG, part of Springer Nature 2018
R. El-Mallawany (ed.), *Tellurite Glass Smart Materials*,
https://doi.org/10.1007/978-3-319-76568-6_6

105

6.1   Introduction

Glasses containing metallic nanoparticles (NPs) attract scientist's interest because NPs may originate changes of the material's luminescence characteristics as well as enhancement of the nonlinear optical properties. Nucleation of metallic NPs inside tellurite dioxide glasses were reported [1]. In all cases, studies of the presence of NPs contribute to enhance the material's luminescence efficiency either due to the energy transfer (ET) mechanism from NP to the rare-earth (RE) ion or through the large induced local field on the RE ions located in the vicinity of the NPs. The contribution of ET processes and the intensified local field are due to the NPs allowed obtaining enhanced luminescence in the green, orange, and dark-red spectral wavelengths. The introduction of the NPs as the second dopants reveals its promising application to develop the properties and characteristics of new functional materials [1].

There are several techniques used for manufacturing of glass material such as thermal evaporation, sputtering, chemical reaction, amorphization, irradiation, melt quenching, sol gel, etc. [2]. Melt quenching, vapor deposition, and sol gel processing solutions are the three common methods that are usually used in industrial process and in researches for producing glasses. The preparing technique of glasses by means of rapid quenching of a melt is historically the most established and it is still the most widely used in the preparation of amorphous metallic oxide materials. Melt quenching technique is the most important conventional way and widely used technique to produce glass in which the glass will not crystallize. Besides, this technique is simple and low cost compared to other methods [2, 3].

The applications of the glass are so wide, so that they can be used in our daily life. For example, glasses are mainly used in packaging housing and buildings, electronics, automotive, optical glass, fiber-optic cables, renewable energy, cookware, chemical laboratory equipment, flat panel devices such as television screen, and dental products such as dental crowns and bridges. A great research interest in the science and technology of glasses has received a great attention. Meanwhile, for scientific research purpose, it is particularly interesting due to the unique properties of glass such as its corrosion resistance, unreactive with other chemical elements, thermal shock resistance, and electrical insulation. New science and technology have dynamically given great improvement in glass manufacturing and their new technological applications. The main advantages of this will be to provide the fundamental base of new optical glasses with a number of applications especially tellurite glass-based solid-state lasers, optical fibers, optical switches, third-order nonlinear optical materials, optical amplifiers, light-emitting diodes, and upconversion glasses [4].

Nowadays, addition of a small amount of transition metal ions into the glass network system becomes more favorable due to their luminescence properties [5]. In modern technology, RE ions are very attractive candidates as active ions in many optical materials because many fluorescence states are chosen among the (4f) electronic configuration and most of them are placed in the visible region [6]. The excited levels of these RE ions emit in the visible region, which exhibit possibly high quantum efficiency and show different quenching mechanisms. Glass is a promising host to investigate the influence of chemical environment on the physical, structural, thermal, and optical properties of the RE ions.

In particular, the investigation of energy transfer (ET) process in glasses having energy gap in the visible region deserves great attention because RE ion-doped glasses may present efficient visible luminescence [6]. In principle, ET process may favor particular applications, such as the operation of lasers, optical amplifiers. Enhancement of optical properties either by ET between two rare earth ions or by influence of the large local field on rare earth ions positioned near the doped elements, is of great technological interest [3, 7]. Therefore, revealing the new approaches to increase the emission cross-section and its intensity, is an important subject of the study which needs more attention due to importance of laser glasses in wide range of the applications.

The luminescence properties of RE ion-doped glasses with addition of metallic NPs have been investigated due to their applications as optical devices. Usually, in order to minimize luminescence quenching and to make the devices with optimized photonic properties, the concentration of the RE has to be kept low. However, an alternative way to avoid the quenching effect is to modify the RE ion environment by embedding metallic NPs [7, 8]. Therefore, glasses containing small amount of RE ions embedded with metallic NPs are of considerable interest, because the luminescence efficiency may increase many times when the pump excitation or emission frequency is in resonance or closely matches with the surface plasmon resonance (SPR) wavelength of metallic NPs.

Turba et al. [9] reported that the thermally induced particle growth either involves migration of adatoms (Ostwald ripening) or migration and coalescence of NPs. It is asserted that the SPR wavelength depends on the host and metal dielectric as well as together it depends on the size and shape of the NPs. The tunability of plasmon band positions at different wavelengths giving an effect to the plasmon resonance facilitates varieties of applications [9]. Accordingly, the electrons modify their collective oscillation with the change in shape or size of the NPs. Variation in the dielectric constant of the surrounding medium affects the oscillation frequency, where the surface is capable in accommodating electron charge density from the NPs. However, the influence of different metallic NPs (growth controlled by thermal annealing) at low concentration of RE ions on the spectral modifications is not established extensively.

Undoubtedly, alike metals the glass material is an integral part of the civilization. The in-depth investigations of glass properties are of great significance due to potential applications in various engineering and technological fields. However, glasses doped with lanthanides (Ln) ions received attention in recent year due to their outstanding optical features. Heavy metal oxide glasses such as tellurite and antimony are superior host than other host matrices like fluoride, bromide, and chalcogenide due to their low phonon energy and high refractive index which increases the radiative emissions in rare earth-doped glasses. It was demonstrated that TeO_2 acts as a very good glass former when a modifier with moderate concentration in range of 10–30 mol% is added.

Lately, researchers synthesized binary glasses TeO_2-M_2O ($M =$ Li, Na, and K) to examine the effects of alkali modifiers on the fluorescence efficiency of Sm^{3+} ions [10]. Interestingly, the fluorescence intensity of Sm^{3+} ions is found to increase with different alkali modifiers in the order of $K_2O < Li_2O < Na_2O$ [11]. In fact, Na_2O is nominated as the best modifier among the other alkali oxides. Additionally, Na_2O presents the most glass-forming ability on the basis of stability against crystallization [12]. Currently, RE-doped optical materials are investigated extensively due to their potential applications in color display, optical data storage, sensor, laser, and optical amplifier for communication. Especially, significant attention is given on Sm^{3+} ions due to their potentiality for blue, green, and red emissions. However, the optical gain can only be increased up to a certain point with the increase of Sm^{3+} ions concentration and decreased or quenched thereafter. Beyond a critical concentration, the RE ion cluster aggregates and quenches the luminescence due to an increasing ion-ion interaction between RE ions. In order to avoid the RE ion concentration quenching, the introduction of second dopant has been suggested [8, 13]. This second dopant can be another RE element and/or metallic nanoparticles.

Lately, vitreous materials containing an appropriate combination of RE ions and metallic NPs are attracting large interest. Surprisingly, the presence of metallic NPs intensifies luminescence and enhances the nonlinear optical properties [9]. Among the three rare metals (silver Ag, gold Au, cooper Cu) that display plasmon resonances in the visible spectrum, Au exhibits the highest plasmon excitation efficiency. Gold NPs are more stable than silver and have a reasonably sharp surface plasmon resonance (SPR) peak. Moreover, optical excitation of plasmon resonances in nanosized Au particles is the most efficient mechanism by which light interacts with matter. Besides, the intense plasmon resonance of Au can be tuned to any wavelength in the visible spectrum by variation of the size of the nanoparticles [9].

6.2   Terminology of the Glassy State

Glasses are often described as supercooled liquids. Almost any solid can be produced in a glassy state if the melt is cooled sufficiently quickly. Most glasses are comprised predominately of silica with an empirical formula of SiO_2. The most common form of natural silica is quartz. However, quartz is a crystalline solid and has a regular repeating crystalline lattice. In contrary, glass has no regular repetition in its macromolecular structure, and therefore it has a disordered structure, much like a substance in the liquid state. The arrangement of these molecular units in a glass is analogous to those in a crystal, but in the former the bond lengths and angles can vary randomly [14, 15].

The oldest established method of producing glass is to cool the molten form of the material sufficiently quickly. An essential prerequisite for glass formation from the melt is that the cooling must be sufficiently fast to preclude crystal nucleation and growth. In order to define a material as a glass, it must have two common characteristics [4]; firstly, there is no long-range order and periodic atomic arrangement. Secondly, all glasses must exhibit time-dependent glass transformation behavior.

Table 6.1 Melting temperature (T_m) and transition temperature (T_g) of glass [16]

Element	T_m (°C)	T_g (°C)
Oxygen	55	37
Sulfur	388	262
Selenium	494	328
Tellurium	725	484

Based on the behavior of glass transformation range, the glassy and crystalline materials can be differentiated. In general, cooling at temperature below melting temperature of the crystal would lead to the formation of crystal; if this happened, the volume and the enthalpy will decrease rapidly to the volume or enthalpy corresponding to the crystal. A slow cooling allows enough time for a viscous liquid to alter its local atomic arrangement to attain the minimum free energy at the corresponding temperature. Conversely, a rapid cooling causes an increase of viscosity that is too quick for the local atomic arrangement to follow and results in a transition into a glass at a higher temperature. However, if the melt liquid could "bypass" or can cool below the melting temperature, then a supercooled liquid or glass-forming liquids are obtained. This could be due to the melt liquid having a high viscosity which makes the atoms not able to completely rearrange to the equilibrium structure or the melt liquid cooled rapidly that it does not have sufficient time to crystallize. The transition from a viscous liquid to a solid glass is called the "glass transition," and the temperature corresponding to this transition is called the "glass transition temperature" denoted by T_g [16]. Table 6.1 represents the typical melting points and transition temperature for few elements in the glass medium.

The reversible transformation from a glass to a viscous liquid also takes place if a glass is heated to a temperature above T_g. The value of T_g is not always the same in spite of same chemical composition. This is because glass transition occurs as a result of the increase of viscosity which in turn depends on the cooling rate. Therefore, T_g is usually different depending on the cooling rate of a liquid. The volume or enthalpy versus temperature diagram is shown in Fig. 6.1 the presence of glass formation behavior can be explained. Figure 6.1 schematically demonstrates the path difference for a specific physical property change such as molar volume in supercooled state. In the process of supercooling, the glass transition range is achieved by avoiding the crystallization. There are two factors which actually controlled the deviation from the equilibrium curve, which are the viscosity of the liquid and the cooling rate of the liquid. The use of slower cooling rate causes the glass transformation region to be shifted to lower temperatures and the glass obtained to have a lower enthalpy.

6.3   Rare-Earth Oxides as Dopants

The role of RE ions as optical activators in glasses has been continuously studied due to their contribution for photonic applications such as solid-state lasers and optical amplifiers. They were used to probe the local structure variations in host glasses

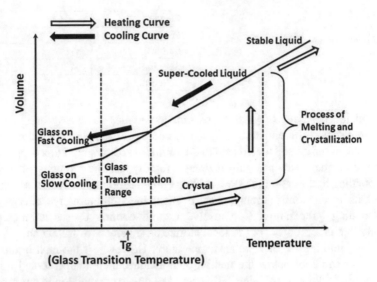

Fig. 6.1 Effect of temperature on the enthalpy of glass-forming melts [17]

because of their unique spectroscopic properties resulting from the optical transitions in the infra $4f$ shell. Glass is a promising host to investigate the influence of chemical environment on the physical, structural, thermal, and optical properties of the RE ions. Among several glasses, tellurite glass containing RE element is attractive for nonlinear optical devices because of its large third-order nonlinear optical susceptibility [18, 19].

In solid-state laser, materials such as doped crystals and glasses, the most interesting rare-earth ions, are those in the lanthanum (Ln) group. These ions usually appear in a trivalent state. The 4f electron shell determines the optical properties of lanthanides; it is almost insensitive to the surrounding atom of the host environment because of screening by 5s and 5p electron shells. The reason is due to the weak interaction between optical centers and the crystalline field (weak electron-phonon coupling). Such weak interactions between the 4f electrons and the crystalline field produce a very well-resolved Stark structure of the levels, which varies slightly from host to host. For the same reason, electronic transitions in trivalent rare-earth ions are very narrow and demonstrate very weak phonon bands [14, 21]. The spectral shape of the optical transitions of lanthanides in glasses is determined mostly by the following factors [22]:

1. Stark splitting of the degenerated energy levels of the free ion, determining the number of Stark levels.
2. Magnitude of the splitting, which is determined by the host material.
3. Different line broadening mechanisms, such as homogeneous and inhomogeneous broadening can occur.

A two-dimensional projection of this model is shown in Fig. 6.2; the model serves only to indicate the feasibility of constructing a large continuous random network [23].

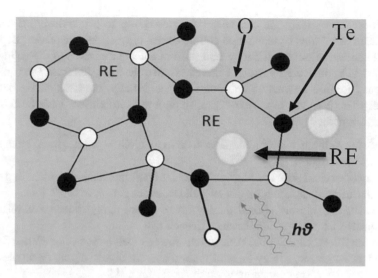

Fig. 6.2 Glass network with RE elements [23]

All Ln ions are characterized by a xenon core, an unfilled 4f shell, and filled $5s^2$ and $5p^6$ shells that screen the 4f electrons from external perturbations. This screening effect protects the optically active electrons from the influence of the crystal field, giving the Ln their characteristic sharp and well defined spectral features [15]. When incorporated in crystalline or amorphous hosts, the RE exists in 3+ or occasionally 2+ ion states. Some of the divalent species exhibit luminescence principally samarium (Sm^{2+}) and europium (Eu^{2+}). However, the trivalent ions are of special interest because the most stable oxidation state for all Ln elements is the 3+ state. The shielding of 4f electrons influences the positions of RE electronic levels via spin-orbit interactions more than due to the applied crystal field. The intra 4f transitions are parity forbidden but partially allowed by crystal field interactions which mix wave functions of opposite parity. Luminescence lifetimes are therefore long (millisecond range), and line widths appear to be narrower. By careful selection of the appropriate dopant ion, an intense narrow-band emission can be achieved in the visible as well as in the near-infrared region.

Particularly, RE ions such as Sm^{3+}, Eu^{3+}, Dy^{3+}, Er^{3+}, and Pr^{3+} are used to develop various active optical devices [24]. Among the lanthanides, samarium ion (Sm^{3+}) is the one that has been studied the most since its laser oscillation is utilized as an optical fiber amplifier. Samarium is a rare-earth element belonging to the group of the lanthanides with an atomic number of 62. Electronic structure of samarium consists of [Xe] $4f^6$ $6s^2$, where the [Xe] represents the electronic structure of xenon. At the atomic number of 57, 5s and 5p shells are full and 4f is unfilled. By increasing the atomic number in lanthanide group, the radius of 4f shell decreases gradually [25, 26]. It is known that Sm^{3+} is one of the most popular and efficient ions since it can perform the lasing process in glasses. A large number of researches worked on Sm^{3+} ions due to their potentiality for yellow, blue, green, and red emissions; the optical gain can only be increased up to a certain point with the increase of Sm^{3+} ions concentration and decreases or quenches thereafter. Beyond a critical concentration, the RE ions tend to form a cluster in most of the solid hosts. Consequently, the RE ions cluster aggregates and quench the luminescence due to an increasing ion-ion interaction between RE ions. Finally, the host matrix becomes optically inactive [24]. Sm^{3+} ion-doped sodium tellurite glasses have been chosen for the following reasons [27]:

1. Sm^{3+} ion exhibits a strong luminescence in the yellow, blue, green, orange, and red spectral region.
2. Decay of excited states of Sm^{3+} involves different mechanisms depending on the composition of the glass matrix as well as concentration of Sm^{3+} ions.
3. The rich and closely spaced energy levels allow many more cross-relaxation channels that result in fluorescence quenching.
4. The Sm^{3+} ions can also act both as sensitizer as well as activator. When Sm^{3+} is used as sensitizer, energy transfer takes place via dipole-dipole or dipole-quadrupole mechanism depending upon the host matrices.
5. The Sm^{3+} ion possesses strong fluorescence intensity, rich energy levels, large emission cross-section, and high quantum efficiency which are suitable to improve fluorescence properties and in turn to develop new optical functions.

Figure 6.3 shows the different arrangements of 4f electrons in generating different energy levels of RE ions. This is due to the unfilled 4f shells of Ln elements. The 4f electron transitions, among the various energy levels, are responsible for the occurrence of numerous absorption and emission wavelengths [28].

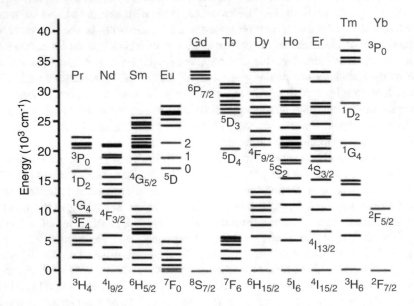

Fig. 6.3 Energy level for some trivalent RE ions [28]

Table 6.2 Energy (cm^{-1}) levels of Sm^{3+} ion in free state and LaCl$_3$ crystal host [29]

Levels	Free ion (cm^{-1})	LaCl$_3$ (cm^{-1})
$^6H_{5/2}$	36	36
$^6H_{7/2}$	1080	1080
$^6H_{9/2}$	2286	2286
$^6H_{11/2}$	3608	3608
$^6H_{13/2}$	5000	5014
$^6F_{5/2}$	7100	7033
$^4G_{5/2}$	17,900	17,860
$^4F_{3/2}$	18,900	18,860
$^4G_{7/2}$	20,050	20,010
$^4I_{9/2}$	20,640	20,600
$^4I_{11/2}$	21,100	20,977
$^4I_{13/2}$	21,600	21,562
$^4F_{5/2}$	22,200	22,129
$^4G_{9/2}$	22,700	22,850

The energy levels of Sm^{3+} ions in the free state and in LaCl$_3$ host crystals are listed in Table 6.2. In different materials, the energy differences for a given level are generally less than 100 cm^{-1} except in few cases of 200 cm^{-1} [29]. The descriptions of energy level position in different hosts are usually referred with respect to LaF$_3$ crystal. The general idea of the energy level positions for all the RE ions offers an aid to the analysis of unknown spectra in determining the exact energy level positions.

6.4 Nanoglass Plasmonic

Nanostructures being small enough manifest noticeably different chemical and physical properties compared to their bulk counterpart. Nanoscience can be defined as the science of objects and phenomena occurring at scale of 1–100 nm. Nanotechnology is the understanding and control of matter at dimensions of roughly 1–100 nm, where unique phenomena enable novel applications. Metal NPs attract strong interest because they open up a new field in fundamental science and because of their technological applications which trigger tunability of their optical properties. They are convenient components of sub-wavelength optical devices, nonlinear optics, and surface-enhanced spectroscopy. Metallic NPs have proven to be the most flexible nanostructures owing to the synthetic control of their size, shape, composition, structure, assembly, and encapsulation [30, 31]. The optical properties of metallic NPs emerge from a complex electrodynamics effect that is strongly influenced by the surrounding dielectric medium. Light impinging on metallic particles enables optical excitations to the electrons causing their collective oscillation in the valence band. Such coherent oscillations occur at the interface of a metal with a dielectric medium are called surface plasmons. Glass network with RE and NPs in the host matrices is shown in Fig. 6.4.

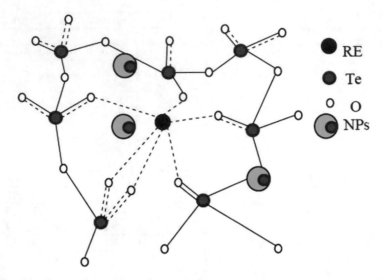

Fig. 6.4 Glass network with RE and NPs element

NPs are used in a variety of applications because of their superior characteristics, such as high specific area of contact, great mechanical properties, high electrical conductivity, and high optical scattering efficiency [32]. The metallic NPs in the vicinity of RE ions inside the dielectric glass host enhance the rate of excitation and emission of RE ions. This effect is attributed to the local field enhancement on excitation and emission of RE ions. Furthermore, the energy transfer (ET) is possibly stimulated by the intense local electric field induced by surface plasmon resonance (SPR). The coupling of RE ions with metallic NPs is developed as a valuable strategy to improve the luminescence yield of RE ions [33].

At this stage, it is important to justify the selection of Au NPs as stimulator in the tellurite glass host. The interaction of light with Sm^{3+}-doped tellurite glasses containing metallic NPs received much attention due to their various applications in optoelectronic devices and optical communications [34]. As RE have lower cross-section area, the small absorption cross section of the Sm^{3+} disqualifies them to be strong emitter as the coupling between centers and the lattice is weak as 4f electrons shielded by $5S^2$ and $5P^6$ electrons.

Bulk gold (Au) has considerably higher chemical stability than silver, although the SPR band of silver is much sharper than gold [35]. For instance, solid gold is very stable under many circumstances, while solid silver is gradually oxidized in air. Actually, in Au the prominent 5d contribution to the conduction electrons plays a key role in protecting gold from oxidation in air [12]. Gold NPs exhibit unique and tunable optical properties owing to the phenomenon of SPR. Incorporation of metallic NPs, often gold and silver, is found to be as fruitful alternative route to enhance the luminescence of RE. However, unlike Ag and Au, other metals such as platinum, palladium, iron, etc. do not show strong SPR effects. Au NPs show strong SPR absorption in the visible region. The special characteristic of Au NPs is

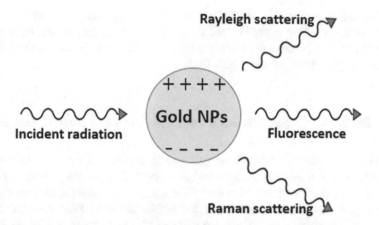

Fig. 6.5 The schematic diagram illustrates the interaction of incident radiation with free electrons in Au NPs

suppressed by the mechanisms such as strong conduction electron relaxation or radiation damping due to scattering. Major processes of interactions of radiation with Au NPs are illustrated in Fig. 6.5. All these processes are enhanced strongly owing to the unique interaction of light with the free electrons in NPs.

Generally, the enhancement in photophysical properties of Au NPs is related to the number of process which occurs when the matter is exposed to light [31]:

(i) The light can be absorbed and metal NPs absorb and scatter light. Their strong interaction with light occurs due to the conduction electrons on the metal surface undergoing a collective oscillation when they are excited by light at specific wavelengths.

(ii) The light can be scattered at the same frequency as the incoming light (Mie or Rayleigh scattering). Mie-scattered light travels in all directions from the particle. Most of the Mie-scattered light intensity travels in the forward direction. The scattered intensity becomes lower with a particular direction at smaller particle size. In elastic scattering or Rayleigh scattering, the energy of the light does not change. However, scattering of polarized light is the strongest in directions perpendicular to the electric field of the light and vanishes in the direction parallel to the electric field.

(iii) The absorbed light can be re-emitted such as fluorescence. PL from asymmetric metal NPs in the visible range was studied by several researchers. PL in noble metal is generally attributed to a three-step process. Firstly, the electrons present in the d-bands are excited to the vacant sp-conduction band to generate electron-hole pair. Secondly, subsequent intra-band scattering processes move the electrons nearer to the Fermi level with partial ET to the phonon (lattice vibrations). Lastly, the electron-hole pair recombination results in photon emission.

(iv) The enhancement of local EM field of the incoming light. Metal NPs exhibit the local electric field caused by the localized SPR behavior. When incident light hits onto the surface, the surface plasmon excitation enhances the EM field in the neighborhood of metal NPs [36].

6.5 Metallic NPs Embedded into RE-Doped Tellurite Glass

Glasses are superior candidate to embed rare-earth ions and metallic nanoparticles due to their high transparency, mechanical strength, and simple preparation in any size and shape. Therefore, metallic glass composites containing nanoparticles become potential substrate materials for significant local field enhancement to increase the luminescence of rare-earth ions [37]. In order to avoid the RE ion concentration quenching, the introduction of second dopant has been suggested. This second dopant can be another RE element and or metallic NPs. Lately, glasses doped with RE ions and metallic NPs are investigated due to their applications in diversified optical devices [38]. It is demonstrated that the metallic NPs enhance the RE luminescence and improve the nonlinear optical properties [7]. The introduction of the NPs as the second dopants reveals its promising application to develop the properties and characteristics of new functional materials. The effect of NPs inside tellurite glasses was investigated [39–41]; they demonstrated that the presence of NPs enhance the material's luminescence efficiency either via energy transfer (ET) from the NPs to the RE ions or by the influence of the large local field on the RE ions positioned in the vicinity of the NPs. However, the ratio between metal NPs and RE concentrations must be less than one to maximize the luminescence intensity.

The coupling of RE ions with metallic NPs is demonstrated to be an alternative route to improve the luminescence yield of RE ions. These efforts are in conjunction with the potential of plasmonic metal nanoparticles to confine the electromagnetic EM energy or optical excitation nanoscale volume [42]. The strong optical interactions within this volume are mediated by local electric field enhancements. The effects of energy transfer between metallic NPs and RE ions are twofold: they may enhance luminescence by energy transfer from particles to ions and/or may quench luminescence by energy transfer from ions to particles. A schematic diagram displaying the interaction mechanisms between the RE and NPs is illustrated in Fig. 6.6. The concepts of energy transfer rely on the wavelength of the incident light beam or luminescent wavelength which is close to the SPR wavelength of the NPs.

The optical properties of tellurite glasses are affirmed to depend strongly on the Au concentration in the dielectric medium [43, 44]. To maximize the luminescence intensity, the ratio between Au and Sm^{3+} concentrations must be less than one or any other else, and the probability of multipole formation increases and contributes to unfavorable properties. Linear and nonlinear optical properties of the RE-doped glasses embedded with metallic NPs have been investigated by different authors in the last few years [42].

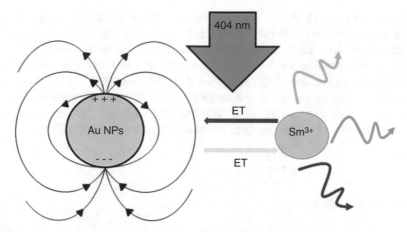

Fig. 6.6 Interaction of RE ions and Au NPs involving the energy transfer process under 404 nm excitations

The interactions in the Sm^{3+}-Au-co-doped heavy tellurite glass can be discussed through three different manners: (i) the interaction of excitation light with Sm^{3+} ion, (ii) excitation light with Au NPs, and (iii) Sm^{3+} ions with Au NPs. In a Sm^{+3}-Au-doped tellurite glass, plasmon-enhanced fluorescence results in increased excitation rate of the Sm^{+3} ions due to the interaction of NPs by the excited Sm^{+3} ions to the surface of Au NPs [45]. Furthermore the probability of energy transfer from RE to metallic NPs is increased resulting in quenching of luminescence. Hence, the concentration of RE must be greater than that of NPs. In these situations, to control the formation of metallic NPs inside the glass matrix is really a challenge with exceptional applications [39, 40].

6.6 UV-Vis-NIR Absorption Spectroscopy

Optical absorption in solid materials occurs in various mechanisms which involve absorption of photon energy either by lattice structures or electrons where the transferred energy can be conserved. The optical absorption spectra are used to determine the transition state in the glass system and also to obtain the optical bandgap values of the prepared glass systems. The absorption energy can be determined by calculating the difference of excited state energy with ground state energy. The transition which resulted in absorption of electromagnetic radiation is the transition between electronic energy levels. In general, the bandgap refers the energy difference between the valence and conduction bands. The optical bandgap is the first starting energy where the photons are being absorbed. The optical bandgap is referring to the energy corresponding to the fundamental absorption edge λ_{edg} in which the absorption edge is the transition from the maximum of valence band to the minimum of conduction band [46]. The fundamental absorption edge is represented by the region in which the glass sample starts to absorb the energy completely at a characteristic wavelength in the UV region. The interacting between the photon and the valence electron across the energy gap which occurs at the fundamental absorption edge of crystalline and non-crystalline materials can be ascribed to either the indirect or direct transition [46]. The amount of light absorbed in a sample when it passes through or gets reflected is estimated by the difference between the incident intensity (I_o) and the transmitted intensity (I_t). The amount of light absorbed is expressed either as transmittance or absorbance. Transmittance is defined as:

$$T_R = \frac{I_t}{I_o} \tag{6.1}$$

Absorbance yields:

$$A_b = -\log T_R \tag{6.2}$$

For most applications, absorbance values are used since a well-defined linear relationship exists between absorbance, concentration, and path length. The mathematical expression for Lambert law is given by:

$$T_R = \frac{I_t}{I_o} = e^{-kb} \tag{6.3}$$

where e is the base of natural logarithms, k is constant, and b is the path length (usually in centimeters). According to Beer's law, the amount of light absorbed by a sample is proportional to the number of absorbing molecules. Combining the two laws, one achieves the following Beer-Lambert law:

$$T_R = \frac{I_t}{I_o} = e^{-kbc_a} \tag{6.4}$$

where c_a is the concentration of the absorbing species. By taking the natural logarithm for both sides of Eq. (6.4), we get:

$$A_b = -\log T_R = -\log\left(\frac{I_t}{I_o}\right) = \log\left(\frac{I_o}{I_t}\right) = \varepsilon_a b c_a \tag{6.5}$$

where ε_a is the molar absorption or extinction coefficient which is the characteristic of a given sample under precisely defined wavelength, solvent, and temperature [47]. The absorption coefficient $\alpha(\nu)$ can be calculated from the transmission spectra by using the Beer-Lambert law:

$$\alpha(\nu) = \frac{1}{t} \ln\left(\frac{I_o}{I_t}\right) \tag{6.6}$$

where I_o and I_t are the intensity of the incident and transmitted light, respectively, and t is the thickness of the sample. The optical absorption coefficient $\alpha(\nu)$ in many amorphous materials reflects the density of states at the band edges and can be estimated from the optical absorption spectrum as follow:

$$\alpha(\nu) = 2.303\frac{A_b}{t} \tag{6.7}$$

where A_b is the absorbance defined by $\log(I_o/I_t)$. In amorphous materials, the absorption due to the electronic transition within the band in relation with the optical bandgap is described by Mott and Davis' [50] given equation as:

$$\alpha(\omega) = \frac{\text{const}}{\hbar\omega}\left(\hbar\omega - E_g\right)^{n_B} \tag{6.8}$$

where $\hbar\omega$ is the photon energy and n_B refers to direct and indirect transitions.

The value of E_g can be obtained by extrapolating the linear part of the $(\alpha\hbar\omega)^{1/2}$ against phonon energy ($\hbar\omega$). The value n_B is 2 for the indirect transitions as for non-crystalline materials and n_B is 1/2 for the direct transition. The determination of bandgap values gives information about the structural changes in the glass. Covalent nature of the glass matrix decreases when bandgap values decrease with the increase of rare-earth contents. Moreover, decrease in the bandgap energy suggests that non-bridging oxygen is increasing. In general, both direct and indirect transitions can occur in crystalline and non-crystalline material where the smallest gap leads to direct transitions; indirect transitions are valid according to the Tauc relations which involve simultaneous interaction with lattice vibrations and the wave vector of the electron. By creating a tangent line at the linear portions of the curve to $(\alpha h\nu)^{1/2} = 0$ and $(\alpha h\nu)^2 = 0$, the optical bandgap values (E_g) can be obtained. The interactions of photons with lattice vibrations take place at this allowed indirect transition. Any change in the oxygen bonding in the glass network will change the characteristic of absorption edge [48].

The band tail is due to the transitions from occupied extended state of the conduction band. Band tailing in the forbidden energy bandgap is observed in glass and amorphous materials. The width of this band tail is extended into the bandgap, and the edge is known as the Urbach energy. When the energy of the incident photon is less than the bandgap, the increase in absorption coefficient is followed with an exponential decay of density of states localized into the gap. The lack of crystalline long-range order in amorphous or glassy materials is associated with a tailing of density of states into forbidden energy. Urbach energy is known as the edge of the exponential decay of density of localized state in the forbidden gap which is causing the increase in absorption coefficient when the energy of the incident is less than the bandgap [49]. The disorderness of the material can be estimated using Urbach rule. At lower values of the absorption coefficient, the extent of the exponential tail of the absorption edge is characterized by the Urbach energy as given by the equation:

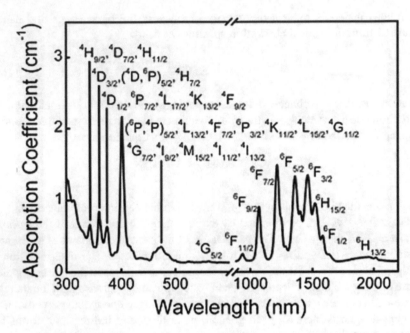

Fig. 6.7 UV-Vis-NIR absorption spectra of tellurite glasses with 1 mol% of Sm^{3+} ions [27]

$$\alpha(\omega) = B\exp\left(\frac{\hbar\omega}{E_u}\right) \quad (6.9)$$

where B is a constant and E_u is the width of the band tail of the electron states. In attempting to determine the Urbach energy E_u, the plot of $\ln\alpha$ against photon energy hv is plotted. The Urbach energy is obtained by extrapolating the linear region of the curve to $(hv = 0)$ followed by taking the reciprocals of the obtained values. Davis and Mott suggested that values of Urbach energy in crystalline and glassy material lie between ~ 0.045 and 0.67 eV [50].

The UV-Vis-NIR absorption spectra of tellurite glasses with different concentrations of Sm^{3+}ions are studied by many researchers, the 1 mol % of Sm^{3+} ion concentration was shown in Fig. 6.7. There are 14 absorption bands appearing at the wavelengths of 1820, 1585, 1527, 1478, 1374, 1223, 1075, 944, 473, 439, 417, 402, 375, and 362 nm. These are the characteristic wavelengths in the absorption spectrum of Sm^{3+} ions, which are attributed to transitions from the $^6H_{5/2}$ ground state to $^6H_{13/2}$, $^6F_{1/2}$, $^6H_{15/2}$, $^6F_{3/2}$, $^6F_{5/2}$, $^6F_{7/2}$, $^6F_{9/2}$, $^6F_{11/2}$, $(^4M_{11/2}, ^4I_{11/2,13/2})$, $(^4M_{17/2}, ^4G_{9/2}, ^4I_{9/2})$, $(^4M_{19/2}, ^6P_{5/2})$, $(^4F_{7/2}, ^6P_{3/2})$, $^6P_{7/2}$, and $^4D_{3/2,5/2}$ excited levels [27].

The UV absorption of a material can be used as a method to obtain the refractive index of the condensed matter which is derived practically by Dimitriv and Sakka [51]. They used the optical bandgap estimated from UV edge of absorption spectrum to calculate the refractive index using [52]:

$$\frac{n^2 - 1}{n^2 + 2} = 1 - \sqrt{\frac{E_g}{20}} \qquad (6.10)$$

The Lorentz-Lorenz equation that relates to molar refractivity (R_M), refractive index, and molar volume (V_M) are given by [52]:

$$R_M = \frac{n^2 - 1}{n^2 + 2}(V_M) \qquad (6.11)$$

This equation gives the average molar refraction of isotropic substances. The molar refractivity has the dimension of a volume (cm³/mol). The molar refraction which is related to the structure of the glass measures the bonding condition which gives the average R_M of isotropic substances. The Lorentz-Lorenz equation presents the polarizability which is the magnitude of the response of the electrons to an electromagnetic field. When Avogadro's number is introduced, the molar refractivity can be expressed as a function of polarizability. The polarizability (α_e) of the glasses is determined from [52]:

$$R_M = \frac{n^2 - 1}{n^2 + 2}(V_M) = \frac{4}{3}\pi N_A \alpha_e \qquad (6.12)$$

where N_A is Avogadro's number.

The room temperature UV-Vis-NIR absorption spectra of Series I and Series II are presented in Figs. 6.8 and 6.9, respectively. The absorption spectra are recorded in the range of 300–1800 nm. It can be observed that the absorption spectra for both Series I and Series II consist of well-defined nine prominent peaks, which are mainly attributed to the transition from ground level $^6H_{5/2}$ to the excited states of Sm^{3+} ion. The absorption peaks are identified and centered around 401, 474, 948, 1086, 1239, 1386, 1490, 1552, and 1597 nm for TNS samples and 402, 473, 941, 1076, 1226, 1368, 1469, 1534, and 1587 nm for TNSA samples corresponding to transition

Fig. 6.8 UV-Vis-NIR absorption spectra of (79-x) TeO$_2$-20Na$_2$O-xSm$_2$O$_3$ glass samples

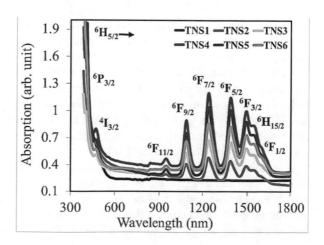

Fig. 6.9 UV-Vis-NIR absorption spectra of $(79-x)$ TeO_2-$20Na_2O$-$1Sm_2O_3$-$xAuCl_3$ glass samples

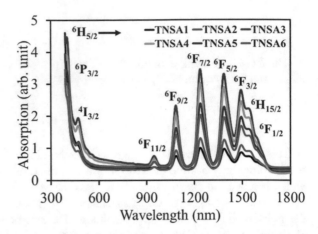

Table 6.3 Assignment of absorption levels and their energies of TNS glass samples for different concentrations of Sm_2O_3

Sample code	$TeO_2\%$	Na_2O %	$Sm_2O_3\%$	Transition band	Energy (cm^{-1})	Wavelength (nm)
TNS2	79.7	20	0.3	$^6P_{3/2}$	25,000	400
				$^4I_{11/2}$	21,097	474
TNS3	79.4	20	0.6	$^6F_{11/2}$	10,537	949
				$^6F_{9/2}$	9208	1086
TNS4	79	20	1	$^6F_{7/2}$	8071	1239
				$^6F_{5/2}$	7209	1387
TNS5	78.8	20	1.2	$^6F_{3/2}$	6706	1491
				$^6H_{15/2}$	6443	1552
TNS6	78.5	20	1.5	$^6F_{1/2}$	6261	1597

bands from ground state $^6H_{5/2}$ to excited states $^6P_{3/2}$, $^4I_{11/2}$, $^6F_{11/2}$, $^6F_{9/2}$, $^6F_{7/2}$, $^6F_{5/2}$, $^6F_{3/2}$, $^6H_{15/2}$, and $^6F_{1/2}$. The occurrence of all these peaks is in agreement with other findings, and the intensities of all absorption peaks vary for all glass samples as previously studied. All the transition bands were originated from the electric dipole contribution ($\Delta J \leq 6$), and these assignments are made by Mawlud et al. [18, 19]. The absorption band positions along their assignments of transitions for the TNS and TNSA glasses are given in Tables 6.3 and 6.4, respectively. For the most glassy oxide materials, no sharp absorption edge can be observed which emphasizes its amorphous nature [31].

It is clear that as the Sm_2O_3 content increased the energy of absorption increases, this is quite understandable since more energy is required to push electrons from the higher level to the upper level of the excited state. The increase of the absorbance with the addition of Au content signifies the increasing concentration of Au NPs. The intensities of all absorption peaks are increased by the addition of $AuCl_3$ ion from

Table 6.4 Assignment of absorption levels and their energies of TNSA glass samples for different concentrations of Au NPs

Sample code	TeO_2%	Na_2O%	Sm_2O_3%	$AuCl_3$%	Transition band	Energy (cm^{-1})	Wavelength (nm)
TNSA1	79	20	1	0			
					$^6P_{3/2}$	24966.5	400.5
TNSA2	78.8	20	1	0.2	$^4I_{11/2}$	21252.3	470.5
					$^6F_{11/2}$	10551.2	947.8
TNSA3	78.6	20	1	0.4	$^6F_{9/2}$	9209.9	1085.8
					$^6F_{7/2}$	8069.6	1239.2
TNSA4	78.4	20	1	0.6	$^6F_{5/2}$	7207.0	1387.5
					$^6F_{3/2}$	6697.2	1493.2
TNSA5	78.2	20	1	0.8	$^6H_{15/2}$	6476.5	1544.0
					$^6F_{1/2}$	6258.3	1597.9
TNSA6	78	20	1	1			

0.2 mol% to 1 mol%. This observation is in agreement with other findings. However, the appearance of plasmon band is not seen in the samples containing both Au NPs and Sm^{3+} ions due to the overlapping of the plasmon band with the Sm^{3+} peaks. Ground state absorption and excited state absorption are the important involved mechanisms. Using the data in Table 6.3, the plot of the partial absorption energy level diagram of Sm^{3+} ion is schematically drawn in order to discuss the downconversion process and the involved mechanisms, and the results are shown in Fig. 6.10 [18, 19].

6.6.1 Energy Bandgap and Urbach Energy

The calculation of optical absorption and the absorption edge play an important role in the electronic structure study of amorphous material. Direct ($n = 1/2$) and indirect ($n = 2$) optical band gaps are defined according to Davis and Mott theory [50]. The absorption coefficient ($\alpha(\omega)$) is plotted as a function of photon energy ($\hbar\omega$) at high values of absorption near the UV region. The values of direct, indirect band gaps and Urbach energy of Series I and Series II are listed in Tables 6.5 and 6.6, respectively, along with the reported optical properties for different tellurite glass compositions. Figures 6.11 and 6.12 for glass sample in Series I and Figs. 6.13 and 6.14 for glass sample in Series II show typical Tauc plot for E_{dir} and E_{ind} bandgap.

Direct and indirect bandgap energy reduced with the increasing of Au NPs. This is due to the energy contributed from Au NPs to NBO in the transition of electron from lower to upper state. The free electron at NBO has extra energy which can easily jump from valance to conduction state, thus producing lower optical energy

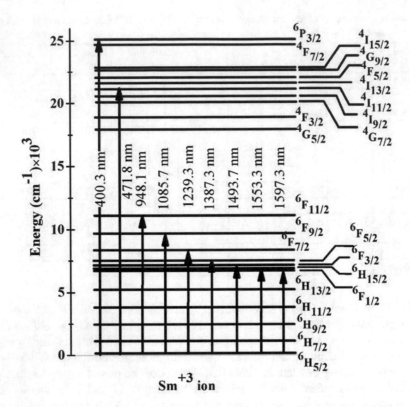

Fig. 6.10 Energy level diagram of absorption transitions of Sm^{3+} doped sodium tellurite glass matrix

bandgap. Another reason is, as in the case of Sm-Cl, as reported by Qi et al. [55], it is expected that the Sm-Cl will be formed as a result of adding $AuCl_3$ to the glass. Since the bonding of Sm-Cl is ionic bond, the release of an electron is easier than that of Sm-O. As a result, the value of E_{dir} and E_{ind} bandgap energy is decreased. The decreasing of E_{dir} and E_{ind} bandgap energy reflects to the creation of NBO. It is reported that NBO is more negatively charged than those of BO. Therefore, more electrons can easily be transferred from the valance band to the conduction band. This reflects a reduction of E_{dir} and E_{ind} optical bandgap energy in the glass. Furthermore, Halimah et al. [12] reported that the decrease in bandgap is mainly due to the addition of metallic NPs which causes less order in glass structure and there is a process of breaking of the regular structure of host glass. The Urbach energy (E_u) or band tail, which is a character of disorder in the amorphous materials, can be defined by a plot of natural logarithm of absorption coefficient ($\alpha(\omega)$) as a function of photon energy ($\hbar\omega$) and determining the inverse of its slope, as described by Eq. (6.9). Figures 6.15 and 6.16 show the plot of $\ln\alpha$ versus photon energy of Series I and Series II glass system, respectively. It can be observed that there is a linear part of the curve near the fundamental band edge. The linear part indicates an

Table 6.5 Glass codes, direct bandgap energy E_{dir} (eV), indirect bandgap energy E_{ind} (eV), and Urbach energy E_U (eV) of the proposed $(80-x)TeO_2-20Na_2O-xSm_2O_3$ glasses (Series I)

Sample code	Sm_2O_3	E_{dir}	E_{ind}	E_U	Reference
TNS1	0	3.044	2.544	0.312	Present work
TNS2	0.3	3.167	2.695	0.256	Present work
TNS3	0.6	3.186	2.769	0.226	Present work
TNS4	1	3.162	2.793	0.267	Present work
TNS5	1.2	3.135	2.712	0.279	Present work
TNS6	1.5	3.114	2.613	0.309	Present work
TMS	0–1.2	2.94–3.10	2.42–2.74	0.19–0.26	[53]
TNS	0.2–1	–	2.36–2.8	0.21–0.38	[54]

Table 6.6 Glass codes, direct bandgap energy E_{dir} (eV), indirect bandgap energy E_{ind} (eV), and Urbach energy E_U (eV) of the proposed $(79-x)TeO_2-20Na_2O-1Sm_2O_3-xAuCl_3$ glasses (Series II)

Sample code	$AuCl_3$	E_{dir}	E_{ind}	E_U	Reference
TNSA1	0	3.321	2.891	0.262	Present work
TNSA2	0.2	3.281	2.853	0.271	Present work
TNSA3	0.4	3.271	2.801	0.299	Present work
TNSA4	0.6	3.241	2.756	0.327	Present work
TNSA5	0.8	3.142	2.721	0.328	Present work
TNSA6	1	3.121	2.638	0.337	Present work
TMSA	0–1	2.88–3.01	2.58–2.82	0.18–0.26	[53]
TZSA	0.2–1	3.11–3.13	2.86–2.93	0.69–0.76	[18, 19]

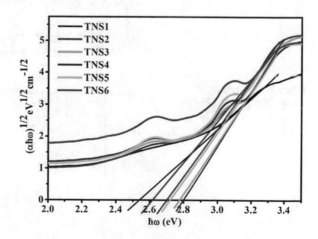

Fig. 6.11 Typical indirect energy bandgap for $(80-x)$ $TeO_2-20Na_2O-xSm_2O_3$ glasses

exponential behavior of the photon absorption process which reflects the formation of defect in the structural network. The fitting of the linear part agrees well with the Urbach tail behavior [18, 19].

The Urbach energy is estimated from the inverse slope of the linear part. The Urbach energy is depicted and listed in Tables 6.5 and 6.6 for glasses in Series I and

Fig. 6.12 Typical direct
energy bandgap for (80-*x*)
TeO$_2$-20Na$_2$O-*x*Sm$_2$O$_3$
glasses

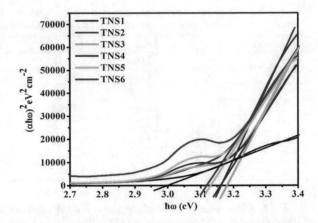

Fig. 6.13 Typical indirect
energy bandgap for (79-*x*)
TeO$_2$-20Na$_2$O-1Sm$_2$O$_3$-
*x*AuCl$_3$ glasses

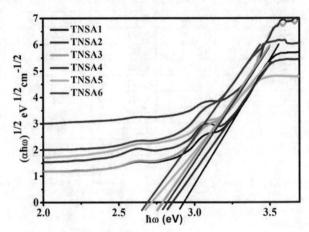

Fig. 6.14 Typical direct
energy bandgap for (79-*x*)
TeO$_2$-20Na$_2$O-1Sm$_2$O$_3$-
*x*AuCl$_3$ glasses

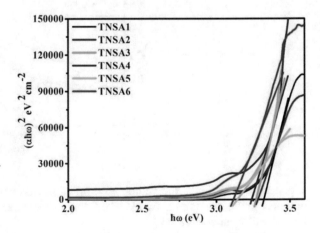

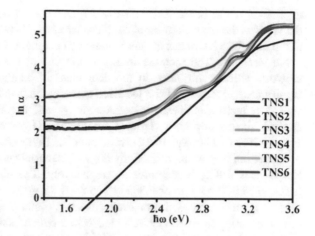

Fig. 6.15 Plot of $\ln\alpha$ versus $\hbar\omega$ for Urbach energy of $(80\text{-}x)$ TeO_2-$20Na_2O$-xSm_2O_3 glass system

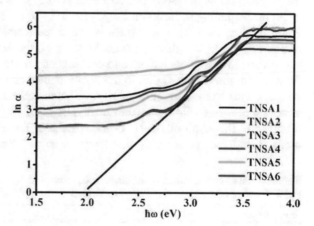

Fig. 6.16 Plot of $\ln\alpha$ versus $\hbar\omega$ for Urbach energy of $(79\text{-}x)$ TeO_2-$20Na_2O$-$1Sm_2O_3$-$xAuCl_3$ glass system

Series II, respectively. It can be seen that the Urbach energy changes inversely with optical energy bandgap value and this totally agrees with the results that were already done by Arunkumar and Marimuthu [46]. Urbach energy is attributed to disorderness of material, and the changes of Urbach energy's value is due to the formation of defects in the prepared glasses. It is observed that Urbach energy decreased up to 0.6 mol% which might be due to the arising of local long-range order in glass matrix, thus causing the decrease in the width of the band tail. However, as the concentration of Sm_2O_3 is beyond 0.6 mol%, higher Urbach energy is observed, possibly due to the arising of local short-range order from the maximum number defects. The Urbach energy at 1.5 mol% is consistent to the increase of optical bandgap energy which is due to the formation of intermediate bond of TeO_{3+1} whereas able to increase or decrease width of band tail. However, it is worth to notice that this is an anomaly phenomenon.

The Urbach energy increases as the concentration of Au NPs increased, this reflects the formation of oxygen vacancies, which closely related to the rising of local short-range order. The value of E_U is found to be increased from 0.262 to 0.337 eV as the concentration of Au NPs increased. E_U which is the scale of disorder in the material system has greater tendency to convert weak bonds into defects as its value increases. The increase in E_U occurs due to the densification of the glass network which contributes to the decrement in bandgap [12]. These behaviors signify more extension of the localized states within the gap [12].

The refractive indices, molar refractivity, and electronic polarizability of the studied glasses for both Series I and Series II were calculated according to Eqs. (6.10), (6.11), and (6.12), respectively. Tables 6.7 and 6.8 summarize the refractive index n, molar refractivity R_M, and electronic polarizabilities α_e for various Sm_2O_3 and $AuCl_3$ concentrations, respectively. The obtained value of refractive index to be in the range of 2.349–2.385 for TNS glass and 2.316–2.365 for TNSA glass are comparable with other tellurite glass systems as stated in Tables 6.7 and 6.8. The variation of refractive index which either increases or decreases can be explained in terms of disorder effect in the glass system [18, 19].

At the beginning of insertion Sm_2O_3 from 0.3 to 0.6 mol%, the refractive index gradually decreases. This decrement is due to the formation of TeO_{3+1} polyhedral from TeO_3 tbp in the glass network. In this part, the Sm atom has less impact of comparison toward glass lattices and consequently resulted in a lower lattice strain. Further addition of Sm_2O_3 up to 1.5 mol% shows that refractive index increased which reflects to the formation of NBO. This increment is due to the alteration of structural network with the transformation of TeO_4 tbp to TeO_{3+1} tp as has been observed in the result of optical energy bandgap. The structure changes toward a more disordered state with the substitution of Sm atoms. The electronic

Table 6.7 Glass codes, refractive indices n, molar refractions R_M (cm³/mol), and polarizabilities α_e ($\times 10^{-18}$ cm³) of the proposed $(80-x)TeO_2$-$20Na_2O$-xSm_2O_3 glasses (Series I)

Sample code	Sm_2O_3 mol%	n	R_M	α_e	Reference
TNS1	0	2.385	17.424	4.839	Present work
TNS2	0.3	2.354	17.233	4.786	Present work
TNS3	0.6	2.349	17.245	4.79	Present work
TNS4	1	2.355	17.224	4.784	Present work
TNS5	1.2	2.362	17.131	4.758	Present work
TNS6	1.5	2.367	17.509	4.863	Present work

Table 6.8 Glass codes, refractive indices n, molar refractions R_M (cm³/mol), and polarizabilities α_e ($\times 10^{-18}$ cm³) of the proposed $(79-x)TeO_2$-$20Na_2O$-$1Sm_2O_3$-$xAuCl_3$ glasses (Series II)

Sample codes	$AuCl_3$ mol%	n	R_M	α_e	Reference
TNSA1	0	2.316	17.159	4.766	Present work
TNSA2	0.2	2.325	17.242	4.789	Present work
TNSA3	0.4	2.328	17.224	4.784	Present work
TNSA4	0.6	2.335	17.444	4.845	Present work
TNSA5	0.8	2.360	17.675	4.909	Present work
TNSA6	1	2.365	19.058	5.293	Present work

polarizability drastically decreased at 1.2 mol%. This could be due to the transformation of BO to NBO being in a small fraction and the mixing of TeO_{3+1} which could not be able to increase the polarizability. Further increasing of Sm_2O_3 concentration up to 1.5 mol% shows that the electronic polarizability increases. At this point Sm atoms tend to break Te-O bonds and create a number of NBO which have higher ability to be polarized [56].

The insertion of Au NPs into the glass structure results in an increasing both of refractive index and electronic polarizability. This increment for some reasons has been clarified which includes the increasing of structural disorder, the increase of internal strain, and more change in stoichiometry [18, 19]. At this point, the additive of Au NPs may be able to change the structural network with the creation of more defects. Au atoms mostly stay at interstitial site, and there is compression toward Te-O bond structure. The interstitial atom Au causes the change in the initial Te-O bond length to a relative length, thus increasing the lattice strain.

6.6.2 Nephelauxetic Ratio and Bonding Parameter

Bonding properties can be investigated from the nephelauxetic ratio β and bonding parameter δ. The nephelauxetic ratio which indicates the properties of the rare earth-oxygen (RE-O) bonding can be calculated by using equation [57]:

$$\beta = \frac{\bar{v}_c}{v_a} \tag{6.13}$$

where v_c^- is the wave number (in cm^{-1}) of a particular transition in the host matrix under selection and v_a is the wave number (in cm^{-1}) of the same transition in the aquo-ion system. The value of $\bar{\beta}$ corresponds to "nephelauxetic effect" or shift in energy, due to metal-ligand orbital overlap [18, 19, 57]. From the mean values of β, the bonding parameter δ can be calculated using the expression:

$$\delta = \frac{\left(1 - \bar{\beta}\right)}{\bar{\beta}} \tag{6.14}$$

The bonding parameters depend on the field environment of the host network [58]. The positive or negative sign of δ is a measure of ionic or covalent nature of the bonding of RE ions and their surrounding ligands. The δ value indicates the bonding between RE ions and their surrounding ligand which depends on the field environment of the host network. Nephelauxetic effect occurs due to metal-ligand bond formation, that is, overlap between the metal and ligand orbitals forming larger molecular orbital leading to delocalization of the electron cloud over a larger area. Bonding properties can be investigated from the nephelauxetic ratio $(\bar{\beta})$ and bonding parameter (δ) which are calculated by using Eqs. 6.13 and 6.14. The observed band

Table 6.9 The transitions, energy levels, and nephelauxetic and bonding parameters for the (80-x) TeO$_2$-20Na$_2$O-xSm$_2$O$_3$ ($x = 0, 0.3, 0.6, 1, 1.2, 1.5$) glasses

Transition	Energy level (cm^{-1})					Aquo-ion [29]
	TNS2	TNS3	TNS4	TNS5	TNS6	
$^6H_{5/2} \rightarrow {}^6P_{3/2}$	24,961	25,021	25,000	25,001	25,022	24,999
$^6H_{5/2} \rightarrow {}^4I_{11/2}$	21,142	21,275	21,200	21,275	21,277	21,096
$^6H_{5/2} \rightarrow {}^6F_{11/2}$	10,550	10,539	10,548	10,542	10,553	10,517
$^6H_{5/2} \rightarrow {}^6F_{9/2}$	9201	9195	9197	9196	9193	9136
$^6H_{5/2} \rightarrow {}^6F_{7/2}$	8070	8070	8070	8068	8068	7977
$^6H_{5/2} \rightarrow {}^6F_{5/2}$	7208	7206	7205	7210	7213	7131
$^6H_{5/2} \rightarrow {}^6F_{3/2}$	6698	6698	6698	6700	6705	6641
$^6H_{5/2} \rightarrow {}^6H_{15/2}$	6440	6452	6447	6450	6448	6508
$^6H_{5/2} \rightarrow {}^6F_{1/2}$	6266	6272	6266	6278	6276	6397
β	9.0110	9.0205	9.0152	9.0212	9.0235	–
$\bar{\beta}$	1.0012	1.0023	1.0017	1.0024	1.0026	–
δ	−0.1216	−0.2278	−0.1681	−0.2345	−0.2602	–

positions (cm^{-1}) which are originating from the ground state $^6H_{5/2}$ to excited states $^6P_{3/2}$, $^4I_{11/2}$, $^6F_{11/2}$, $^6F_{9/2}$, $^6F_{7/2}$, $^6F_{5/2}$, $^6F_{3/2}$, $^6H_{15/2}$, and $^6F_{1/2}$ transitions, nephelauxetic ratio, and bonding parameters for Series I and Series II are summarized in Tables 6.9 and 6.10, respectively. It is observed that the bonding parameter δ possesses a negative sign for all glass samples in Series I and Series II which means that Sm^{3+} and ligand are ionic bonding in nature. It can be seen that the δ is increasing and decreasing with the increasing of Sm$_2$O$_3$ concentration. The increasing of bonding parameter could be explained from the value of electronegativity difference [59]. As the electronegativity between two atoms becomes higher, the polar covalent bond also becomes higher. This means that one of the atoms is more positive and the other has more negative charge. The atom which attracts electrons toward itself becomes more negative. The phenomenon of electron pairs shifting toward one atom creates the formation of covalent bonds. The electronegativity of Sm is 1.17, oxygen is 3.610, and Te is 2.158 which shows that the electronegativity difference between Sm and O is higher than those of Te-O bond. Thus, the addition of Sm$_2$O$_3$ at the expense of TeO$_2$ may increase the number of covalence bond. The electron pair that is Sm-O bonding together will shift toward the oxygen atom because of the larger electronegativity value of oxygen. It is observed that the value of bonding parameter shifted toward a higher direction with the increase of Au NPs concentration up to 0.4 mol%. The participation of Au has changed the site symmetry of Sm^{3+} and oxygen by moving it toward a higher asymmetry site which results in the asymmetrical distribution of electrons. Then the asymmetrical electrons possess a different electron density at different sites of oxygen atom where one site has more electrons (negativity) and the other site has less electrons (positivity) [18, 19, 59]. Such electrons must have a greater potential to be attracted by Sm^{3+} ion which leads to increase the number of covalence bond as well as decrease the number of ionic bonds. With

Table 6.10 The transitions, energy levels, and nephelauxetic and bonding parameters for the $(79-x)TeO_2$–$20Na_2O$–$1Sm_2O_3$–$xAuCl_3$ ($x = 0, 0.2, 0.4, 0.6, 0.8, 1$) glasses

Transition	Energy level (cm^{-1})						Aquo-ion [29]
	TNSA1	TNSA2	TNSA3	TNSA4	TNSA5	TNSA6	
$^6H_{5/2} \rightarrow \, ^6P_{3/2}$	24,995	24,922	24,946	24,946	24,995	24,995	24,999
$^6H_{5/2} \rightarrow \, ^4I_{11/2}$	21,323	21,270	21,078	21,323	21,270	21,252	21,096
$^6H_{5/2} \rightarrow \, ^6F_{11/2}$	10,553	10,551	10,542	10,564	10,547	10,551	10,517
$^6H_{5/2} \rightarrow \, ^6F_{9/2}$	9215	9205	9205	9218	9212	9205	9136
$^6H_{5/2} \rightarrow \, ^6F_{7/2}$	8074	8066	8069	8074	8069	8066	7977
$^6H_{5/2} \rightarrow \, ^6F_{5/2}$	7212	7204	7202	7212	7206	7204	7131
$^6H_{5/2} \rightarrow \, ^6F_{3/2}$	6701	6695	6694	6701	6699	6694	6641
$^6H_{5/2} \rightarrow \, ^6H_{15/2}$	6492	6466	6461	6487	6486	6466	6508
$^6H_{5/2} \rightarrow \, ^6F_{1/2}$	6263	6260	6240	6269	6261	6257	6397
β	9.0318	9.0177	9.0046	9.0315	9.0253	9.0190	–
$\bar{\beta}$	1.00353	1.00197	1.00051	1.00350	1.00281	1.00212	–
δ	−0.3517	−0.1966	−0.0514	−0.3485	−0.2804	−0.2111	–

Table 6.11 Comparison of average nephelauxetic ratio $\bar{\beta}$ and bonding parameter δ with different glass hosts

Sample code	$\bar{\beta}$	δ	Glass host	Reference
TNSA3	1.00051	−0.0514	Sodium tellurite	Present work
S3	1.0123	−0.121	Magnesium tellurite	[53]
PKAMZF-Sm10	1.0076	−0.7601	Sodium fluoroborate	[13]
B4TS	1.0040	−0.3984	Fluorophosphates	[60]
BLNS	1.0072	−0.7137	Borotellurite	[61]
1SmPbFb	1.0078	−0.7784	Fluoroborate	[46]

further addition of Au NPs up to 0.6 mol%, the bonding parameter will shift toward a lower value. This leads to increase the number of ionic bond.

It has been confirmed that the sample containing 1 mol% Sm_2O_3 and 0.4 mol% Au NPs possesses the highest covalence bond character compared to others. At this composition, Au NPs contents are sufficient to change the symmetrical site of Sm^{3+} because they have higher strength to attract electron. Since that, this composition possessed the lowest inter-electronic repulsion and the highest covalency between Sm^{3+} and the ligand in composition with other composition. Using this argument, a comparison with the other glass containing 1 mol% Sm_2O_3 in terms of its nephelauxetic ratio and bonding parameter can be made, and the results are listed in Table 6.11. It is observed that the covalence bond increases as 1SmPbFb<BLNS<B4TS<PKAMZFSm10<S3<TNSA3. Thus, it can be concluded that TNSA3 has lower inter-electronic repulsion of Sm-ligand which results in a strong covalence bond [18, 19].

6.7 Surface Plasmon Resonance

Transmission electron microscopy (TEM) permits the direct particles' morphology image inside the glass structure and provides information about the particles in terms of their shape and size distribution. The existence of nanocrystalline phase of Au in the presence of regular morphologies of the resultant microstructures after crystallization is verified by TEM analysis. The black spots verify the occurrence of Au NPs having different sizes and shapes of non-spherical NPs and are visibly dispersed homogeneously in the glass matrix. Figure 6.17a shows the TEM micrograph of the glass with 0.4 mol% of $AuCl_3$ (TNSA3) [18, 19].

The shapes of the NPs exert prominent effects in the optical absorption spectra of metal NPs compared to the size and surrounding effect [43]. The fusion of Au NPs within glass matrix restricted the electromagnetic energy or optical excitation in a nanoscale volume, and the local field enhancements mediated strong optical interactions within this volume. The occurrence of broad particle size distribution is shown in Fig. 6.17b which displays the histogram fitted to a Gaussian curve. From Fig. 6.17b, the average diameter of Au NPs inside the glass matrix is

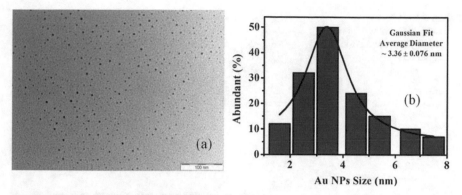

Fig. 6.17 (**a**) TEM image of glass containing 0.4 mol% of Aucl₃ (TNSA3 glass) displays the distribution of spherical and non-spherical gold NPs. (**b**) Histogram of size distribution of gold NPs with average diameter of ~ 3.36 nm for respective glass

Fig. 6.18 HRTEM image of a part of Au NPs

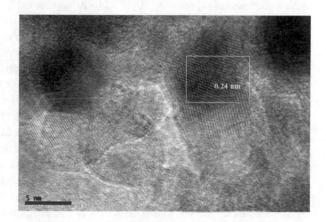

~3.36 ± 0.076 nm for AuCl₃ 0.4 mol% (TNSA3). The different size distributions of the NPs caused by the diffusion limit of the NPs growth [37].

Figure 6.18 shows the high-resolution transmission electron microscopy (HRTEM). The fringe spacing of gold was 0.24 nm which corresponded well to the spacing between (111) planes of face-centered cubic gold (JCPDS card No. 04-0784).

It is well-known that the absorption in the UV-Vis spectrum is the most reliable tool to analyze the function of metallic nanoparticles and to observe the existence of SPR band. The absorption spectra of Sm^{3+} ions exhibiting several bands in the visible region overshadow the plasmon band of gold NPs [18, 19, 38]. No plasmon peak is observed in co-doped Au NPs and Sm^{3+} ions with sodium tellurite glasses due to the overlapping of the plasmon band with the Sm^{3+} peaks. In order to study and to locate the plasmon resonance band clearly, Series III (TNA) of glass samples

Fig. 6.19 UV-Vis
absorption spectra of glass is
showing SPR band positions
of gold NPs

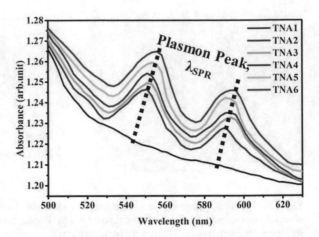

with a chemical composition $(80-x)\text{TeO}_2\text{-}20\text{Na}_2\text{O-}x\text{AuCl}_3$ ($x = 0$, 0.2, 0.4, 0.6, 0.8, 1 mol%) and without Sm_2O_3 is prepared. The clear evidence of intensive SPR absorption band around 550–590 nm in the UV-Vis spectrum is shown in Fig. 6.19; it confirms the existence of nanosized gold particles in the glass matrix.

Figure 6.19 presents the absorption spectra of TNA glass system which indicates the presence of two SPR bands centered at about 550 and 590 nm. The distinct plasmon bands are revealed from the elongated NPs which are related to transverse and longitudinal SPR oscillations of the Au NPs, respectively. These peaks are in a good agreement with the previous study (530 nm). In the previous study, it was reported that the observation of SPR band of isolated gold NPs is in the region of 500–600 nm [45]. With further addition of Au NPs in the glass matrix, a slight significant red shift in the plasmon peak is observed. The plasmon bands shift from 550 to 556 nm (red shift). This shift is related to the higher refractive index possessed by most tellurite glasses which normally lies approximately in the range of 2–2.5 [20, 21]. This is particularly reliable since the higher refractive index will cause the SPR to shift to a longer wavelength. The red shift in the SPR absorption band can be also ascribed to the increase in mean diameter of gold NPs. However, the position of plasmon band is comparable and positioned in the visible region. The intensity of both plasmon peaks is not so immense. This is due to the existence of spherical and non-spherical shapes of Au NPs in the glass matrix. It is established that NPs diameter and the features of SPR band can be tuned by changing the concentration of Au [31].

6.8 Photoluminescence Spectral Analysis

Photoluminescence (PL) is the emission of light from a material under optical excitation [37]. Luminescence is the phenomenon involving the energy absorption by material and subsequent re-emission of light. Excitation of the luminescent substance occurs when an electron returns to the electronic ground state (valence band) from an excited state (conduction band) and loses its excess energy as a photon. In the metal, the energy of the emitted photon is very small, since the valence and conduction bands overlap and the wavelength is longer than the visible spectrum. Therefore, luminescence does not occur. However, in certain glass materials, the energy bandgap between the valence and conduction band is such that an electron dropping through this gap produces a photon in the visible range [18, 19, 37].

Luminescence occurs from electronically excited states; and this phenomenon is divided into two categories including fluorescence and phosphorescence. The type of luminescence is decided by the nature of the excited state. Fluorescence occurs when the stimulus agent is removed. As a result, all excited electrons drop back to valence band, and the photon is emitted within about 10^{-8} s. The radiative lifetimes of the excited electronic states, which are responsible for the time constants for fluorescence, vary from 10^{-10} s to 10^{-1} s [57]. Phosphorescence occurs when the stimulated electron first drops into the donor level and is trapped, and then the electron must escape the trap before returning to the valence band [18, 19, 37]. The visible down-conversion of the luminescence spectra for the prepared glass in Series I and Series II are shown in Figs. 6.20 and 6.21, respectively. The emission spectra were recorded at a room temperature (25 °C) under 404 nm excitation wavelengths. The emission spectra exhibit four emission bands centered at 563, 599, 644, and 703 nm for TNS glass system and 577, 614, 658, and 718 nm for TNSA glass system. The emission peaks were assigned to the transition of $^{4}G_{5/2} \rightarrow {}^{6}H_{5/2}$, $^{4}G_{5/2} \rightarrow {}^{6}H_{7/2}$, $^{4}G_{5/2} \rightarrow {}^{6}H_{9/2}$, and $^{4}G_{5/2} \rightarrow {}^{6}H_{11/2}$, respectively. From these emission bands, a possible of moderate green, moderate yellow, intense orange, and feeble red color could be expected for all samples in Series I and Series II.

Obviously, the relative intensity in transition bands $^{4}G_{5/2} \rightarrow {}^{6}H_{7/2}$ is higher compared to the other transition bands, and $^{4}G_{5/2} \rightarrow {}^{6}H_{11/2}$ is found to be weak in intensity which is in a good agreement with the previous research [62]. The obtained emission peaks of glass system confirm the presence of samarium trivalent state that results in large luminescence efficiency. A change in the emission profile with the

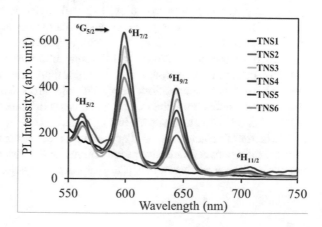

Fig. 6.20 Downconversion photoluminescence spectra of Sm^{3+}-doped tellurite glass with different concentrations of Sm^{3+} ions

Fig. 6.21 Downconversion
photoluminescence spectra
of Sm^{3+}-doped tellurite
glass embedded with Au
NPs with different
concentrations of AuCl$_3$
ions

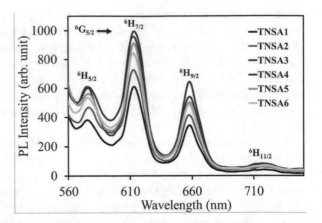

introduction of Au NPs is clearly noticeable. Furthermore, the glasses containing Au
NPs possess higher intensities for all emission bands.

Figure 6.22 shows the simplified emission energy diagram of Sm^{3+}-doped
TeO$_2$-Na$_2$O glass embedded with Au NPs. Thus, the possible mechanism for
Sm^{3+} emission can be explained through the partial energy diagram. It can be
seen that, as the Sm^{3+} ions are pumped with 404 nm excitation wavelength, they
excite to ^{6}P$_{3/2}$ state by ground state absorption (GSA) after getting the extra energy
from SPR of Au NPs. An energy transfer enhancement (ETE) may occur which
promotes the higher number of photon emitted which can be detected on PL
intensity. Then some of Sm^{3+} ions were found relaxed NR to a relaxation lower
level ^{4}G$_{5/2}$ since the ^{6}P$_{3/2}$ level has comparatively a shorter lifetime. Meanwhile, in
the intermediate energy level, there is an energy transfer between Sm^{3+} ions called
cooperative energy transfer (CET) between ^{4}G$_{5/2}$ and ^{6}P$_{3/2}$. A similar mechanism
in other systems containing Sm^{3+} ions embedded with Ag NPs has also been
established elsewhere. The Sm^{3+} ions populated ^{4}G$_{5/2}$, and then they decay to
^{6}H$_{5/2}$, ^{6}H$_{7/2}$, ^{6}H$_{9/2}$, and ^{6}H$_{11/2}$ by emitting moderate green, moderate yellow,
intense orange, and feeble red color. This is depicted in the energy level diagram
in Fig. 6.22. The ^{4}G$_{5/2}$ energy level possesses purely radiative relaxation due to
sufficient energy gap of ~7200 cm^{-1} with respect to the next lower level ^{6}H$_{11/2}$.
It is very important to mention that if the concentration of Sm^{3+} is too high, energy
transfers between them can occur and dominate the energy transfer process
[18, 19]. At this concentration, energy transfer from Sm^{3+} to Au NPs will take
place, and the quenching phenomenon is observed. The energy transfer quenching
(ETQ) of this kind is also shown in Fig. 6.22.

The factor of enhancement and quenching of the emission intensity is determined
from the ratio between intensity of the samples in the series and intensity of the first
sample in the respective series. The behavior of quenching or enhancement of

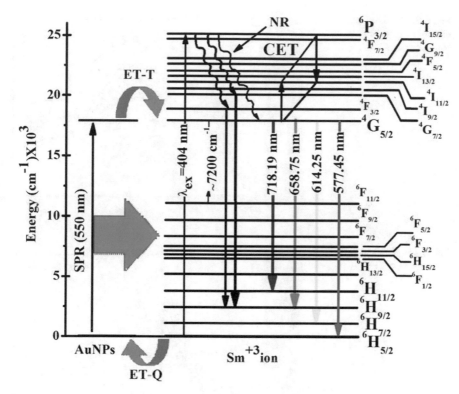

Fig. 6.22 Schematic energy level diagram of Sm^{3+} ion shows ground state absorption and excited state absorption mechanisms

fluorescence is attributed to the geometry-dependent variation in nanometal surfaces or particles [7]. In gold NPs, plasmons can be excited at frequencies that can be tuned through the visible range [56]. A plot of integrated intensity versus concentration of Sm_2O_3 mol% and $AuCl_3$ mol% are drawn as shown in Figs. 6.23 and 6.24 for Series I and Series II glass system, respectively. Enhancement and quenching effect also occur in glass with different Sm_2O_3 as shown in Fig. 6.23. It is observed that the PL intensity shows the increment with the addition of Sm_2O_3 up to 1 mol%. This enhancement indicated that Sm^{3+} leads to increase the probability of photon emitting. However, beyond 1 mol% Sm_2O_3, the intensity is quenched by a factor of 1.63:1.53:1.61:1.76 times for all the transitions. This quenching is due to the decrease in distance between Sm^{3+} ions which imply the energy transfer of Sm^{3+} ions due to the existence of inter-atomic interaction [63]. A similar approach has also previously been done in the Sm^{3+}-doped magnesium tellurite glasses embedded with Ag NPs.

From Fig. 6.24, it is observed that the PL intensity is enhanced as Au NPs concentration increased up to 0.4 mol % for 1.90:1.82:1.97:2.25 times to the one without Au NPs for all transition bands $^4G_{5/2} \rightarrow {}^6H_{5/2}$, $^4G_{5/2} \rightarrow {}^6H_{7/2}$, $^4G_{5/2} \rightarrow {}^6H_{9/2}$, and $^4G_{5/2} \rightarrow {}^6H_{11/2}$, respectively. This enhancement is attributed by two possible

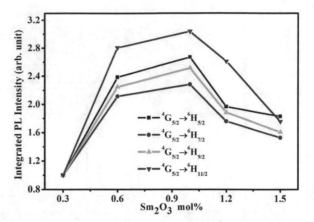

Fig. 6.23 PL-integrated intensity versus Sm_2O_3 concentration

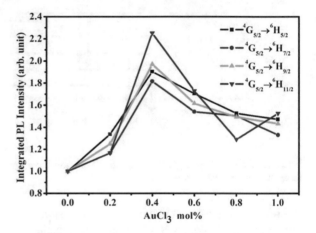

Fig. 6.24 PL-integrated intensity versus $AuCl_3$ concentration

reasons, firstly due to the local field effect of Au NPs in the vicinity of Sm^{3+} ions. This is due to the difference between the relative permittivity of the metal and the surrounding glass matrix, which would allow the SPR to produce surface waves that move along the metal dielectric interface [18, 19, 64]. As a result, the electromagnetic energy will be concentrated more tightly; thus, it produces the giant electric field around the particle especially at a sharp metallic tip [38]. Hence, with respect to the incident field, a large local electric field is induced in the vicinity of lanthanide ions. Secondly, it is due to an energy transfer between the Au NPs and Sm^{3+} ions [33]. A similar enhancement has also been observed in the glass containing Sm^{3+}-doped magnesium tellurite embedded with Ag NPs. Further addition of Au NPs, the quenching effect can be observed in sample with concentration more than 0.4 mol% of Au NPs. The emission intensity is decreased which is due to the energy transfer from Sm^{3+} to Au species. Since the largest enhancement for most of the sample was observed at 0.4 mol% Au NPs, it is assumed that this is the best concentration for the Au NPs contribution in achieving orange-red laser production.

References

1. R. De Almeida, D.M. da Silva, L.R. Kassab, C.B. de Araujo, Eu^{3+} luminescence in tellurite glasses with gold nanostructures. Opt. Commun. **281**(1), 108–112 (2008)
2. D.A. Drabold, X. Zhang, S. Nakhmanson, in *Properties and Applications of Amorphous Materials*, NATO ASI series, ed. by M. F. Thorpe, L. Tichy (Kluwer Academic Publishers, Dordrecht, 2001)
3. I.Z. Hager, R. El-Mallawany, Preparation and structural studies in the (70-x) TeO_2-20WO_3-10Li_2O-xLn_2O_3 glasses. J. Mater. Sci. **45**(4), 897 (2010)
4. I.Z. Hager, R. El-Mallawany, A. Bulou, Luminescence spectra and optical properties of TeO_2-WO_3-Li_2O glasses doped with Nd, Sm and Er rare earth ions. Phys. B Condens. Matter **406**(4), 972–980 (2011)
5. N. Kiran, C.R. Kesavulu, A.S. Kumar, J.L. Rao, Spectral studies on Mn^{2+} ions doped in sodium-lead borophosphate glasses. Phys. B Condens. Matter **406**(20), 3816–3820 (2011)
6. H. Li, Y. Wang, T. Wang, Z. Li, Ionic liquids and rare earth soft luminescent materials, in *Application of Ionic Liquids on Rare Earth Green Separation and Utilization*, (Springer, Berlin Heidelberg, 2016), pp. 157–178
7. A.P. Carmo, M.J.V. Bell, V. Anjos, R. De Almeida, D.M. Da Silva, L.R.P. Kassab, Thermo-optical properties of tellurite glasses doped with Eu^{3+} and Au nanoparticles. J. Phys. D. Appl. Phys. **42**(15), 155404 (2009)
8. N.S. Hussain, G. Hungerford, R. El-Mallawany, M.J.M. Gomes, M.A. Lopes, N. Ali, J.D. Santos, S. Buddhudu, Absorption and emission analysis of RE^{3+} (Sm^{3+} and Dy^{3+}): Lithium boro tellurite glasses. J. Nanosci. Nanotechnol. **9**(6), 3672–3677 (2009)
9. T. Turba, M.G. Norton, I. Niraula, D.N. McIlroy, Ripening of nanowire-supported gold nanoparticles. J. Nanopart. Res. **11**(8), 2137 (2009)
10. S. Thomas, R. George, S.N. Rasool, M. Rathaiah, V. Venkatramu, C. Joseph, N.V. Unnikrishnan, Optical properties of Sm^{3+} ions in zinc potassium fluorophosphate glasses. Opt. Mater. **36**(2), 242–250 (2013)
11. L. Balachandera, G. Ramadevudub, M. Shareefuddina, R. Sayannac, Y.C. Venudharc, IR analysis of borate glasses containing three alkali oxides. ScienceAsia **39**, 278 (2013)
12. M.K. Halimah, W.M. Daud, H.A.A. Sidek, A.W. Zaidan, A.S. Zainal, Optical properties of ternary tellurite glasses. Mater. Sci. Pol. **28**, 173–180 (2010)
13. M.R. Dousti, M.R. Sahar, S.K. Ghoshal, R.J. Amjad, A.R. Samavati, Effect of AgCl on spectroscopic properties of erbium doped zinc tellurite glass. J. Mol. Struct. **1035**, 6–12 (2013)
14. M.M. El-Zaidia, A.A. Ammar, R.A. El-Mallwany, Infra-red spectra, electron spin resonance spectra, and density of (TeO_2) 100-x-(WO_3)x and (TeO_2) 100-x-($ZnCl_2$) x glasses. Phys. Status Solidi(a) **91**(2), 637–642 (1985)
15. N.S. Hussein, J.D. da Silva Santos, *Physics and Chemistry of Rare-Earth Ions Doped Glasses* (Trans Tech Publications, Stafa-Zuerich, 2008)
16. R. El-Mallawany, *Tellurite Glasses Handbook: Physical Properties and Data* (CRC Press, Boca Raton, 2011)
17. J.E. Shelby, *Introduction to Glass Science and Technology, Thomas Graham House, UK,* (Royal Society of Chemistry, 2005)
18. S.Q. Mawlud, M.M. Ameen, M.R. Sahar, K.F. Ahmed, Plasmon-enhanced luminescence of samarium doped sodium tellurite glasses embedded with gold nanoparticles: Judd-Ofelt parameter. J. Lumin. **190**, 468–475 (2017)

19. S.Q. Mawlud, M.M. Ameen, M.R. Sahar, Z.A.S. Mahraz, K.F. Ahmed, Spectroscopic properties of Sm^{3+} doped sodium-tellurite glasses: Judd-Ofelt analysis. J. Opt. Mater. **69**, 318–327 (2017)

20. P.G. Pavani, K. Sadhana, V.C. Mouli, Optical, physical and structural studies of boro-zinc tellurite glasses. Phys. B Condens. Matter **406**(6), 1242–1247 (2011)

21. P.G. Pavani, S. Suresh, V.C. Mouli, Studies on boro cadmium tellurite glasses. Opt. Mater. **34** (1), 215–220 (2011)

22. V. Ter-Mikirtychev, *Fundamentals of Fiber Lasers and Fiber Amplifiers, Springer International Publishing Switzerland, vol 181* (Springer, 2014)

23. B. Astuti, *Study of Optical Properties of P_2O_5-Sm_2O_3-MnO_2 Glass System*, Doctoral dissertation, Universiti Teknologi Malaysia, Faculty of Science, 2005

24. Selvaraju K. and Marimuthu K., Structural and spectroscopic studies on concentration dependent Sm^{3+} doped boro-tellurite glasses. J. Alloys Compd. **553**, 273–281 (2013)

25. A. Abdel-Kader, R. El-Mallawany, M.M. Elkholy, Network structure of tellurite phosphate glasses: Optical absorption and infrared spectra. J. Appl. Phys. **73**(1), 71–74 (1993)

26. S. Hufner, *Optical Spectra of Transparent Rare Earth Compounds, Academic Press INC. New York, USA,* (Elsevier, 2012)

27. V.B. Sreedhar, C. Basavapoornima, C.K. Jayasankar, Spectroscopic and fluorescence properties of Sm^{3+}-doped fluorophosphate glasses. J. Rare Earths **32**(10), 918–926 (2014)

28. C.H. Huang, *Rare Earth Coordination Chemistry: Fundamentals and Applications, John Wiley & Sons (Asia) Pte Ltd,* (Wiley, 2011)

29. W.T. Carnall, P.R. Fields, K. Rajnak, Electronic energy levels in the trivalent lanthanide aquo ions. I. Pr^{3+}, Nd^{3+}, Pm^{3+}, Sm^{3+}, Dy^{3+}, Ho^{3+}, Er^{3+}, and Tm^{3+}. J. Chem. Phys. **49**(10), 4424–4442 (1968)

30. R. El-Mallawany, M. Sidkey, A. Khafagy, H. Afifi, Ultrasonic attenuation of tellurite glasses. Mater. Chem. Phys. **37**(2), 197–200 (1994)

31. X. Huang, P.K. Jain, I.H. El-Sayed, M.A. El-Sayed, Gold nanoparticles: Interesting optical properties and recent applications in cancer diagnostics and therapy. Nanomedicine **2**(5), 681–693 (2007)

32. A. Rastar, M.E. Yazdanshenas, A. Rashidi, S.M. Bidoki, Theoretical review of optical properties of nanoparticles. J. Eng. Fabrics Fibers (JEFF) **8**(2), 85–96 (2013)

33. Z. Pan, A. Ueda, R. Aga, A. Burger, R. Mu, S.H. Morgan, Spectroscopic studies of Er^{3+} doped Ge-Ga-S glass containing silver nanoparticles. J. Non-Cryst. Solids **356**(23), 1097–1101 (2010)

34. N.M. Yusoff, M.R. Sahar, Effect of silver nanoparticles incorporated with samarium-doped magnesium tellurite glasses. Phys. B Condens. Matter **456**, 191–196 (2015)

35. A. Shakiba, S. Shah, A.C. Jamison, I. Rusakova, T.C. Lee, T.R. Lee, Silver-free gold nanocages with near-infrared extinctions. ACS Omega **1**(3), 456–463 (2016)

36. T.S.T. Amran, M.R. Hashim, N.K.A. Al-Obaidi, H. Yazid, R. Adnan, Optical absorption and photoluminescence studies of gold nanoparticles deposited on porous silicon. Nanoscale Res. Lett. **8**(1), 1 (2013)

37. L.R. Kassab, C.B. de Araújo, R.A. Kobayashi, R. de Almeida Pinto, D.M. da Silva, Influence of silver nanoparticles in the luminescence efficiency of Pr^{3+}-doped tellurite glasses. J. Appl. Phys. **102**(10), 103515 (2007)

38. Rivera V.A.G., Osorio S.P.A., Ledemi Y., Manzani D., Messaddeq Y. , Nunes L.A.D.O., Marega E., Localized surface plasmon resonance interaction with Er^{3+}-doped tellurite glass. Opt. Express **18**(24), 25321–25328 (2010)

39. L.R. Kassab, R. de Almeida, D.M. da Silva, T.A. de Assumpção, C.B. de Araújo, Enhanced luminescence of Tb^{3+}/Eu^{3+} doped tellurium oxide glass containing silver nanostructures. J. Appl. Phys. **105**(10(2009)), 103505 (2009)

40. L.R. Kassab, D.S.D. Silva, R.D. Almeida, C.B.D. Araujo, Photoluminescence enhancement by gold nanoparticles in Eu^{3+} doped GeO_2-Bi_2O_3 glasses. Appl. Phys. Lett. **94**(10), 101912 (2009)

41. H.M. Moawad, H. Jain, R. El-Mallawany, T. Ramadan, M. El-Sharbiny, Electrical conductivity of silver vanadium tellurite glasses. J. Am. Ceram. Soc. **85**(11), 2655–2659 (2002)
42. J. Jana, M. Ganguly, T. Pal, Enlightening surface plasmon resonance effect of metal nanoparticles for practical spectroscopic application. RSC Adv. **6**(89), 86174–86211 (2016)
43. A. Awang, S.K. Ghoshal, M.R. Sahar, M.R. Dousti, R.J. Amjad, F. Nawaz, Enhanced spectroscopic properties and Judd–Ofelt parameters of Er-doped tellurite glass: Effect of gold nanoparticles. Curr. Appl. Phys. **13**(8), 1813–1818 (2013)
44. D.S. Da Silva, T.A. de Assumpção, L.R. Kassab, C.B. de Araújo, Frequency upconversion in Nd^{3+} doped $PbO-GeO_2$ glasses containing silver nanoparticles. J. Alloys Compd. **586**, S516–S519 (2014)
45. S.P. Osorio, V.A.G. Rivera, L.A.O. Nunes, E. Marega Jr., D. Manzani, Y. Messaddeq, Plasmonic coupling in Er^{3+}: Au tellurite glass. Plasmonics **7**(1), 53–58 (2012)
46. Arunkumar S. and Marimuthu K., Concentration effect of Sm^{3+} ions in B_2O_3 - PbO - PbF_2 - Bi_2O_3 - ZnO glasses-structural and luminescence investigations. J. Alloys Compd. **565**, 104–114 (2013)
47. T. Owen, *Fundamentals of Modern UV-Visible Spectroscopy, Hellwet -Packard, Palo Alto, CA,* (Agilent Technologies, 2000)
48. L.S. El-Deen, M.S. Al Salhi, M.M. Elkholy, IR and UV spectral studies for rare earths-doped tellurite glasses. J. Alloys Compd. **465**(1), 333–339 (2008)
49. C.R. Kesavulu, R.P.S. Chakradhar, C.K. Jayasankar, EPR, optical, photoluminescence studies of Cr^{3+} ions in $Li_2O-Cs_2O-B_2O_3$ glasses-An evidence of mixed alkali effect. J. Mol. Struct. **975**(1), 93–99 (2010)
50. Mott N. and Davis E., Conduction in non-crystalline systems V. Conductivity, optical absorption and photoconductivity in amorphous semiconductors. Philos. Mag. **22**(179), 0903–0922 (1970)
51. V. Dimitriv, S. Sakka, Electronic oxide polarizability and optical basicity of simple oxides. J. Appl. Phys. **79**, 1736–1741 (1996)
52. M. Born, E. Wolf, *Principles of Optics, Electromagnetic Theory of Propagation, Interference and Diffraction of Light*, 2nd edn. (Macmillan Company, New York, 1964), p. 325
53. Y. Nurulhuda, *Physical and Spectroscopic Characterization of Samarium Doped Magnesium Tellurite Glass Embedded Silver Nanoparticle*, PhD Thesis, Universiti Teknologi Malaysia, Skudai, 2015
54. Saman Q. Mawlud, M.M. Ameen, R. Md. Sahar, K.F. Ahmed, Influence of Sm_2O_3 ion concentration on structural and thermal modification of TeO_2-Na_2O glasses. J. Appl. Mech. Eng. **5**, 5 (2016)
55. J. Qi, T. Xu, Y. Wu, X. Shen, S. Dai, Y. Xu, Ag nanoparticles enhanced near-IR emission from Er^{3+} ions doped glasses. Opt. Mater. **35**(12), 2502–2506 (2013)
56. B. Eraiah, Optical properties of lead-tellurite glasses doped with samarium trioxide. Bull. Mater. Sci. **33**(4), 391–394 (2010)
57. F. Wooten, *Optical Properties of Solids, Academic Press INC. New York, USA,* (Academic Press, 2013)
58. P.V. Do, V.P. Tuyen, V.X. Quang, N.T. Thanh, V.T.T. Ha, N.M. Khaisukov, Y. Lee, B.T. Huy, Judd-Ofelt analysis of spectroscopic properties of Sm^{3+} ions in K2YF5 crystal. J. Alloys Compd. **520**, 262–265 (2012)
59. A. Awang, S.K. Ghoshal, M.R. Sahar, R. Arifin, F. Nawaz, Non-spherical gold nanoparticles mediated surface plasmon resonance in Er^{3+} doped zinc-sodium tellurite glasses: Role of heat treatment. J. Lumin. **149**, 138–143 (2014)
60. M. Abdel-Baki, F. El-Diasty, Role of oxygen on the optical properties of borate glass doped with ZnO. J. Solid State Chem. **184**(10), 2762–2769 (2011)
61. K. Maheshvaran, K. Linganna, K. Marimuthu, Composition dependent structural and optical properties of Sm^{3+} doped boro-tellurite glasses. J. Lumin. **131**(12), 2746–2753 (2011)

62. K. Swapna, S. Mahamuda, A.S. Rao, S. Shakya, T. Sasikala, D. Haranath, G.V. Prakash, Optical studies of Sm^{3+} ions doped zinc alumino bismuth borate glasses. Spectrochim. Acta A Mol. Biomol. Spectrosc. **125**, 53–60 (2014)
63. S. Sailaja, C.N. Raju, C.A. Reddy, B.D.P. Raju, Y.D. Jho, B.S. Reddy, Optical properties of Sm^{3+}-doped cadmium bismuth borate glasses. J. Mol. Struct. **1038**, 29–34 (2013)
64. R. El-Mallawany, *Tellurite Glasses, Handbook: Physical Properties and Data* (CRC Press, Boca Raton, 2002)

Chapter 7
Lanthanide-Doped Zinc Oxyfluorotellurite Glasses

M. Reza Dousti and Raja J. Amjad

Abstract In this chapter, we present the versatile examples and optical properties of zinc oxyfluorotellurite glasses doped with rare earth ions, and current challenges in this field are discussed. Zinc-tellurite glasses are among the most important heavy metal glass compositions with a wide range of excellent structural, thermal, chemical, and optical properties. When doped with rare earth ions, zinc-tellurite glasses show superior properties than other glass compositions, such as wide broadband luminescence and efficient upconversion emissions of Er^{3+} ions, as well as high rare earth solubility, which facilitate the incorporation of sensitizers such as Ce^{3+} and Yb^{3+} ions. When modified with some fluoride components, the optical and thermal stability of rare-earth-doped zinc-tellurite glasses does not change drastically, while the average phonon energy stays in a low-range energy, and the excited state lifetime of the rare-earth ions increases due to the different site symmetry provided by F^{-1} ions rather than O^{-2} ions. The recent developments in the oxyfluorotellurite glasses doped with rare earth ions are given in this chapter, which will be compared to those achieved with the zinc-tellurite oxide glasses.

7.1 Introduction

Material science is a branch of knowledge which involves in classification of solid substances by their versatile structural properties and exterior figure. It is the combination of three major branches of science: chemistry, physics, and engineering, which aim to build useful and economical composites. In a chemical taxonomy, materials are classified in four different classes categorized by their electronic and bonding structures: metallic, ionic, covalent, and van der Waals. In an engineer's

M. Reza Dousti (✉)
Grupo de Nano-Fotônica e Imagens, Instituto de Física, Universidade Federal de Alagoas, Maceió, AL, Brazil
e-mail: mrdphysics@gmail.com

R. J. Amjad
Department of Physics, COMSATS Institute of Information Technology, Lahore, Pakistan

© Springer International Publishing AG, part of Springer Nature 2018
R. El-Mallawany (ed.), *Tellurite Glass Smart Materials*,
https://doi.org/10.1007/978-3-319-76568-6_7

143

point of view, a compound can be set in one of four, metals, semiconductors, polymers, and ceramics, groups. Physicists divide the solid materials to either crystalline or noncrystalline (amorphous) solids.

Composites like ceramics and glasses are complimentary materials with diverse applications from solar cells, optical amplifiers, solid-state lasers, and army applications to development of medical devices. Over the years, necessity of research on glasses is increased to understand their structure, properties, and functionality. An exponential increase in number of publications in the field of glasses in general and tellurite glass in particular is an evident of such needs [1]. According to the Scopus database [1], more than 180,000 documents are indexed on glasses in the last 50 years, where more than 108,000 documents out of total (60%) are published since 2000 to now (end of 2017). In this regard, 50.6% of documents are indexed in the field of materials science; 39.9%, 33.4%, and 15.6% of documents are indexed by physics (and astronomy), engineering, and chemistry, respectively.

On the other hand, special optical properties of the rare-earth (RE) ions and their photonic applications in addition to non-radiative energy transfer (ET) processes led to a wide study on REs. In principle ET processes may favor particular applications (such as the operation of anti-Stokes emitters), but it may be detrimental as in the case of RE-based lasers because interactions among the active ions contribute for the increase of the laser threshold. In particular, the study of ET processes in glasses having frequency gap in the visible region deserves large attention because when doped with RE ions, some glasses may present efficient visible luminescence. The efficient emission intensity of REs embedded in glasses is limited to particular concentration so that further introduction of REs results in clustering of dopants or inefficient energy transfers. The formation of such clusters rises to increasing the non-radiative (NR) transfer rates between REs; therefore "concentration quenching" phenomena will occur [2, 3]. Several methods have been proposed to avoid the quench phenomena by increasing the efficiency of RE emissions. Enhancements through the effect of co-dopants on local symmetry of RE by adjusting the local interaction with suitable crystal structure, reduction of quench centers, semiconductor-induced energy transfer, and introduction of noble metallic nanoparticles (NPs) are presented by Zheng et al. [4]. Glasses doped with RE ions and metallic nanoparticles have been investigated due to their applications as optical devices. It has been demonstrated that metallic nanoparticles may enhance the RE luminescence and improve nonlinear optical properties.

Among the oxide glasses, tellurite glasses are in much interest due to their potential applications; they have attracted interest for optoelectronic and photonic applications because of promising properties such as high refractive index (~ 2.0), large transparency in the infrared region (up to 5 μm), low cutoff phonon energy (~ 800 cm^{-1}), high solubility of rare-earth ions, chemical durability, and thermal stability. However, it has been shown that the tellurium oxide can be used as glass former unaccompanied. Modifier oxides and fluorides such as zinc oxide, tungsten oxide, lead oxide, and some alkaline oxide such as Li, Mg, and K can improve the glass formability of tellurium oxide. Zinc oxides which modified tellurite glasses have been studied mostly because of their large refractive index, thermal stability, and improvement of tellurite structure. The heavy metal oxide glass formers (e.g., tellurite and antimony) and modifiers (e.g., zinc and lead) showed more polarizability and covalency properties. Therefore, they are promising to increase the emission cross sections of RE ions in the glassy hosts. On the other hand, the metallic NPs in the system aim to enhance the optical properties of Er^{3+}-doped glasses. However, quenches are also observed in different studies.

In this chapter, firstly, a short introduction on importance of material science and glass technology has been explained. It will be followed by describing the importance of the glasses and RE-doped glasses. This will be continued by a noteworthy review on tellurite glasses giving a chronological study on the history of glasses, the first-time preparation experience of tellurite glass. Next, the structural, thermal, and optical properties of tellurite glass modified by different oxides will be discussed. Next, the spectroscopic properties of trivalent rare-earth ions will be explored, and probable mechanism, such as energy transfer and relaxation processes, will be given. The rest of this chapter is devoted to summarize the recent results of the spectroscopic studies of rare-earth ion-doped tellurite and some oxyfluorotellurite glasses. The influence of metallic NPs on the optical behavior of these glassy systems will also be reviewed. The new achievements and importance of the research are ascertained in this chapter. However, there are numerous works in recent years in this area, which could not be covered.

7.2 Glasses: From the Art to the Science

Currently, glasses exist all around the world and in our daily life, starting from drinking cups to the dressing mirrors, from electric lamps to the communication fibers, and from window glass to wine bottle and many attractive decorative jams. There are magnificent collections and museum of glasses which inspire the human mind by those timeless and limitless colorful and shaped objects of art. The first glasses on the earth, indeed, have been made by nature. The ashes of overflowed volcanoes get cooled down slowly and make natural glasses, containing aluminum silicon, sodium, potassium, calcium, and iron.

The history of glasses made by man is started since 4000 B.C. in Mesopotamia, western Asia. These glasses were glazed due to application of copper compounds. Colorful glasses have been prepared between fourteen and sixteen century B.C. in Egypt. However, the art of glass transferred to Syria, Cyprus, and Palestine at eleventh century B.C. All nations have used same technique to prepare the glasses; the glasses were melted, drown out, and winded around a clay core which was kept by an iron rod. The art of glasses has many centers through the history from Greece and Macedonia, Roman Empire, and Muslim countries, especially in Syria and Palestine at seventh century A.D., China (second century B.C.) and India (fifth century B.C.), Baghdad (ninth century), and Persia. Later, large tax reliefs and social recognition from European governments (starting from Italy and Germany) pushed the glass-makers to develop the techniques of glass blowing. Addition of calcia to make shinier and brilliant glasses (Crystallo glasses) and "ice glasses" methods were two biggest achievements of that time. Later, crystals became very popular, and then fluorescent glasses, opaline, HF-etched glasses, and cobalt blue glasses were prepared. In the twentieth century, by developments of chemistry and physics of inorganic materials, glasses were also grown, not only for decorative objects, and daily life facilities, but for scientific purposes. Silica glasses are the first production of this century, which was mixed with other oxides. By understanding the amorphous nature of glasses, study of the local structure and bonding in glasses was acknowledged as important factors. The cutting, polishing, and coating methods led to develop glass pieces potential for microscope, telescope, glass fibers (cables), electronic bulbs, and beverage bottles.

More recently, zinc phosphate and lead borate glasses were used as solders. Phosphate and fluorophosphate glasses were applied as laser hosts. Memory panels and switching materials were prepared by chalcogenide glasses. Superior transparency of halide glasses suggests them good candidate to be substitute for silica fibers. Modern glasses are versatile. They showed different physical and optical properties due to their specific structures, which make them excellent candidate for application in different branches of science and technology. In a scientific approach, tellurite glasses attracted large interest due to its significant optical, thermal, and physical properties. On the following section, a review on history and properties of tellurite glasses will be presented.

7.3   Preparation of Tellurite Glass: Physical and Structural Properties

The most stable oxide of tellurium is tellurium dioxide (TeO_2). This property of TeO_2 introduced it as a suitable host for crystalline solids and glasses. Some of the physical properties of TeO_2 are listed in Table 2.2. The first tellurite glass was synthesized by Cheremisinov and Zalomanov, at 1962 [5]. They used almost pure tellurium dioxide (99–99.5%) to melt in an alumina crucible and melt at 800–850 °C for 30–40 min. Then the melts were cooled with a rate of 100 °C/h and kept at 400 °C for 1 h. Next, the furnace was turned off and glass was left to be cooled naturally. They yielded a greenish hue cylinder of transparent glass which contains 6% Al_2O_3 contamination. Lambson and co-workers (1984) also made them efforts to prepare tellurite glass by high-purity crystalline tetragonal powder. They used different methods of melting, casting, and cooling; however, they failed [6]. At the same time, they finally succeed to synthesize a greenish hue tellurium glass with 1.6% Al_2O_3 impurity by (1) avoiding the formation of the polycrystal and (2) removing any crystal center in the melt. Barady [7] also reported on tellurite oxide glasses. He noticed that the tellurium dioxide is a poor glass former, meaning that it is necessary to add a modifier to start TeO_2 powder to form the glass. Ten mol% concentration of any modifier is required to form the tellurite glass. In 1978,

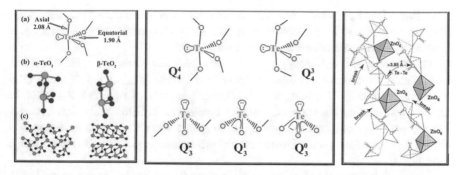

Fig. 7.1 (Left) Different forms of pure TeO_2 in its crystallized phase. (Center) Different polyhedral groups of modified TeO_2 system. (Right) Atomic arrangement illustration of $(TeO_2)_{80}$-$(ZnO)_{20}$ glass system [13]

Mochida et al. [8] and Kozhokarov et al. [9] were reported on binary tellurite glass modified by monovalent and divalent cations and binary tellurite glasses containing transition metal oxides, respectively. The physical and optical of these systems are reported by El-Mallawany [10].

Two crystalline phases exist in pure TeO_2: paratellurite (α-form) [11] and tellurite (β-form) [12]. In both forms, four coordinate tellurium and four bridging oxygen consist a completely interconnected network. Figure 7.1a shows the TeO_4 trigonal bipyramid (tbp) which is the basic polyhedron in both tellurite crystalline forms. This group is known as Q^4_4 in which a lone pair of electrons filled an equatorial position. Axial and equatorial bond lengths of Te-O are 2.08 A and 1.90 A, respectively. Figure 7.1b shows the difference between α- and β-forms, which is the sharing behavior of polyhedral by corners and edges, respectively. Another difference of these two crystalline phases is shown in Fig. 7.1c, where α-TeO_2 consists of a three-dimensional network, but β-TeO_2 is an infinite two-dimensional sheet.

Addition of modifiers breaks the connected network of TeO_2. For example, addition of only 20 mol% of Na_2O [14] breaks some of the Q^4_4 groups in TeO_2 glass to form the Q^3_4 and Q^2_3 polyhedra ($Na_2Te_4O_9$). Tellurium is bonded to three bridging oxygen and one non-bridging oxygen in Q^3_4 group, while it is connected to two bridging oxygen atoms and one doubly bonded non-bridging oxygen. Addition of 33 mol% of Na_2O [15] breaks all the bounds in Q^4_4 group while forming new group in tellurite glass, Q^1_3. An addition of 50 mol% Na_2O (Na_2TeO_3) leads to the appearance of the TeO_3^{2-} trigonal pyramid (tp) (or Q_3^0 groups) [16]. As shown in Fig. 7.1, tellurium in Q_3^0 group is bonded to three non-bridging oxygen atoms.

The first neutron diffraction study on $20Li_2O$-$80TeO_2$ by Neov [17] showed that coordination number of Te-O is around 4 in 1.9–2 A range. The ^{23}Na NMR studied by Zwanziger et al. [18, 19] revealed that there is no Te-O-Na bond in sodium-tellurite glass. Suzuki et al. [20] studied the neutron diffraction on barium-tellurite glass, and existence of two different Te-O distances was revealed at 1.9 and

2 A. Three short bonds and one longer bond were observed in pure TeO_2 due to the presence of distorted TeO^{3+}_1 tetrahedron. The short Te-O bond distance declined to 1.86 by increasing the concentration of barium oxide.

The first Raman spectroscopy studies on tellurite glass by Sakiya et al. [21] led to assignments of different groups in tellurite network, Q^4_4, Q^3_4, Q^2_3, and Q^1_3. Tatsumisago et al. [22] investigated the Raman spectrum of lithium-tellurite oxide and showed that introduction of modifier breaks the TeO_4tbp and constructs the TeO_3tp groups, consisting of non-bridging oxygen. By the NMR technique, it is also shown that abundance of TeO_4tbp and TeO_3tp is independent of the added alkali modifiers [23, 24].

Chemical durability of tellurite glasses was studied by Stanworth (1952) [25]. Tellurite glass shows large durability against water, acids, and alkaline. The weight loss of the order of 5×10^{-7} and 20×10^{-7} $g/cm^2/day^{-1}$ was measured for 18%PbO-82TeO$_2$ and 22PbO-78TeO$_2$ glasses, respectively. The water durability of lead-tellurite glass reduces by introduction of modifiers such as lithium oxide and sodium oxide, while this effect is more rigorous by adding the boron oxide in tellurite glass. Tellurite glasses containing BaO, Li_2O, Na_2O, and As_2O_5 showed heavy attacks and reduction of water durability.

El-Mallawany et al. [26, 27] has recently studied the structural and optical properties of some lanthanide-doped tellurite glasses based on the glass composition. They showed that the rare-earth ions were incorporated in the network of the present glasses and act as a network intermediate. They suggested that these glasses are promising host for the rare-earth ions and suitable for optical properties and optical applications. El-Mallawany et al. also studied optical properties [28, 29], elastic moduli and ultrasonic attenuations [30–32], heat capacity [33], and electrical [34, 35] and structural properties [26, 29, 36] of some new tellurite glasses, in the last few years.

Figure 7.1 (right) shows the structure of α-TeO_2 tellurite glass as reported by Kozhokarov et al. 1986 [13]. The shortest Te-Te distance in this system shows the presence of weak metal bonds between two cations. Zinc-tellurite glasses are also investigated by Sidek [37], Hoppe [38], and Sahar [39] and [40]. Some of the physical, structural, and optical properties of zinc-tellurite glasses studied by Sidek are listed in Table 7.1. Sidek et al. showed that the introduction of ZnO to TeO_2 network shifts the band at 626 cm^{-1} to the band at 669 cm^{-1} from pure TeO_2 to 15ZnO-85TeO$_2$ glassy system, respectively. The addition of zinc oxide to TeO_2 matrix increases the density, refractive index, and Urbach energy, while polarizability and molar volume and optical band gap are decreased gradually. Hoppe et al. [38] showed that introduction of ZnO in TeO_2 glass system decreases the total Te-O coordination number, while TeO_4 tbp groups transform to TeO_3 units. The bond distances for two existing species of Te-O band were measured to be 1.9 Å and 2.1 Å. Hoppe et al. [38] interpreted the change of coordination number as follows: addition of second oxide to corner-connected TeO_4 in TeO_2 network, breaks the Te-O-Te bridging, and converts two TeO_4 units to TeO_3 units with one non-bridging oxygen for each. By means of the neutron diffraction method, the atomic arrangement of a zinc-tellurite glass with composition 80%TeO$_2$ + 20%ZnO has been

Table 7.1 Physical, optical, and structural properties of $ZnO_{(x)}TeO_{2(1-x)}$ glass system [37]

Glass samples	Composition		Density (Kg m^{-3})	V_m (cm^3 moL^{-1})	n (± 0.01)	R_M (± 0.01)	α ($\times 10^{-24}$) (± 0.01)	(E_{opt}) $(\alpha h\nu)^{1/2}$ (± 0.01) eV	E_U (± 0.01) eV	IR bands
	X	$1-x$								
TZ0	0.00	1.00	4806	33.21	1.99	16.50	6.54	2.34	0.85	444; 626; 760
TZ1	0.10	0.90	5098	29.77	1.96	14.52	5.75	2.61	0.60	443; 668; 745
TZ2	0.15	0.85	5102	28.98	1.97	14.20	5.63	2.56	0.71	457; 669; 726
TZ3	0.20	0.80	5136	28.03	1.98	13.83	5.48	2.51	0.82	441; 668; 684
TZ4	0.25	0.75	5194	26.96	1.99	13.39	5.31	2.49	0.80	428; 675;
TZ5	0.30	0.70	5211	26.13	2.00	13.14	5.21	2.20	1.45	431; 668; 730
TZ6	0.35	0.65	5280	25.04	2.07	13.32	5.28	2.18	1.50	429; 665; 675
TZ7	0.40	0.60	5283	24.29	2.10	12.72	5.04	1.88	1.91	428; 675; 685

studied [41]. Approximately 35% of the Te atoms conserve the same coordination state as in the TeO crystal structure. The remaining Te atoms are of lower coordination. The observed variations of the Te-Te and Te-O distances greater than 3.0 Å have an essential contribution to tellurite melt amorphization [41].

7.4   Thermal Properties

The melting, crystallization, and glass transition temperatures, T_g, are well-known thermal properties of amorphous glass networks, which are related to structural arrangements of the system. Glass transition temperature, crystallization temperature, and melting temperature are the three major thermal characteristics, which can be evaluated by differential thermal analyzer (DTA) and differential scanning calorimeter (DSC) measurements. Here, we review on some previously reported results of thermal measurements on tellurite glasses. Three thermal properties (T_g, T_c, and T_m) of different tellurite glasses are listed in Table 7.2. Tellurite glasses show high thermal expansion coefficients ($150\text{--}200 \times 10^{-6}\,^\circ\text{C}^{-1}$) and low deformation temperature, around $250\text{--}350\,^\circ\text{C}$.

Table 7.2   Thermal properties of different tellurite glasses

Year [Ref]	Glass	T_g (°C)	T_c (°C)	T_m (°C)	$T_c - T_g$ (°C)	H
2009 [42]	TZLF-Er	292	465		173	0.59
2007 [43]	TeO$_2$-Na$_2$O-ZnO	285	482	–	197	0.69
2010 [44]	TeO$_2$-Li$_2$CO$_3$-Er$_2$O$_3$-AgNO$_3$	250	330, 357, 380	427	80, 107, 130	0.32
2009 [45]	TeO$_2$-ZnF$_2$-PbO-Nb$_2$O$_5$	263	380	–	117	0.44
2008 [46]	TeO$_2$-ZnO-BaO	340	550	–	210	0.62
2007 [47]	TeO$_2$-ZnO-Eu$_2$O$_3$	326	434	663	108	0.47
	TeO$_2$-ZnO-LiF-Eu$_2$O$_3$	264	420	583	156	0.95
	TeO$_2$-ZnO-Na$_2$O-Li$_2$O-Eu$_2$O$_3$	248	418	615	170	0.73
	TeO$_2$-ZnO-Na$_2$O-Li$_2$O-Nb$_2$O$_5$-Eu$_2$O$_3$	277	401	670	124	0.46
2009 [48]	TeO$_2$-ZnO-Nb$_2$O$_5$-MoO$_3$-Pr$_6$O$_{11}$	407	548	–	141	0.34
2008 [49]	TeO$_2$-ZnO-Er$_2$O$_3$	335	475	–	140	0.42
2007 [50]	TEGNAZO10	293	416	445	123	0.42
2006 [51]	GeO$_2$-PbO-PbF$_2$-CdF$_2$-Er$_2$O$_3$	406	543	828	137	0.34
	TeO$_2$-GeO$_2$-PbO-PbF$_2$-CdF$_2$-Er$_2$O$_3$	318	432	838	114	0.36

Stabilization of the glass is possible by keeping the glass at temperature slightly below glass transition temperature. It is worth to mention that at T_g, the viscosity is extremely large ($\sim 10^{13}$ poise). The stability factor, S, can be calculated by $S = T_c - T_g$. The time of annealing the glass at around T_g changes the properties of the glass, while volume of the glass also depends on cooling rate. Another factor to determine the tendency of the system to become a glass rather than a crystal can be defined by

$$K_g = \frac{T_c - T_g}{T_m - T_c} \tag{7.1}$$

where K_g is known as glass-forming tendency and T_m is the melting point of the glass [52]. Equation (7.1) is mostly referred as Hruby's formula, (H).

DTA and DSC techniques are usually used to investigate the glass transformation mechanism which could be studied by observing the kinetic of glass transition. In this regard, the glass transition temperature is defined as a function of heating rate. Also, activation energy (E_η) which is one of the important kinetic thermal parameters can be determined by thermal analyzing techniques [53]. Strong glasses show high resist, small structural change, and small heat capacity change around the T_g with the changing temperature [54–56]. On the contrary, fragile glasses are known as the materials with large structural and heat capacity change in glass transition temperature. El-Mallawany [10] presented in his book that tellurite glasses follow the "two-thirds rule," which were previously established for silicate, phosphate, borate, and germanate glasses. Two-thirds rule states that the ratio of glass transition temperature (T_g) to the melting temperature (T_m) of the glass is around 2/3 (~ 0.66).

Physical and elastic properties of the ternary zinc oxyfluorotellurite glass system have been studied by Sidek et al. [57]. Rapid melt quenching technique was used to synthesized a series of glasses $(AlF_3)_x(ZnO)_y(TeO_2)_z$ with $x = 0$–19, $y = 0$–20, and $z = 80$, 85, and 90 mol%. The composition dependence of the physical, mainly density and molar volume, and elastic properties is discussed in terms of the AlF_3 modifier additions that are expected to produce quite substantial changes in their physical properties. No crystalline peaks are observed in the X-ray diffraction (XRD) patterns of the present glass indicating the amorphous nature of prepared materials. The addition of AlF_3 modifiers into the zinc-tellurite causes substantial changes in their density, molar volume, as well as their elastic properties. Increasing the amount of AlF_3 decreases the densities of this glass system. A MBS8020 ultrasonic data acquisition system was used to measure the propagation of longitudinal and transverse waves in each glass sample. Velocity data were collected at room temperature and a frequency value of 5 MHz. Velocity data and their respective density were then used to obtain the longitudinal modulus (L), Young's modulus (E), shear modulus (G), bulk modulus (K), and Poisson's ratio (σ). It was found from experimental data that the density and elastic moduli of each AlF_3-ZnO-TeO$_2$ series strongly depend upon the glass composition.

7.5 Spectroscopic Properties of Rare-Earth Ion-Doped Glasses

The theory of spontaneous and stimulated emission was first reported by Einstein [58]. However, it is just at 1961 that Snitzer [59] reported the first fiber prepared by Nd^{3+} ions in solid-state flash lamp-pumped laser, which operates at 1061 and 1062 nm. In 1969, Koester and Snitzer reported on near single-Nd^{3+}-doped fiber laser. Kao and Hockham [60] developed the theory of propagation in core-clad fibers and studied the structured optical fiber. Judd [61] and Ofelt [62] independently and simultaneously formulate the theory of absorption and emission of rare earths. Interestingly, approaches, assumptions, and results of both theories were one and the same; however, there are some differences in definitions. Judd defines the theory as the optical absorption, while Ofelt referred to crystal spectra of rare-earth ions [63]. In 1972, Sandoe et al. [64] reported the first phosphate glass containing Er $^{3+}$ ions which emits at 1530–1560 nm regions. In addition, Stone and Burrus [65] studied the 800 nm emission of Nd^{3+} fibers (CW laser). Mears et al. presented the first tunable and Q-switched fiber laser, which operates in two regions: 1528–1542 and 1544–1555 nm. In 1987, Desurvier et al. [66] explored the dynamic range of Er^{3+}-doped fiber amplifier by model investigation. First transatlantic fiber-optic TAT-8 cables were developed in 1990. Svendsen developed the optical network WDM system, and he installed this system for the first time on long-haul routes in Norway on 1997 [67].

Up to now, there are a lot of reports on photoluminescence, thermoluminescence, and electroluminescence of rare-earth ions in different hosts, since 1976 when Reisfeld et al. [68] reported the Judd-Ofelt intensity, radiative transition, and non-radiative relaxation of Ho^{3+} in different tellurite glasses modified by BaO, ZnO, or Na_2O. Transition probability, branching ratio, radiative lifetime, quantum efficiency, and multiphonon relaxation rates were calculated by collected data. Weber et al. (1981) [69] measured the absorption and luminescence spectrum of Nd^{3+}-doped tellurite glass modified by alkali and phosphates.

7.5.1 Energy Levels of Lanthanide Ions

Energy levels of rare earth ions could be explained through the Hamiltonian operator approach [70]:

$$H = H_{\text{free}-\text{ion}} + H_{\text{"Crytal}-\text{field"}} \tag{7.2}$$

where

$$H_{\text{free-ion}} = -\frac{\hbar^2}{2} \sum_{i=1}^{N} \nabla_i^2 - \sum_{i=1}^{N} \frac{Z^* e^2}{r_i} + \sum_{i<j}^{N} \frac{e^2}{r_{ij}} + \sum_{i=1}^{N} \zeta(r_i) \vec{s}_i . \vec{l}_i + \text{etc.} \qquad (7.3)$$

Here, N is the number of electrons in *4f*; Z^* is the effective nuclear charge, including the inner electrons and nuclei; $\zeta(r_i)$, s_i, and l_i are the spin-orbit coupling efficiency and spin and orbit angular momentum, respectively.

The terms in the above equation define the kinetic energy, Coulomb interaction, mutual Coulomb repulsion, and spin-orbit interaction of the *4f* electrons. The last two terms are responsible for the broadening of energy level structure of Er^{3+} ions in a host, which lifts the degeneracy of the $4f^N$ electron configuration. A non-spherical symmetric crystal field in solids splits the energy levels of ion, which is frequently called as "Stark splitting."

Due to shielding by 5s and 5p electrons, the crystal field Hamiltonian is 100 times weaker than electrostatic and spin-orbit interactions, in 4f electrons [70]. It is worth to mention that electric dipole intra-$4f^N$ transitions are forbidden due to matching parity of all levels; however, they became allowed as a result of mixture into the $4f^N$ configuration of a small amount of excited opposite parity configuration.

In the presence of magnetic field of an incoming light, and considering the ion-ion interaction, two more terms will be added to the equation (above):

$$H = H_{\text{f-ion}} + H_{\text{CF}} + V_{\text{EM}} + V_{\text{ion-ion}} \qquad (7.4)$$

where V_{EM} is the Hamiltonian of interaction of light by ion, and the last term represents the interaction of two neighboring ions. V_{EM} is responsible for absorption transitions, when frequency of incoming magnetic field is in resonance or near resonance with transition between different energy levels of Er^{3+} ions in particular and RE ions in general.

7.5.2   Lanthanide Ions in Different Hosts

Heavy metal oxide glasses such as tellurite and antimony are superior host than fluoride, bromide, chalcogenide, and iodide host matrices due to their low phonon energy, high resistance, and high refractive index which increase the radiative emissions in rare-earth-doped glasses. The stimulated emission cross section of RE-doped glasses is related to refractive index by

$$\sigma_{\text{emi}} \propto \frac{(n^2 + 2)^2}{n} \qquad (7.5)$$

for electric dipole transitions and

Table 7.3 Full width at half maximum, emission cross section, and lifetime of 1.5 μm broadband of Er^{3+} ion in different hosts (AFP stands for alumina fluorophosphates)

Glass	Reference	FWHM (nm)	σ_e ($\times 10^{-20}$ cm^2)	FWHM$\times\sigma_e$	Lifetime (ms)
AFP	[71]	53	0.60	3.18	7.6–8.4
Silicate	[71]	40	0.55	2.2	5–8
Phosphate	[71]	37	0.64	2.37	6–10
Bismuth glasses	[71]	79	0.70	5.54	1.6–2.7
Tellurite	[71]	65	0.75	4.88	2.5–4
Zinc-tellurite (0.5–2.5% Er_2O_3)	[72]	60–80 nm	1.21–0.88	–	1–2 ms
$80TeO_2$-15(BaF_2-BaO)	[72]	91	0.68		2.69
$75TeO_2$ 20ZnO $5Na_2O$	[73]	56	0.88		3.15
$70TeO_25BaF_210ZnBr_215PbF_2$	[74]	56	0.83		4.05
$70TeO_25BaF_210ZnO15PbCl_2$	[74]	52	0.82		4.16
$65TeO_215B_2O_320SiO_2$	[75]	71	0.75		2.38
TeO_2-PbF_2-AlF_3 TG01	[76]	71	–	–	3.06
TeO_2-PbF_2-AlF_3 TG1	[76]	72	0.84		2.99
TeO_2-PbF_2-AlF_3 TG25	[76]	84	0.81		1.76

$$\sigma_{emi} \propto n^3 \tag{7.6}$$

for magnetic dipole transitions. Therefore, tellurite glasses are good candidates to enhance the emission cross section by their large refractive index ($n > 2$). For example, broad and flat stimulated emission cross section in communication band and large amplification gain in L-band of Er^{3+}-doped tellurite glass are introduced as promising materials for broadband applications. The cross section of this broadband in tellurite glass is larger than those phosphate, silicate, and fluoride glasses. The emission cross section, lifetime, and FWHM of the broadband near 1.5 μm spectral region of Er^{3+} ions in some glasses are listed in Table 7.3.

Low phonon energy characteristic is the significant property to enhance the quantum efficiency of emissions from different RE-doped tellurite glasses. Having about ~750 cm^{-1} cutoff phonon energy, tellurite glasses are able to increase the emission cross section by decreasing the non-radiative decay rates and assisting the energy transfer processes. These two last mechanisms are discussed in the next sections. Phonon energy of different glassy hosts is listed in Table 7.4.

Solubility of rare-earth ions is another factor to select the suitable host matrix. The solubility of a dopant directly depends on the strength of the structural bonding. For instance, in silicate glass four oxygen atoms are tightly bonded to silicon atom by strong covalent linkages. Therefore, the incorporation of rare-earth ions in silicate glass is weak, and uniform distribution is not possible. Due to this fact, remarkable amount of modifiers (usually alkalis) is required to break the covalent bonds, and to form the non-bridging oxygens, and to weaken the network structure [78]. Phosphate glasses are also based on tetrahedral structure. However, their covalency is five. The double bond between phosphorus and oxygen increases the number of NBOs.

Table 7.4 Non-radiative transition parameters for various host matrices [10, 77] and references within

Glass host	B (s^{-1})	A (cm)	$\hbar\omega$ (cm^{-1})
Silicate	1.4×10^{12}	4.7×10^{-3}	1100
Phosphate	5.4×10^{12}	4.4×10^{-3}	1280
Tellurite	6.3×10^{10}	4.7×10^{-3}	700
Borate	2.9×10^{12}	3.8×10^{-3}	1400
Germanate	3.4×10^{12}	4.9×10^{-3}	900
Fluoride	9.33×10^{7}	5.19×10^{-3}	500–600
Sulfide	1.26×10^{5}	2.9×10^{-3}	350
Oxyfluorophosphate	–	–	621
Chalcogenide	–	–	300–450
Antimony	–	–	602
Bismuth-gallate	–	–	673
Lead	–	–	1000–1300

Therefore phosphate glasses show better spectroscopic properties and emissions than silicates when doped with REs [79]. On the other hand, tellurite glass shows a two-dimensional system, where the Te-O linkage is significantly weaker than Si-O bond in silicate glass. Thus, it is much easier to break the atomic network of tellurite. Moreover, the atomic/ionic diameter of Te is larger than Si; therefore, the network is not tightly closed. The open and weak network in tellurite glass facilitates the incorporation of RE ions easier and more uniform than silicate and phosphate glasses [80].

7.5.3 Lifetime and Non-radiative Decays

The lifetime of excited states of lanthanide ions is the important factor that defines the possibility of achieving the population inversion and efficiency of pumping in amplifiers and laser applications. Transition rate of an excited state is the inverse of lifetime of that energy level. Transitions from a state include both radiative and non-radiative transitions. Radiative transitions are due to absorption and emission of a photon, while non-radiative transitions correspond to the interaction of ion with lattice quantized network, the phonons. Therefore, total lifetime of a level can be written as

$$\frac{1}{\tau} = \frac{1}{\tau_{\mathrm{r}}} + \frac{1}{\tau_{\mathrm{nr}}} \tag{7.7}$$

where τ, τ_{r}, and τ_{nr} represent the total, radiative, and non-radiative lifetimes of an excited state. Non-radiative transition rate at a certain temperature, T, is given by [81]

$$\left(\frac{1}{\tau_r}\right)_T = B\exp(-\alpha\Delta E)[1 - \exp(-\hbar\omega/kT)]^{-n} \qquad (7.8)$$

where ΔE is the energy gap, $\hbar\omega$ is the phonon energy, and n is the number of phonons required to bridge the gap ($\Delta E/\hbar\omega$). B and α are two positive phenomenological parameters, which vary on glass host. This non-radiative transition is also known as "multiphonon relaxation" process.

Non-radiative decay rates from excited states play a significant role to choose the suitable host for different applications. For example, multiphonon relaxation rate in borate and phosphate glasses in $^4I_{11/2}$ excited states is large which reduces the radiative emissions and quantum efficiency of $^4I_{13/2}$ excited state. On the other hand, tellurite glass possesses low phonon energy which is favorable to enhance the lifetime of $^4I_{11/2}$ level in Er^{3+}; therefore, the efficiency of the tellurite glass is large enough to develop the erbium-doped fiber amplifiers (EDFA), pumping at 980 nm.

7.5.4 Rare Earth's Ion-Ion Interactions

The main ion-ion interactions between the lanthanide ions consist of energy transfer process, excited state absorption, up- and downconversion processes, and cross-relaxation process. The neighboring active ions in short distances can interact with each other in two different approaches:

- Summing up the photon energy by energy transfer (ET)
- Cooperative effects due to emission, absorption, or sensitization

The energy transfer processes are firstly defined as shown in Fig. 7.2, where activator ion is in its ground state and sensitizer ion is in an excited state.

Auzel introduced the situations when activator ion is in an excited state, considering the exchanging of energy due to the difference between ion energy, not only

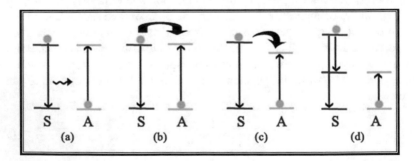

Fig. 7.2 Different energy transfer processes from a sensitizer (S) to an activator (A) in its ground state. Resonant radiative transfer (**a**), resonant energy transfer (**b**), energy transfer assisted by phonons (**c**), and example of quenching of the fluorescence of S by energy transfer to A (**d**)

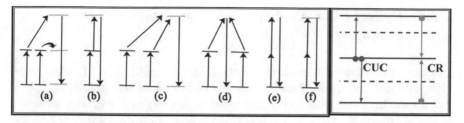

Fig. 7.3 Schematic energy transfer mechanism for different two-photon upconversion processes. APTE effect in YF_3:Yb:Er (**a**), two-step absorption in SrF_2:Er (**b**), cooperative sensitization in YF_3: Yb:Tb (**c**), cooperative luminescence in Yb:PO_4 (**d**), second harmonic generation in KDP (**e**), and two-photon absorption excitation in CaF_2:Eu^{2+} (**f**). The right-hand side figure shows the cooperative upconversion and cross-relaxation process in a schematic energy level diagram of Er^{3+} ion

absolute energy [82]. He also presented and discussed different two-step absorption, two-photon excitation, cooperative luminescence, second harmonic generation (SHG), and cooperative sensitization, as shown in Fig. 7.3. Cooperative upconversion (CUC) includes both cooperative sensitization and cooperative luminescence mechanisms. In high concentration of rare-earth ions, the interaction between two electric dipole moments of ions is effected by short-distance neighboring. CUC depends greatly on pumping intensity and concentration of ions. At low laser intensity, the CUC is not efficient. The required concentration of Er^{3+} ions in a system to show a significant CUC process is larger than 1×10^{-21} cm^{-3} [83]. Excited state absorption or two-step absorption is shown in Fig. 7.3. An ion in its metastable level can absorb a second photon and be excited to higher energy levels. Normally, non-radiative decays through multiphonon relaxations will help to populate lower metastable levels where the absorption of second or third photons may excite the ion, gradually. The so-called "energy transfer" upconversion [82, 84] is the general form of Dexter energy transfer [85] when the activator is in a metastable excited state. In a Dexter ET process, two ions, or two molecules or two parts of a molecule, mutually swap their electrons. Increased distance between two particles results in an exponential decrease in the rate of process. This exchange mechanism is also named as "short-range" energy transfer. In addition, the interaction of activator and sensitizer should be weaker than the vibronic interaction of two parties. The latest circumstance suggests the coupling of single-ion level to host network. By and large, this situation is more probable in high concentration of REs in glass or crystal, where the splitting of pairs' level is as small as 0.5 cm^{-1} [86]. Besides, the transfer probability of such ET processes must be faster than radiative and non-radiative transitions from the metastable level. Therefore, two or three photons lead to generation of only one metastable state ion. The excited state absorption depends strongly on the pump intensity. Cross-relaxation process is the reverse mechanism of CUC as shown in Fig. 7.3. The transition of two electrons in high- and low-energy levels leads to populate the middle-energy state. Second-order cross-relaxation of excited levels is negligible due to low concentration of ions, in this excited state.

7.5.5   Optical Fiber Amplifiers Based on Rare-Earth Ion Doping

Telecommunication devices are one of the immense applications of optical materials, with large transparency and less signal loss. Long-haul communication systems can transfer the optical data (or signals) up to few thousands of kilometers. However, it is still crucial to overcome the loss of such optical fiber amplifiers. TeO_2- and GeO_2-based erbium-doped fiber amplifiers (EDFAs) showed losses between 0.5 and 1.5 dB m^{-1} near 1300 nm [87]. However, the loss in mid-IR range is extremely high which complicates the design of mid-IR lasers. Erbium-doped fiber amplifiers (EDFA) are motivating materials to improve the telecommunication systems.

Figure 7.4 presents the two types of amplifiers. The pumping process from ground state to the highest excited state followed by a non-radiative decay to the next lower state is same for both systems. However, in a case of three-level fiber amplifiers, the laser transition takes place from E_2 excited state to ground state (E_1), while in four-level system, the ion will be relaxed non-radiatively to its ground state after a laser emission (E_2 to E_1).

The incident light will be amplified by stimulated emission from stored energy in mid-energy state. The three-level pumping system is applied in EDFA. By using a suitable design, optical fibers with 35 nm can be developed with a flat gain over the broadband region. WDM systems need broadband as width as 50 nm to transmit more than 50 channels. Therefore, designs with higher gains recently attracted the research interest by applying a two-stage approach. Wysocki et al. [88] reported a flat gain (~0.5 dB) in the 1544–1561 nm regions through the fabrication of two amplifiers. The EDFA co-doped with ytterbium, and phosphorous also is reported [89] to develop a power amplifier in a fluoride host under the 1481 nm pumping excitation wavelength. The combination of Raman amplification and one or two EDFAs to obtain a flat gain with 65 nm bandwidth (from 1549 to 1614 nm) is also introduced as another approach for high-gain flat broadband amplifiers [90].

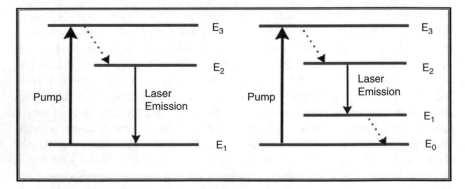

Fig. 7.4 Three-level (**a**) and four-level (**b**) pumping process in a schematic fiber amplifier. Wavy arrows show the fast relaxation of the level population by non-radiative process

7.5.6 Rate Equations

A rate equation system models the optical and electrical performance of a lasing device. Therefore, a set of ordinary differential equations relates the density or number of charge carriers (or ions) to the photons. In 2002, Santos et al. [91] reported the infrared to visible upconversion in Pr^{3+}-Yb^{3+} and Er^{3+}-Yb^{3+}-co-doped tellurite glasses. By generating the intense green and red fluorescence emission through the excitation, and investigation on the temperature dependence of upconversion efficiency, they consider a three-level system to explain the luminescence processes in their samples. By an analytical solution of the rate equations derived from Santos model, light intensity emitted from the $^3P_0$ was deduced. Afterward, Nie et al. [92] by developing a five-level model for transitions in Tm^{3+}-Yb^{3+}-co-doped tellurite glass and extracting the rate equations, a numerical solution has been done to determine the population of excited and ground states. Although most of the parameters applied in the rate equation were obtained by direct experimental data, energy transfer coefficients have been evaluated through a best-fit procedure.

Another three-level model is also reported by Shen et al. [93] to study the temperature quenching of upconversion luminescence in erbium-doped tellurite glass under 970 nm excitation. Three intense upconversion emissions around 530, 545, and 657 were observed. By increasing the temperature, the intensity of green emission around 530 nm ($^2H_{11/2}$) increases continuously, while the emission around 545 nm ($^4S_{3/2}$) increases from 20 to 80 K and then shows a quenching behavior after its largest value around 80 K. The temperature dependent of intensity was analyzed through a simple three-level system, and the temperature dependence on the multiphonon relaxation rates was fitted with four-phonon process. In another work, Cherif et al. [94] developed a four-level model to explain the red upconversion in Er^{3+}-doped TeO_2-ZnO glasses under 980 nm. By considering the rate equations of growth and decay in the populations of $^4I_{15/2}$, $^4I_{13/2}$, $^4I_{11/2}$, and $^4F_{9/2}$ levels, and simplifying presented model, the ground-state absorption probability (GSA), excited state absorption (ESA) probabilities, energy transfer upconversion coefficients (ETU), branching ratio of the transitions, and cross-relaxation rates were measured by fitting and calculation of Judd-Ofelt theory parameters.

7.6 Recent Research on Rare-Earth-Doped Tellurite and Fluorotellurite Glasses

7.6.1 Er^{3+}-Doped Glasses

Er^{3+} singly doped, Er^{3+}/Yb^{3+}-co-doped glasses, and Er^{3+}-doped glasses have been studied extensively. The major interest lies on the upconversion emissions of Er^{3+} ions in the green spectral region (520–560 nm), and a broad near-infrared band at

Fig. 7.5 Intensity of the upconverted green emission from Er^{3+} versus power excitation for a 0.4 at.% Er^{3+} in $70TeO_2$-$30ZnO$ glass using a 797 nm excitation. Measurements were performed at 300 K. (Taken from [95])

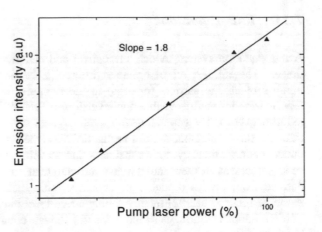

around 1530 nm. Zinc-tellurite glasses are one of the most studied examples. Optical absorption and photoluminescence properties of Er^{3+}-doped $70TeO_2$-$30ZnO$ and $80TeO_2$-$20ZnO$ glasses are investigated by Jaba et al. [95] and Dousti et al. [72], respectively. An infrared to visible upconversion is usually observed at room temperature in tellurite glass system using a 797 nm excitation line or 980 nm laser source. A study of the $^4S_{3/2} \rightarrow {}^4I_{15/2}$ transition (554 nm) versus power excitation provides evidence for a two-step upconversion process under theses excitations (see Fig. 7.5). Moreover, a red emission (663 nm) originating from the $^4F_{9/2} \rightarrow {}^4I_{15/2}$ transition is usually observed. It is shown that the efficiency of the red upconversion line is enhanced considerably with the Er^{3+} concentration relative to the green emission (554 nm). This behavior could be explained in terms of an energy transfer between excited ions. The temperature-dependent upconversion intensity has been investigated in the temperature range 40–310 K [95]. It was found that the thermal quenching of the green emission ($^4S_{3/2} \rightarrow {}^4I_{15/2}$) is large enough compared with those of the red transition ($^4F_{9/2} \rightarrow {}^4I_{15/2}$). Jaba et al. has proposed an additional decay rate probably due to a non-radiative energy transfer and/or a charge transfer through trapping impurities to explain the observed phenomenon, and a good agreement has been achieved between measured and computed data.

Dousti et al. [72] revisited Er^{3+}-doped zinc-tellurite glasses prepared by the melt quenching technique, and they found some favorable effects of increasing dopant concentration on chemical durability, water resistivity, and thermal stability (up to 140 °C). The conventional photophysical properties of the glasses were investigated by absorption and luminescence spectroscopic techniques. The Stokes and anti-Stokes emissions of erbium were analyzed, and it was verified that the width of the emission band at 1532 nm strongly depends on Er^{3+} concentration varying from 60 to 82 nm for 0.5 and 2.5 mol% of Er_2O_3, respectively, as listed in Table 7.3. The intensity of green and red upconversion emissions was evaluated, and the increased efficiency of red emission with increasing concentration is attributed to energy transfer mechanisms between infrared energy levels.

Er^{3+}-doped oxyfluorotellurite glasses have also attracted a large attention as the presence of fluorine ions has improved the optical properties of RE ions. For example, time-resolved and steady-state spectroscopies were used to study Er^{3+} ions in a novel glass based on TeO$_2$-PbF$_2$-AlF$_3$ oxyfluoride tellurites. The broad ^{4}I$_{13/2}$ → ^{4}I$_{15/2}$ emission transition at around 1.53 mm had a wide broadening of around 70 nm and a relative long lifetime of around 3 ms compared to other glass hosts, which shows potential applications of this glass in the design of erbium-doped fiber amplifiers. The stimulated emission, gain cross-sections, and absorption of this transition are given in Table 7.3. The lifetime of the tellurite glasses is less than those phosphate and silicate glasses, while FWHM of the Er^{3+} ion broadband in tellurite glasses is larger than those silicate and phosphate. While adding some fluorine ions to tellurite oxide glasses increase the lifetime of the ^{4}I$_{13/2}$ excited state. The lifetime of the NIR broadband of some tellurite glasses is given in Table 7.3. It can be concluded that the FWHM of this band strongly depends on the glass host choice as well as other parameters such as concentration of ions, thickness of the sample, and the intensity of the excitation beam [96].

The upconverted near-infrared emissions of Er^{3+}-doped fluorotellurite transparent glass ceramics obtained by thermal treatment of the precursor Er-doped TeO$_2$-ZnO-ZnF$_2$ glass are also reported by Miguel et al. [97]. ErF$_3$ nanocrystals with an average size of 45 ± 10 nm were homogeneously nucleated in the glass-ceramic samples. The comparison of the fluorescence properties of Er^{3+}-doped precursor glass and glass-ceramic confirms the successful incorporation of the rare earth into the nanocrystals. An enhancement of the red upconversion emission is observed under excitation at 801 nm in the glass-ceramic sample. The enhancement of the red emission is attributed to the energy transfer processes, analyzing the temporal evolution of the red emission together with the excitation upconversion spectrum.

Er^{3+}-doped oxyfluorotellurite glasses with a molar composition of (60-x)TeO$_2$-20ZnO-20LiF-xEr$_2$O$_3$, where $x = 0.1, 1.0, 2.0$, and 3.0 (referred as TZLFEr), were prepared by melt quenching technique by Babu et al. [42]. The ΔT ($\Delta T = T_x - T_g$) values for the present glasses are increased with concentration up to 2.0 mol% and then decrease marginally for 3.0 mol% Er^{3+}-doped glass. The increase of ΔT mostly resulted from the increase of T_x. This suggests that the thermal stability of the glass becomes higher up to 2.0 mol% of Er^{3+} concentration. The calculated Judd-Ofelt intensity parameters have been used to determine radiative properties for the important luminescent levels of Er^{3+} ions in 1.0 mol% Er^{3+}-doped TZLFEr glass, to study the radiative properties for the important luminescent levels of Er^{3+}ions' calculated Judd-Ofelt intensity parameters. The strength of covalent bonding between Er-O and/or asymmetry of Er^{3+} sites can be assessed by the greater Ω_2 and lesser Ω_6 JO parameters of the present glass. Enhanced spectroscopic quality factor shows that major transitions of Er^{3+} ions in this host are most likely to be considerably stronger. Under normal excitation, TZLFEr glasses radiate strong green and weak red luminescence with 451 nm. Concentration quenching by energy transfer processes between Er^{3+} ions results in the continuous decrease in emission intensities and lifetimes of the ^{4}S$_{3/2}$ level. Different parameters of the ^{4}S$_{3/2}$ and ^{4}I$_{13/2}$ levels like stimulated emission cross section and quantum efficiencies have been measured and compared with already reported results.

Optical properties of Er^{3+} ions are investigated using steady-state and time-resolved spectroscopies as a function of rare-earth-doped concentration, in a novel oxyfluoride tellurite-based glass (TeO_2-PbF_2-AlF_3) [68]. After basic optical characterizations, different parameters like absorption and emission spectra, cross-sections, Judd-Ofelt parameters, the radiative probabilities, and fluorescence decays and lifetimes were determined. Particular attention has been paid toward the broad $^4I_{13/2}$–$^4I_{15/2}$ transition at around 1.53 μm, having bandwidth of around 70 nm and a relatively longer life of ~3 ms as compared to other host glasses. This shows its potential applications in designing erbium-doped fiber amplifiers. The absorption, stimulated emission, and the gain cross sections of this transition have been obtained and evaluated in comparison with different hosts. At last, infrared-to-visible upconversion processes taking place at ~800 nm have been investigated, and different processes involved in energy conversions have been propounded.

Using standard experimental and theoretical techniques, spectroscopic investigations on Er^{3+}/Yb^{3+}-co-doped (Ba-La)-fluorotellurite glass compositions have been performed [90]. Quantitative study of absorption and emission spectra as well as emission lifetimes at room temperature yields various significant spectroscopic parameters like radiative decay rates, emission/absorption cross sections, and fluorescence branching ratios. Furthermore, internal radiative quantum yields have been evaluated for infrared at 1571 nm and at 547 nm for the upconversion emission. Quantitative estimation of impact of several non-radiative properties such as concentration quenching, quenching by hydroxyl radicals, and multiphonon relaxation has been carried out and correlated with the observed spectral properties. Comparison with other glasses and different tellurite compositions revealed that current glass could be used as a potential candidate for broadband amplifiers.

Study in [98] reported triply doped low phonon glass systems and their enhanced MIR emission bandwidth from Yb^{3+}/Er^{3+}/Dy^{3+}. NIR emission band can be increased from ~2600 to 3100 nm (~ 500 nm) with significant gain cross section. This band broadening is practically not possible to obtain from Er^{3+} or Dy^{3+} ions, i.e., single-doped glass systems. Co-doped Dy^{3+} ion not only helps in quenching the undesired visible upconverted emissions from Er^{3+} ions, but it also diminishes the prominent ~1.5 μm emission. Upon efficient energy transfer (ET) Yb^{3+} → Er^{3+} → Dy^{3+} at ~980 nm excitation, a broad NIR emission was obtained on superimposition of Er^{3+} (~ 2.76 μm) and Dy^{3+} (~ 2.95 μm) emissions. Present glasses could be a potential candidate for developing compact and tunable NIR solid-state fiber laser sources.

The temperature-dependent luminescence for various concentration of Er^{3+} ions in fluorotellurite glass was studied [99] in the 100–573 K range. Such studies open the window for interesting application of glasses as optical temperature sensors, as discussed in Chap. 8 of this book. Under commercial CW 488 nm laser excitation, the temperature-induced changes in the Er^{3+} green emissions were calibrated by doping the fluorotellurite glasses with three concentrations (0.01, 0.1, and 2.5 mol% of Er^{3+} ions). It is shown that concentration of Er^{3+} ions plays a role on the defined intensities, quantum efficiencies, line profiles, and fluorescence intensity ratio of the $^2H_{11/2}$- and $^4S_{3/2}$-thermalized green emitting levels, due to the existence of radiative and non-radiative energy transfer processes between Er^{3+} ions. The maximum value for thermal sensitivity was obtained for the fluorotellurite glass doped with the lowest concentration of Er^{3+} ions, 79×10^{-4} K^{-1} at 541 K, one of the highest values found in the literature, consistent with the theoretical thermal sensitivity analysis by Judd-Ofelt theory. It is concluded that the optical sensing calibrated based on Er^{3+} ions is strongly affected by radiative energy transfer processes, suggesting the utilization of a relatively low concentration of Er^{3+} ions to avoid the mentioned energy transfer processes. Similar studies have been also carried in other works under a CW laser diode excitation at 800 nm. The fluorescence intensity ratio between the thermally coupled emitting levels as well as the temperature sensitivity has been experimentally obtained up to 540 K. A maximum sensitivity of 54×10^{-4} K^{-1} at 540 K is reported for the less Er^{3+} concentrated glass [100].

Many efforts have been done to enhance the intensity and gain of major emissions of rare-earth ions. First, it was reported that increment of concentration of REs in the system could intensify the up- and or downconversion luminescence. However, a quench is observed often, after the introduction of 1–2 mol% of the emitting ions. This happens due to further energy transfer among the RE ions, which results to increase the lifetime. Another approach was demonstrated to increase the absorption cross section of REs by introduction of second dopant, commonly trivalent Yb. The trivalent ytterbium ion shows large absorption cross section; therefore, the large concentration of Yb^{3+} in vicinity of Er^{3+} provides larger absorption and emission gain, through energy transfer from $\{2F_{5/2}: Yb^{3+}\}$ to $\{^4F_{7/2}:^4F_{9/2}:Er^{3+}\}$ [101]. On the other hand, enormous studies are focused on glasses doped with RE ions and metallic nanoparticles (NPs). The major objectives of these investigations are to characterize the thermal, structural, and optical properties of RE-doped glasses with and without metallic NPs and to examine the influence of nanoparticles in optical properties of such glasses. The introduction of metallic nanoparticles into the glassy hosts was first reported by Malta et al. [102]. A large enhancement of the order of 5.6 for luminescence of Eu^{3+} in borate glass (emission at 612 nm under 312 nm excitation wavelength) has been shown due to the presence of silver particles by a concentration around 7.5 wt%. The absorption peak of small silver particles in their study showed a sharp peak at 312 nm. In 2002, Strohhofer et al. [103] reported on the enhanced emission of Er^{3+} in a borosilicate glass by an ion-exchange process. They observed 70 and 220 times enhancements in broadband line under 488 and 360 nm excitation wavelength.

Recently, many efforts have been done to improve the luminesce properties of rare-earth ions in tellurite glasses. For example, Rivera et al. [104] showed that by exciting the Er^{3+} ions doped in tellurite glass containing silver NPs upon 980 nm laser, a blue shift occurs in the peaks of broadband emission (~1.55 μm). The modification of Stark energy levels (blue shift) was attributed to oscillator strengths of NPs, which results in the energy transfer from NP to Er^{3+} ions. The small plasmon absorption peak was revealed in an Er^{3+}-free Ag NP-doped tellurite glass centered at

479 and 498 nm for 3 and 6 h annealed samples, respectively. Slight increase in FWHM and intensity of broadband were observed by increasing the annealing time interval. The lifetime of 1.55 µm decreases by introduction of Ag NPs in their system comparing to Er^{3+} single-doped tellurite glass. Many authors have studied the effect of silver or gold NPs on the emissions of RE-doped tellurite glasses [105–115].

The effect of silver NPs on the optical and structural properties of Er^{3+}-doped zinc-tellurite glass is investigated by Dousti et al. [98, 116, 117]. The absorption bands are located at 445, 488, 522, 654, 800, 976, and 1526 nm and ascribed to the electric transitions from the $^4I_{15/2}$ ground state of Er^{3+} to its excited states $^4F_{3/2}/^4F_{5/2}$, $^4F_{7/2}$, $^2H_{11/2}$, $^4F_{9/2}$, $^4I_{9/2}$, $^4I_{11/2}$, and $^4I_{13/2}$, respectively. The introduction of Ag NPs enhanced the upconversion emissions of green and red bands centered at 520, 550, and 640 nm by 4–6 times. The enhancement is attributed to the presence of silver NPs with an average size ~ 10 nm and with SPR localized at 522 nm. The effect of heat treatment is also reported by same authors [98, 117, 118], where it was shown that further annealing above the glass transition temperature improves the optical properties of Ag-embedded Er^{3+}-tellurite glasses. By introduction of 0.5 mol% of Ag NPs with an average size of 12 nm, 3.5-fold enhancement was observed for green emission ($^2H_{11/2} \rightarrow {}^4I_{15/2}$) due to formation and growth of NPs by 8 h annealing [98]. The SPR for 2 h annealing time is also reported. Two peaks in UV-Vis-IR spectra were attributed to SPR, centered at 550 and 580 nm. For 8 h annealing, however, they observed three SPR peaks which are ascribed to different modes of oscillating particles. In the case of 1 mol% of Ag NPs in the same glassy system, UC emissions were enhanced up to 6.5-folds after 4 h annealing above the T_g. The average size of manipulated Ag NPs was 14 nm, with two SPR bands at 560 and 594 nm due to non-spherical NPs.

Dousti et al. [117] also observed same behavior of Ag NPs in tellurite glass, while broad emission band centered at 500 nm originated due to 786 nm excitation wavelength. SPR band of the Ag NPs in this glass is reported around 564 and 594 nm. Besides the investigations on the effect of metallic NPs on the optical properties of RE-doped glasses, one may also study the effect of noble nanometal on the structural and thermal properties of the glassy systems, or even glass-ceramics and crystals. Recently, Dousti et al. [116] reported the effect of Ag NPs synthesized by AgCl addition in Er^{3+}-doped zinc-tellurite glass, on the structural, band gap properties and thermal properties of this common glass sample. As listed in Table 5.3, the addition of AgCl decreases the density of tellurite samples and increases the molar volume and refractive index. Moreover, a decrease in both direct and indirect band gaps was determined by increasing the Ag NPs concentration, while Urbach energy varies erratically (Table 7.5).

The thermal properties of the tellurite glass also are changed due to participation of Ag NPs. Onset and peak of glass transition temperature, onset and peaks of crystallization, and melting temperature are alerted as reported by Dousti et al. [116]. The DTA characteristics of Er^{3+}-doped zinc-tellurite glass without and with

Table 7.5 Density (ρ, g.cm^{-3}), photon cutoff wavelength (W_{ph}, nm), direct ($E_{d.r}$, eV) and indirect (E_{ind}, eV) band gaps, Urbach energy (E_U, eV), concentration of dopants (N, $\times 10^{20}$ ions cm^{-3}), molar volume V_m(cm^3 mol^{-1}), and average molecular weight M_{av}(gr mol^{-1}) of proposed glass samples. Internuclear distances for Er-Er, Er-Ag, and Ag-Ag couples are given by d_{Er}, d_{EA}, and d_{Ag} (nm), respectively

Glass	ρ	W_{ph}	E_{dir}	E_{ind}	E_U	n	N (Er^{3+}/Ag)	V_m	M_{av}	d_{Er}	d_{Ag}	d_{EA}
TZE	5.293	356	3.139	2.938	0.762	2.360	2.18/0	27.62	146.19	1.661	NA	NA
TZEA	5.251	354	3.138	2.934	0.693	2.361	2.16/1.08	27.83	146.11	1.667	2.009	1.456
TZEA1	5.217	348	3.127	2.909	0.775	2.364	2.1$_{5/2}$.15	27.99	146.03	1.669	1.669	1.325
TZEA2	5.188	347	3.139	2.901	0.724	2.360	2.14/4.28	28.12	145.89	1.672	1.327	1.159
TZEA3	5.156	373	3.112	2.861	0.694	2.367	2.13/6.39	28.26	145.70	1.674	1.161	1.005

Table 7.6 Thermal properties of prepare samples (T_g, T_c, and T_m are assigned for transition, crystallization, and melting temperatures, respectively)

Glass	T_g (°C)	T_{c1} (°C)	T_{c2} (°C)	T_m (°C)	$T_c - T_g$ (°C)	H
TZE	324	421	452	607	97	0.52
TZEA	322	430	447	610	108	0.60
TZEA1	321	436	–	607	115	0.67
TZEA2	320	392	419	594	72	0.36
TZEA3	320	399	417	595	79	0.40

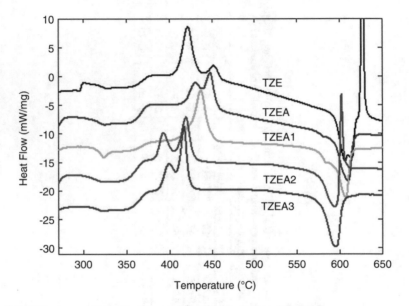

Fig. 7.6 DTA curves of tellurite glasses with and without silver NPs [116]

different concentration of Ag NPs are shown in Fig. 5.2 and listed in Table 7.6. The glass transition temperature is decreased by introduction of Ag NPs due to the decrement in the rigidity of glass network. The crystallization peaks in zinc-tellurite glass that refers to $ZnTeO_3$ and $Zn_2Te_3O_8$ phases [119] are suppressed and sharpened, respectively, because of new heterogeneous nucleation sites by Ag NPs in the host [120]. The formation of latest environment prevents the crystallization process by declining the rate of diffusion. Two different factors, $\Delta T = T_c - T_g$ and Hruby parameter, were used to estimate the glass thermal stability. The results are summarized in Table 5.4 which indicates the enhanced stability is provided by addition of Ag NPs up to 1 mol%. Fragility of the glasses can be estimated taking into account by determining the difference between onset glass transition temperature T_g and the temperature in which the transition completes (Figs. 7.6 and 7.7).

Oxyfluorotellurite glasses co-doped with Er^{3+}/Yb^{3+} and containing silver species and nanoparticles (NPs) were prepared using the conventional melt quenching technique [72]. The X-ray diffraction patterns obtained in this work do not reveal

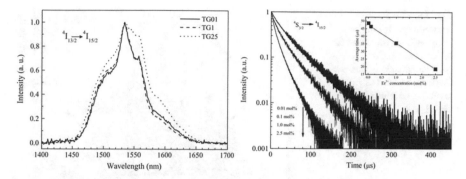

Fig. 7.7 (Left) NIR emission spectra as a function of Er^{3+} concentration (0.1, 1, and 2.5%) exciting at 488 nm. (Right) Decay curves of the green transition as a function of Er^{3+} concentration. The inset figure shows the calculated average lifetimes. (Results are taken from [76] for the TeO_2-PbF_2-AlF_3 oxyfluoride tellurite glasses at room temperature)

any crystalline phase in the glass. Heat treatment at 290 °C, for about 6 h, of an Er^{3+}/Yb^{3+} co-doped oxyfluorotellurite glass containing 1.0 mol% $AgNO_3$ yielded well-dispersed, nearly spherical Ag NPs, as verified by transmission electron microscopy. The observed surface plasmon resonance (SPR) band was observed around 490 nm for heat-treated glasses containing only silver. Upon off resonance (375 nm) and in resonance (486 nm) excitation of SPR band, the intense visible (520, 540, and 650 nm) and NIR (0.98 and 1.53 μm) downconversion emission bands were observed and further enhanced with the increase of Ag concentration and heat-treatment duration up to 6 h. An intensity drop was observed by further heat treatment (up to 9 h). In addition, enhancement of upconversion emissions, in the green and red, as a function of Ag concentration and heat-treatment time, has also been verified. The intensity was found to be enhanced by increasing the Ag concentration and heat-treatment time up to 3 h. However, further increase in the heat-treatment time duration (up to 9 h) resulted in reduced intensities. Enhancement in the luminescence intensity was discussed in terms of the SPR, whereas the quenching was attributed to the intense absorption of photons, emitted from RE ions, by larger Ag NPs.

7.6.2 Nd³⁺-Doped

Neodymium/erbium ions co-doped in the system of zinc-tellurite with the composition of $(70-2x)TeO_2$-$30ZnCl_2$-xNd_2O_3-xEr_2O_3 concentration from 1.0 to 3.0 mol% ($x = 1$, 2, and 3) glasses were prepared by using conventional melt quenching technique [121]. The amorphous nature of the glass was confirmed from X-ray diffraction technique. The UV absorption spectra recorded several bands and the values of the optical band gaps found around 3 eV, while the Urbach energy values are between 0.27 and 1.01 eV. The optical energy gap for indirect transition and

M. Reza Dousti and R. J. Amjad

Fig. 7.8 Steady-state emission spectra of the $^4F_{3/2} \to {}^4I_{11/2}$ transition for different excitation wavelengths along the low Stark component of the $^4F_{3/2}$ level for the sample doped with 1 wt% of NdF_3. Data correspond to 10 K. (Ref. [122])

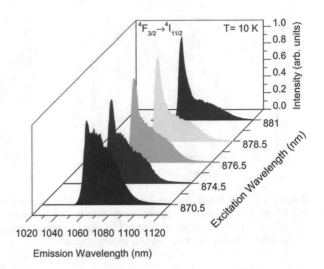

Urbach energy had minimum value for Nd^{3+}/Er^{3+} at 1% mol. The varying concentration of Nd^{3+} and Er^{3+} ions found to have a strong effect on optical and structural properties of the glass.

In this work Miguel et al. [122] present, for the first time to our knowledge, laser emission under wavelength-selective laser-pulsed pumping in Nd^{3+}-doped TeO_2-ZnO-ZnF_2 bulk glass for two different Nd^{3+} concentrations. The fluorescence properties of Nd^{3+} ions in this matrix which include Judd-Ofelt calculation, stimulated emission cross section of the laser transition, and lifetimes are also presented. The site-selective emission and excitation spectra along the $^4I_{9/2} \to {}^4F_{3/2}$ absorption band show the inhomogeneous behavior of the crystal field felt by Nd^{3+} ions in this fluorotellurite glass which allows for spectral tuning of the laser output pulse as a function of the pumping wavelength. The emission cross section obtained from the Judd-Ofelt analysis and spectral data (4.9×10^{-20} cm^2) is in fairly good agreement with the value obtained from the analysis of the laser threshold data (4×10^{-20} cm^2).

In another work, Miguel et al. [122] presented TeO_2-ZnO-ZnF_2 bulk glass for two different Nd^{3+} concentrations in laser-pulsed pumping in Nd. The site-selective steady-state emission spectra of the laser transitions were obtained at low temperatures for different excitation wavelengths along the low-energy component of the $^4F_{3/2}$ level. The room temperature properties of Nd^{3+} were checked too. The Nd^{3+} ion was doped as NdF_3 by the ratio of 1% by weight. Refractive index was 2.604, λp (nm) was 1059, $\Delta\lambda_{eff}$ was 25 nm, σ_p was 4.9×10^{-20} cm^2, τ_R was 145 µs, and τ_{exp} was 128 µs. As can be seen in Fig. 7.8, the shape, crest position, and linewidth of the discharge band change with the excitation wavelength. The spectra obtained under excitation at the low-energy side of the $^4I_{9/2} \to {}^4F_{3/2}$ absorption band narrow and red shift. The wavelength of the fluorescence peak shifts from 1055 to 1064 nm by varying the excitation wavelength from 870.5 to 881 nm, whereas the effective linewidth is reduced from about 24 to 13 nm. Similar variations in fluorescence profile, peak position, and linewidth have been observed in phosphate glasses. These results show the inhomogeneous behavior of the crystal field felt by Nd^{3+} ions in this fluorotellurite glass. As can be seen in Fig. 7.8, only when we excite at the high-energy side of the absorption band it is possible to cover the full spectral range of the Nd^{3+} emission which is probably helped by vibronic transitions. This inhomogeneous behavior is related to the effect of the pumping wavelength on the spectral behavior of the laser emission.

The effect of silver nanoparticles (NPs) on the optical properties of Nd^{3+}-doped sodium-lead-tellurite glass has been also studied [123]. The average size of silver NPs increases from 7 to 18 nm by addition of $AgNO_3$ content. The surface plasmon resonance band of the silver NPs was recorded at 522 nm for 0.5 mol% $AgNO_3$. Large upconversion enhancements (10–16-folds) were attributed to the enlarged local field in vicinity of silver NPs. Feasible interactions between the excitation light and Nd^{3+}/Ag NP-co-doped tellurite glass are discussed. Similar results were reported by various authors.

7.6.3 Pr^{3+}-Doped

Different characteristics of Pr^{3+}-doped tellurite glasses, such as optical absorption, luminescence, and upconversion, have been brought under investigation, and their corresponding optical properties have been presented in Ref. [1]. In upconversion process, the first NIR photon is absorbed in order to populate $^1G_4$ level with the help of phononic excitations. After excitation, Pr^{3+} ion captivates another NIR photon, transferring upward to ^3P_J levels, that emits fluorescence from $^3P_{2,1,0}$ and $^1D_2$ to the ground ($^3H_4$) and other lower excited states. Several transitions are observed in upconverted luminescence including $^3P_2 \rightarrow {}^3H_4$ which usually lacks in one photon process. Doping concentration of Pr^{3+} ions does not significantly contribute to upconversion luminescence, which confirms that most probable channel for upconversion is ESA. Studies also revealed temperature dependence of upconversion luminescence. For 0.5 mol% Pr^{3+} ion concentration in tellurite glasses, luminescence quantum efficiency and upconversion efficiency are found to be ~3.3% and ~1.7%, respectively, for $^3P_0 \rightarrow {}^3F_2$ transition. Moreover, Kassab et al. [115] reported that the luminescence characteristics of Pr^{3+} in lead-tellurite glass enhance due to the presence of silver NPs with average size around 3.5 nm, by annealing at 350 °C for 7 h. Rai and Kassab have also studied the Stokes emission and the effect of silver NPs on the luminescence of Pr^{3+} ions in different glass compositions [114].

Work of Rajesh et al. [124] has been quantified in the new Pr^{3+}/Yb^{3+}-co-doped transparent oxyfluorotellurite glasses with chemical composition TeO_2-ZnO-YF_3-NaF-$0.5Pr_2O_3$-xYb_2O_3 (x = 0.25, 0.5, 0.75, and 1.0.) mol%. Pr^{3+}-Yb^{3+}-co-doped samples showed emission in the visible (Pr^{3+}: $^3P_0 \rightarrow {}^3F_2$ transition, 640 nm) and in the infrared (Yb^{3+}: $^2F_{5/2} \rightarrow {}^2F_{7/2}$,980 nm) spectral regions when it was excited at 440 nm during the downconversion process. The luminescence decay time of the emitting levels was obtained in co-doped samples as a function of Yb^{3+}

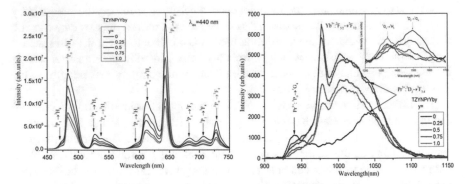

Fig. 7.9 (Left) Visible emission spectra for Pr^{3+} and Pr^{3+}-Yb^{3+}-co-doped glass samples as a function of Yb^{3+} concentration and upon excitation at 440 nm and (right) near-infrared emission spectra for Pr^{3+} singly doped and Pr^{3+}-Yb^{3+}-co-doped samples as a function of Yb^{3+} concentration. (Ref. [124, 125])

concentration, and the results confirmed the occurrence of energy transfer from Pr^{3+} to Yb^{3+} via a combination of two different cross-relaxation processes: (1) Pr^{3+}: $^1D_2$ → $^3F_4$ to Yb^{3+}: $^2F_{7/2}$ → $^2F_{5/2}$ and (2) Pr^{3+}: $^3P_0$ → $^1G_4$ to Yb^{3+}: $^2F_{7/2}$ → $^2F_{5/2}$. Furthermore, due to the reverse-energy transfer mechanism from Yb^{3+}:$^2F_{5/2}$ level to Pr^{3+}:$^1G_4$ level, the intensity at 980 nm for Yb^{3+} concentrations higher than 0.5 mol% is quenched. The energy transfer efficiency was assessed from the intensity ratios and rot times related to the $^3P_0$ → $^3F_2$ progress, and it achieved 66% for a glass co-doped with 0.5 mol% of Pr^{3+} and 1.0mo% of Yb^{3+}. The outcomes show that these glasses are potential contender for control of the solar spectrum, through upconversion and downconversion forms, with a specific end goal to increase the absorption efficiency of right now utilized c-Si photovoltaic solar cells.

The photoluminescence emission spectra for TZYN/Pr and Pr^{3+}-Yb^{3+}-co-doped glasses as a function of Yb^{3+} concentration in the visible region upon excitation at 440 nm are demonstrated in Fig. 7.9. It is evident that for different Yb^{3+} concentrations, the spectra of line shapes and wavelengths are almost identical when Pr^{3+} transition assignment is near the bands. It shows excitation by non-radiative decay from the upper excited levels due to the energy gaps between $^3P_0$ and $^1D_2$ levels, so red luminescence was also observed. The separately doped glasses with Pr^{3+} show high emission band intensity when compared to co-doped glasses. The Yb^{3+} concentration has a negative effect on the emission intensity of Pr^{3+} bands, which indicated energy transfer from Pr^{3+} to Yb^{3+}.

The NIR emission spectra of all samples with same excitation (440 nm) can be observed in Fig. 7.9. $^3P_0$ → $^1G_4$, $^1D_2$ → $^3F_{3,4}$ and $^1D_2$→$^1G_4$ transitions in Pr^{3+} singly doped samples showed spectrum bands at 940, 1048, and 1480 nm, respectively. $^2F_{5/2}$ → $^2F_{7/2}$ transition of Yb^{3+} ion was shown when Yb^{3+} was added indicated by an additional band in the range of 950–1100 nm. A decrease in the intensity of emissions by Pr^{3+} (band at 940 nm) was also observed upon increasing the Yb^{3+}concentrations which was also observed in the visible spectra. The Yb^{3+}

emission band intensity was observed at 977 nm. These results also confirm the hypothesis of energy transfer from Pr^{3+} to Yb^{3+}. Two processes have been proposed for this energy transfer.

1. Cross-relaxation Pr^{3+}: $^1D_2 \rightarrow {}^3F_4$;$Yb^{3+}$: $^2F_{7/2} \rightarrow {}^2F_{5/2}$ followed by the emission of one photon by Yb^{3+}: $^2F_{5/2} \rightarrow {}^2F_{7/2}$
2. Cross-relaxation Pr^{3+}: $^3P_0 \rightarrow {}^1G_4$;$Yb^{3+}$: $^2F_{7/2} \rightarrow {}^2F_{5/2}$ resulting in one photon emitted by Yb^{3+}: $^2F_{5/2} \rightarrow {}^2F_{7/2}$

The back-energy transfer process from Yb^{3+}:$^2F_{5/2}$ level to Pr^{3+}: $^1G_4$ level is the reason of the decrease in the emission intensity upon the increase in Yb^{3+} concentration. The emission intensity of transition from $^1D_2$ to $^1G_4$ also decreases with increase of Yb^{3+} concentration which is also due to the Pr^{3+} to Yb^{3+} energy transfer which was explained in the above two points. The transition of $^1D_2$ to $^1G_4$ which is also ion-ion energy transfer adds a new band at 1340 nm, and transition of $^1G_4$ to $^1H_5$ in Pr^{3+} also increased by increase in Yb^{3+} concentration due to the increase in concomitant population of $^1G_4$.

$AgNO_3$ containing TeO_2-ZnO-YF_3-NaF glasses doped with Pr_2O_3 was synthesized for the first time in a single-step melt quenching technique [125]. Silver nanoparticles were grown at different intervals of time for up to 10 h by putting glasses under controlled heat treatment at a temperature under the glass transition temperature, which was at 300 °C. TEM was used for the indication of silver nanoparticles, and images showed the presence of Ag NP of 1046 nm diameter. The surface plasmon resonance band (SPR) of the NPs in Pr^{3+}-free glasses also resulted in the formation of absorption band which peaked at 492 nm. The characterization of the effects of nanoparticles on the emission properties of Pr^{3+} was done by multiple techniques such as Vis and near-infrared fluorescence techniques in steady-state and time-dependent fluorescence and UV-Vis absorbance. The samples excited off resonance (440 and 470 nm) and in resonance (480 nm) with the SPR band showed increase of Pr^{3+} Stokes luminescence on raising the nanoparticle concentration. The Pr^{3+}:$^3P_0$ level showed increase in lifetime value of excited state when any Ag species or NP was present. This indicates an energy transfer from Ag to Pr^{3+} ions, which increases the enhancement of luminescence intensity which is also attributed to the possible local field effect of the NPs. These glasses have desirable properties for use in the photonic device applications.

The NIR emission at 1.23 μm from Er^{3+}/Pr^{3+} was identified for the investigation of the understudied regions of 1.2 μm [126]. Er^{3+}/Pr^{3+} co-doped with water-free fluorotellurite glass having a chemical composition of $60TeO_2$-$30ZnF_2$-$10NaF$ (TZNF60, mol. %) was used for this study. The directly measured lifetime (τ_f) at 1.23 μm in Er/Pr-co-doped fluorotellurite glasses was about 111.2 μs when pumping was done at 488 nm using optical parametric oscillator (OPO) laser system. The measured lifetime was longer than Er-doped fluorotellurite glass (80.1 μs).When appropriate Pr^{3+} ions were added, then the stimulated emission cross section (σ_{em}) and quantum efficiency (η) for Er^{3+}:$^4S_{3/2} \rightarrow {}^4I_{11/2}$ transition were significantly increased. The increase can be attributed to the low phonon energy and lack of hydroxyl group, which resulted due to the inclusion of high concentration of fluorides into oxide-based host glasses. Er^{3+}/Pr^{3+}-co-doped TZNF60 glass is considered a potential and promising material application in optical amplification and laser operation at comparatively unexplored 1.2 μm region due to high quantum efficiency (56.2%) and a large stimulated cross section (4.03×10^{-21} cm^2).

7.6.4 Sm^{3+}-Doped

Using melt quenching method, glasses with chemical composition of $x = 0$-1.5 mol % in (80-x)TeO$_2$-20ZnO-xSm$_2$O$_3$ were prepared [127]. For observing the structural modification of units of trigonal bipyramid, FTIR spectroscopy has been used. Transition peaks of glass shift, crystallization, and melting temperature can be seen properly in DTA traces. Samples made by this doping are made in such a way that they are stable as compared to the desired Hruby parameters and superior glass-forming ability. The edge of UV-Vis-NIR spectra was used for the measurement of direct band gap (2.75–3.18) eV, indirect band gap (3.22–3.40) eV, and Urbach energy (0.20–0.31) eV. An increase of 0.02 point in the refractive index was observed which was attributed to the generation of non-bridging oxygen atoms by the transition of TeO$_4$ into TeO$_3$ units. UV-Vis-NIR absorption spectra give peaks at 9 centered positions 470, 548, 947, 1085, 1238, 1385, 1492, 1550, and 1589 nm which result in transitions at $^6H_{5/2} \rightarrow {}^4I_{11/2}$, $^4G_{5/2}$, $^6F_{11/2}$, $^6F_{9/2}$, $^6F_{7/2}$, $^6F_{5/2}$, $^6F_{3/2}$, $^6H1_{5/2}$, and $^6F_{1/2}$, respectively. On the other hand, a decrease of 0.06 points in refractive index was also observed which was explained by lower ionic radii (1.079 Å) of Sm^{3+}. PL spectra under the excitation of 452 nm display 4 emission bands centered at 563, 600, 644, and 705 nm equivalent to $^4G_{5/2} \rightarrow 6H_{5/2}$, $6H_{7/2}$, $^6H_{9/2}$, and $^6H_{11/2}$ transitions of samarium ions. These properties make the sample excellent candidate for novel applications. With increase up to 1.2 mol of Sm^{3+}, the emission intensity is found to be increasing, and after further increases the emission intensity showed concentration quenching. Higher energy transfer among Sm^{3+} ions is ascribed to the decrease interionic separation, which was a result of increase in Sm^{3+} concentration and concentration quenching.

Physical and spectroscopic properties of Er^{3+}, Sm^{3+}, and Er^{3+}:Sm^{3+} ion-co-doped barium fluorotellurite (BFT) glasses are also studied elsewhere [128]. The laser excitations at 532 nm and 976 nm were used for the observance of different Stokes and anti-Stokes emissions. Luminescence intensity variation and decay curve analysis confirmed the energy transfer from Er^{3+} ion to Sm^{3+} ion in both results. In Er/Sm-co-doped samples, the emission intensity of Sm^{3+} bands was increased when excited with green light of 532 nm; in contrast upon NIR excitation at wavelength of 976 nm, new emissions bands of Sm^{3+} were observed.

The study of spectroscopic properties Sm^{3+}, Tb^{3+}-doped, and Tb^{3+}:Sm^{3+}-co-doped lead-fluorotellurite glasses has been done vigorously over the years [129]. The excitation of singly and doubly doped by laser at wavelengths of 355 and 532 nm has been done to study the luminescence properties. The presence of Tb^{3+} ions significantly increases the intensity of Sm^{3+} emission bands which is attributed to the energy transfer from Tb^{3+} to Sm^{3+} ions. The efficient energy transfer from Tb^{3+} to Sm^{3+} ions has also been confirmed by calculation of different energy transfer parameters.

7.7 Summary

In this study, various characteristics of tellurite glass without and containing fluoride and rare-earth ions with and without metallic nanoparticles have been determined. Based on the results, some conclusions can be drawn which are summarized below.

The knowledge of the glass and glass-ceramic technology is growing rapidly through the years. Such an increase in the interest on these materials is due to their potential applicability and easy-preparation method as low cost.

Tellurite and oxyfluorotellurite glasses show good optical properties, such as high-transparency window and relatively higher refractive index than other traditional glasses, such as silica and phosphate glasses. Moreover, the low phonon energy characteristics of these glasses facilitate the upconversion process and increase the quantum efficiency of the emitting levels of rare-earth ions, thanks to a drastic decrease in the rate of non-radiative relaxations.

Tellurite and oxyfluorotellurite glasses present good thermal stability and chemical durability. Moreover, the rare-earth solubility in tellurite glasses is comparable to those oxyfluorotellurite glasses and higher than the traditionally used silica glasses.

Adding fluoride to prepare fluorotellurite glasses aims to improve some optical properties such as excited state lifetime. Some recent examples of the tellurite glasses and oxyfluorotellurite glasses doped with various rare earth ions are presented in this chapter. The influence of the metallic nanoparticles is also discussed in both the studied glass compositions.

The transparent tellurite and oxyfluorotellurite glasses are promising materials in a wide range of technological applications and are capable hosts to incorporate rare-earth ions for optical applications and photonics as well as noble metallic nanoparticles.

Acknowledgment M.R. Dousti is grateful to the Institute of Physics at Universidade Federal de Alagoas for the given infrastructure for the research and teaching. R. J. Amjad would like to thank Ms. Sumera Yasmeen and Ms. Asma Muhammad for helping in editing and English revision.

References

1. V.K. Rai, K. Kumar, S.B. Rai, Opt. Mater. (Amst). **29**, 873–878 (2007)
2. K. Maheshvaran, K. Marimuthu, J. Lumin. **132**, 2259–2267 (2012)
3. S. Dai, C. Yu, G. Zhou, J. Zhang, G. Wang, L. Hu, J. Lumin. **117**, 39–45 (2006)
4. H. Zheng, D. Gao, Z. Fu, E. Wang, Y. Lei, Y. Tuan, M. Cui, J. Lumin. **131**, 423–428 (2011)

5. V. Chereminsinov, V. Zalomanov, Opt. Spectrosc. **12**, 110 (1962)
6. E.F. Lambson, G.A. Saunders, B. Bridge, R.A. El-Mallawany, J. Non-Cryst. Solids **69**, 117–133 (1984)
7. G. Barady, J. Chem. Phys. **27**, 1957 (1957)
8. M. Mochida, K. Takashi, S. Shibusawa, J. Ceram. Assoc. Japan **86**, 317 (1978)
9. V. Kozhukharov, M. Marinov, G. Grigorova, J. Non-Cryst. Solids **28**, 429–430 (1978)
10. R.A.H. El-Mallawany, *Tellurite Glasses Handbook: Physical Properties and Data* (CRC Press, Boca Raton, 2002)
11. O. Lindqvist, Acta Chem. Scand. **22**, 977–982 (1968)
12. H. Beyer, Z. Krist. **124**, 228 (1967)
13. V. Kozhukharov, H. Burger, S. Neov, B. Sidzhimov, Polyhedron **5**, 771–777 (1986)
14. S.L. Tagg, J.C. Huffman, J.W. Zwanziger, Chem. Mater. **6**, 1884–1889 (1994)
15. S.L. Tagg, J.C. Huffman, J.W. Zwanziger, Acta Chem. Scand. **51**, 118–121 (1997)
16. R. Masse, J.C. Guitel, I. Tordjman, Mater. Res. Bull. **15**, 431–436 (1980)
17. S. Neov, V. Kozhukharov, I. Gerasimova, K. Krezhov, B. Sidzhimov, J. Phys. C Solid State Phys. **12**, 2475–2485 (1979)
18. S.L. Tagg, R.E. Youngman, J.W. Zwanziger, J. Chem. Phys. **99**, 5111–5116 (1995)
19. J.W. Zwanziger, S.L. Tagg, J.C. Huffman, Science (80-.) **268**, 1510 (1995)
20. K. Suzuki, J. Non-Cryst. Solids **95–96**, 15–30 (1987)
21. T. Sekiya, N. MOchida, A. Ohtsuka, A. Soejima, J. Non-Cryst. Solids **151**, 222–228 (1992)
22. T. Tatsumisago, T. Minami, Y. Kowada, H. Adachi, Phys. Chem. Glasses **35**, 89 (1994)
23. S. Sakida, S. Hayakawa, T. Yoko, J. Non-Cryst. Solids **243**, 13–25 (1999)
24. H. Koller, G. Engelhardt, A.P.M. Kentgens, J. Sauer, J. Ceram. Assoc. Japan **98**, 1544–1551 (1994)
25. J. Stanworth, Soc. Glas. Technol. **36**, 217T–241T (1952)
26. I.Z. Hager, R. El-Mallawany, J. Mater. Sci. **45**, 897 (2010)
27. I. Hager, R. El-Mallawany, A. Bulou, Phys. B Condens. Matter **406**, 972–980 (2011)
28. N. Soorj Hussain, G. Hungerford, R. El-Mallawany, M.J.M. Gomes, M.A. Lopes, N. Ali, J.D. Santos, J. Nanosci. Nanotechnol. **9**, 3672–3677 (2009)
29. A. Abdel-Kader, R. El-Mallawany, M.M. Elkholy, J. Appl. Phys. **73**, 71–74 (1993)
30. R. El-Mallawany, M. Sidkey, A. Khafagy, H. Afifi, Mater. Chem. Phys. **37**, 197–200 (1994)
31. A. El-Adawy, R. El-Mallawany, J. Mater. Sci. Lett. **15**, 2065–2067 (1996)
32. R. El-Mallawany, A. Abd El-Moneim, Phys. Status Solidi **166**, 829–834 (1998)
33. R. El-Mallawany, Phys. Status Solidi(a) **177**, 439–444 (2000)
34. H.M.M. Moawad, H. Jain, R. El-Mallawany, T. Ramadan, J. Am. Ceram. Soc. **85**, 2655–2659 (2002)
35. M.M. Elkholy, R.A. El-Mallawany, Mater. Chem. Phys. **40**, 163–167 (1995)
36. M.M. El-Zaidia, A.A. Ammar, R.A. El-Mallwany, Phys. Status Solidi **91**, 637–642 (1985)
37. H.A.A. Sidek, S. Rosmawati, Z.A. Talib, M.K. Halimah, W. Daud, J. Appl. Sci. **6**, 1489–1494 (2009)
38. U. Hoppe, E. Yousef, C. Russel, J. Neuefeind, A.C. Hannon, J. Phys. Condens. Matter **166**, 1645–1663 (2004)
39. M.R. Sahar, A.K. Jehbu, M.M. Karim, J. Non-Cryst. Solids **213–214**, 164–167 (1997)
40. M.R. Sahar, N. Noordin, J. Non-Cryst. Solids **184**, 137–140 (1995)
41. C. Technology, S. Neov, N. Energy, B. Academy **5**, 771–777 (1986)
42. P. Babu, H.J. Seo, C.R. Kesavulu, K.H. Jang, C.K. Jayasankar, J. Lumin. **129**, 444–448 (2009)
43. M.D. O'Donnell, K. Richardson, R. Stolen, A.B. Seddon, D. Furniss, V.K. Tikhomirov, C. Rivero, M. Ramme, R. Stegeman, G. Stegeman, M. Couzi, T. Cardinal, J. Am. Ceram. Soc. **90**, 1448–1457 (2007)
44. S.K. Singh, N.K. Giri, D.K. Rai, S.B. Rai, Solid State Sci. **12**, 1480–1483 (2010)
45. G. Liao, Q. Chen, J. Xing, H. Gebavi, D. Milanese, M. Fokine, M. Ferraris, J. Non-Cryst. Solids **355**, 447–452 (2009)

46. Y. Yang, B. Chen, C. Wang, G. Ren, Q. Meng, X. Zhao, W. Di, X. Wang, J. Sun, L. Cheng, T. Yu, Y. Peng, J. Non-Cryst. Solids **354**, 3747–3751 (2008)
47. S. Surendra Babu, K. Jang, E. Jin Cho, H. Lee, C.K. Jayasankar, J. Phys. D. Appl. Phys. **40**, 5767–5774 (2007)
48. G. Lakshminarayana, J. Qiu, J. Alloys Compd. **478**, 630–635 (2009)
49. M.R. Sahar, K. Sulhadi, M.S. Rohani, J. Non-Cryst. Solids **354**, 1179–1181 (2008)
50. V.A.G. Rivera, E. Rodriguez, E.F. Chillcce, C.L. Cesar, L.C. Barbosa, J. Non-Cryst. Solids **353**, 339–343 (2007)
51. Z. Pan, A. Ueda, M. Hays, R. Mu, S.H. Morgan, J. Non-Cryst. Solids **352**, 801–806 (2006)
52. A. Hruby, Czechoslov. J. Phys. **22**, 1187–1193 (1972)
53. A.A. Abu-Sehly, M.A. El-oyoun, A.A. Elabbar, Thermochim. Acta **472**, 25–30 (2008)
54. C.A. Angell, J. Non-Cryst. Solids **73**, 1–17 (1985)
55. C.A. Angell, J. Non-Cryst. Solids **131–133**, 13–31 (1991)
56. C.A. Angell, J. Phys. Chem. Solids **49**, 863–871 (1988)
57. H.A.A. Sidek, S. Rosmawati, M.K. Halimah, K.A. Matori, Z.A. Talib, Materials (Basel). **5**, 1361–1372 (2012)
58. A. Einstein, Verh Deut Phys Ges **18**, 318–323 (1916)
59. E. Snitzer, Phys. Rev. Lett. **7**, 444 (1961)
60. K.C. Kao, Proc. Inst Electr Eng-London **113**, 1151 (1966)
61. B.R. Judd, Phys. Rev. **127**, 750–761 (1962)
62. G.S. Ofelt, J. Chem. Phys. **37**, 511–520 (1962)
63. B.M. Walsh, Advances in Spectroscopy for Lasers and Sensing, Springer, 403–433 (2006)
64. J.N. Sandoe, P.H. Sarkies, S. Parke, J. Phys. D. Appl. Phys. **5**, 1788 (1972)
65. J. Stone, C.A. Burrus, Appl. Phys. Lett. **23**, 388–389 (1973)
66. E. Desurvire, J.R. Simpson, P.C. Becker, Opt. Lett. **12**, 888–890 (1987)
67. S.M. Svendsen, Telektronikk **3**, 115 (1997)
68. R. Reisfeld, J. Hormadaly, A. Muranevich, Chem. Phys. Lett **38**, 188–191 (1976)
69. M.J. Weber, J.D. Myers, D.H. Blackburn, J. Appl. Phys. **52**, 2944–2946 (1981)
70. P.C. Becker, N.A. Olsson, J.R. Simpson, *Erbium-Doped Fiber Amplifiers* (Academic Press, San Diego, 1999)
71. Z. Li-Yan, H. Li-Li, Chinese Phys. Lett. **20**, 1836–1837 (2003)
72. M.R. Dousti, R.J. Amjad, M.R. Sahar, Z.M. Zabidi, A.N. Alias, A.S.S. De Camargo, J. Non-Cryst. Solids **429**, 70–78 (2015)
73. I. Jlassi, H. Elhouichet, M. Ferid, C. Barthou, J. Lumin. **130**, 2394–2401 (2010)
74. J. Yang, L. Zhang, L. Wen, S. Dai, L. Hu, Z. Jiang, J. Appl. Phys. **95**, 3020 (2004)
75. T. Xu, X. Zhang, S. Dai, Q. Nie, X. Shen, X. Zhang, Phys. B Condens. Matter **389**, 242–247 (2007)
76. U.R. Rodríguez-Mendoza, E.A. Lalla, J.M. Cáceres, F. Rivera-López, S.F. León-Luís, V. Lavín, J. Lumin. **131**, 1239–1248 (2011)
77. T. Som, B. Karmakar, Spectrochim. Acta Part A Mol. Biomol. Spectrosc. **79**, 1766–1782 (2011)
78. E. Snitzer, Ceram. Bull. **52**, 516–525 (1973)
79. T. Izumitani, H. Toratani, H. Kuroda, J. Non-Cryst. Solids **47**, 87–99 (1982)
80. J.S. Wang, E.M. Vogel, E. Snitzer, Opt. Mater. (Amst). **3**, 187–203 (1994)
81. C.B. Layne, W.H. Lowdermilk, M.J. Weber, Phys. Rev. B **16**, 10–20 (1977)
82. F. Auzel, J. Lumin. **45**, 341–345 (1990)
83. P.G. Kik, A. Polman, J. Appl. Phys. **93**, 5008–5012 (2003)
84. H. Dong, L.D. Sun, C.H. Yan, Basic understanding of the lanthanide related upconversion emissions. Nanoscale **5**, 5703–5714 (2013)
85. D.L. Dexter, J. Chem. Phys. **21**, 836–851 (1953)
86. A. A. Kaplyanskii, R. M. MacFarlane (eds.), *Spectroscopy of Solids Containing Rare-Earth Ion* (North-Holland, Amsterdam, 1987)

87. A. Jha, B. Richards, G. Jose, T. Teddy-Fernandez, P. Joshi, X. Jiang, J. Lousteau, Prog. Mater. Sci. **57**, 1426–1491 (2012)
88. P.F. Wysocki, N. Park, D. Digiovanni, Opt. Lett. **21**, 1744–1746 (1996)
89. H. Ono, M. Yamada, T. Kanamori, Y. Ohishi, Electron. Lett. **33**, 1477–1479 (1997)
90. H. Masuda, K.I. Suzuki, S. Kawai, K. Aida, Electron. Lett. **33**, 753–754 (1997)
91. P.V. dos Santos, M.V.D. Vermelho, E.A. Gouveia, M.T. De Araujo, A.S. Gouveia-Neto, F.C. Cassanjes, S.J.L. Ribeiro, Y. Messaddeq, J. Alloys Compd. **344**, 304–307 (2002)
92. Q. Nie, X. Li, S. Dai, T. Xu, Z. Jin, X. Zhang, J. Lumin. **128**, 135–141 (2008)
93. X. Shen, Q. Nie, T. Xu, S. Dai, X. Wang, J. Lumin. **130**, 1353–1356 (2010)
94. A. Cherif, A. Kanoun, N. Jaba, Opt. Appl. **XL**, 109–118 (2010)
95. N. Jaba, A. Kanoun, H. Mejri, A. Selmi, S. Alaya, H. Maaref, J. Phys. Condens. Matter **12**, 4523–4534 (2000)
96. M.R. Dousti, J. Mol. Struct. **1100**, 415–420 (2015)
97. A. Miguel, R. Morea, J. Gonzalo, J. Fernandez, R. Balda **9359**, 93590X (2015)
98. M. Reza Dousti, M.R. Sahar, S.K. Ghoshal, R.J. Amjad, R. Arifin, J. Mol. Struct. **1033**, 79–83 (2013)
99. S.F. León-Luis, U.R. Rodríguez-Mendoza, I.R. Martín, E. Lalla, V. Lavín, Sensors Actuators B Chem. **176**, 1167–1175 (2013)
100. S.F. León-Luis, U.R. Rodríguez-Mendoza, E. Lalla, V. Lavín, Sensors Actuators B Chem. **158**, 208–213 (2011)
101. M. Pokhrel, G.A. Kumar, S. Balaji, R. Debnath, D.K. Sardar, J. Lumin. **132**, 1910–1916 (2012)
102. O.L. Malta, P.A. Santa-Cruz, G. De Sa, F. Auzel, J. Lumin. **33**, 261–272 (1985)
103. C. Strohhöfer, A. Polman, Appl. Phys. Lett. **81**, 1414–1416 (2002)
104. V.A.G. Rivera, S.P.A. Osorio, Y. Ledemi, D. Manzani, Y. Messaddeq, L.A.O. Nunes, E. Marega, Opt. Express **18**, 25321–25328 (2010)
105. L.R.P. Kassab, C.B. de Araújo, R.A. Kobayashi, R. de Almeida Pinto, D.M. da Silva, J. Appl. Phys. **102**, 103515 (2007)
106. L.R.P. Kassab, R. de Almeida, D.M. da Silva, C.B. de Araújo, J. Appl. Phys. **104**, 93531 (2008)
107. T.A.A. de Assumpção, D.M. da Silva, M.E. Camilo, L.R.P. Kassab, A.S.L. Gomes, C.B. de Araújo, N.U. Wetter, J. Alloys Compd. **536**, S504–S506 (2012)
108. L.R.P. Kassab, K.J. Plucinski, M. Piasecki, K. Nouneh, I.V. Kityk, A.H. Reshak, R.D.A. Pinto, Opt. Commun. **281**, 3721–3725 (2008)
109. L.P.R. Kassab, L. Ferreira Freitas, K. Ozga, M.G. Brik, A. Wojciechowski, Opt. Laser Technol. **42**, 1340–1343 (2010)
110. A.P. Carmo, M.J.V. Bell, V. Anjos, R. de Almeida, D.M. da Silva, L.R.P. Kassab, J. Phys. D. Appl. Phys. **42**, 155404 (2009)
111. L.R.P. Kassab, T.A.A. Assumpção, P. Czaja, E. Gondek, K.J. Plucinski, J. Mater. Sci. Mater. Electron. **23**, 1122–1125 (2011)
112. V.P.P. de Campos, L.R.P. Kassab, T.A.A. de Assumpção, D.S. da Silva, C.B. de Araújo, J. Appl. Phys. **112**, 63519 (2012)
113. R. De Almeida, D.M. da Silva, L.R.P. Kassab, C.B. de Araujo, Opt. Commun. **281**, 108–112 (2008)
114. V.K. Rai, L.D.S. Menezes, C.B. De Araújo, L.R.P. Kassab, M. Davinson, J. Appl. Phys. **103**, 93526 (2008)
115. L.R. Kassab, C.B. De Araújo, R.A. Kobayashi, R.D.A. Pinto, D.M. Silva, J. Appl. Phys. **102**, 103515–103514 (2007)
116. M. Reza Dousti, M.R. Sahar, S.K. Ghoshal, R.J. Amjad, A.R. Samavati, J. Mol. Struct. **1035**, 6–12 (2013)
117. M. Reza Dousti, M.R. Sahar, S.K. Ghoshal, R.J. Amjad, R. Arifin, J. Non-Cryst. Solids **358**, 2939–2942 (2012)

118. M.R. Dousti, M.R. Sahar, R.J. Amjad, S.K. Ghoshal, A. Khorramnazari, A.D. Basirabad, A. Samavati, Eur. Phys. J. D **66**, 1–6 (2012)
119. A. Nukui, T. Taniguchi, M. Miyata, J. Non-Cryst. Solids **293–295**, 255–260 (2001)
120. S. Banijamali, B.E. Yekta, A.R. Aghaei, J. Non-Cryst. Solids **358**, 303–309 (2012)
121. K.F. Ahmed, S.O. Ibrahim, R. Sahar, S.Q. Mawlud **28**, 351–355 (2016)
122. A. Miguel, J. Azkargorta, R. Morea, I. Iparraguirre, J. Gonzalo, J. Fernandez, R. Balda, Opt. Express **21**, 9298 (2013)
123. M. Reza Dousti, J. Appl. Phys. **114**, 3–8 (2013)
124. D. Rajesh, M.R. Dousti, R.J. Amjad, A.S.S. De Camargo, J. Non-Cryst. Solids **450**, 149–155 (2016)
125. D. Rajesh, R.J. Amjad, M. Reza Dousti, A.S.S. de Camargo, J. Alloys Compd. **695**, 607–612 (2017)
126. H. Zhan, A. Zhang, J. He, Z. Zhou, J. Si, A. Lin, Appl. Opt. **52**, 9–11 (2013)
127. Y.A. Tanko, M.R. Sahar, S.K. Ghoshal, Results Phys. **6**, 7–11 (2016)
128. A. Bahadur, Y. Dwivedi, S.B. Rai, Spectrochim. Acta Part A Mol. Biomol. Spectrosc. **77**, 101–106 (2010)
129. A. Bahadur, Y. Dwivedi, S.B. Rai, Spectrochim. Acta Part A Mol. Biomol. Spectrosc. **118**, 177–181 (2014)

Chapter 8
Optical Sensing Based on Rare-Earth-Doped Tellurite Glasses

M. Reza Dousti, Weslley Q. Santos, and Carlos Jacinto

Abstract Tellurite glasses are among the most interesting host matrices for various optical applications. The excellent optical and thermal properties of tellurite glasses are due to their high linear and nonlinear refractive indices, good transparency window, low phonon energy, high rare-earth solubility, and thermal stability. One of the promising applications of tellurite glasses is in optical thermometry. In this chapter, we review the fundaments of thermometry by luminescence spectroscopy, the theoretical background to calculate the thermal sensibility of rare-earth ion-doped materials, and some examples of the thermometry done on tellurite glasses as well as some other host matrices. For example, Er^{3+}-doped tellurite glasses are given as thermal sensors applied in the visible spectral region, where the fluorescent intensity ratio of two green emission bands plays the role to optically determine the temperature. On the other hand, Nd^{3+} ion-doped glasses could be used to measure the temperature in the near-infrared region, using the intensity ratio variations of principal emissions in the 800–1400 nm spectral range. Thermal sensibility of each case is compared to various glass host compositions.

8.1 Introduction

Temperature is one of the most fundamental properties of thermodynamics. From the first attempt (by Galileo Galilei in 1593) to develop a numerical scale for the thermometer, precise temperature measurement with high spatial resolution has been a rather challenging research topic. Advances in nanotechnology and biotechnology require a precise thermometry, capable of evaluating systems in the regime of micro- and nanoscale (nanothermometry), where conventional methods are not able to perform measurements [1–3]. The development of a nanothermometer (NTh) is not only about the miniaturization of size but also about the manipulation of new physical and chemical properties of materials, because these properties are

M. Reza Dousti (✉) · W. Q. Santos · C. Jacinto
Grupo de Nano-Fotônica e Imagens, Instituto de Física, Universidade Federal de Alagoas, Maceió, AL, Brazil

© Springer International Publishing AG, part of Springer Nature 2018
R. El-Mallawany (ed.), *Tellurite Glass Smart Materials*,
https://doi.org/10.1007/978-3-319-76568-6_8

drastically altered on such a small scale. In fact, nanothermometry seeks to extract knowledge of the local temperature of a given system with a sub-micrometric spatial resolution [4–6]. As a first approximation, we can identify three areas that can clearly use nanothermometry and benefit significantly from the increasing development of NThs with respect to higher sensitivities and resolutions: (1) micro-/nanoelectronics, (2) integrated photonics, and (3) biomedicine [7].

It is well known that in any biosystem temperature plays a fundamental role determining its properties and dynamics. For example, temperature is one of the critical and determinant parameters in rates of cell division, enzymatic reactions, changes in metabolic activities, and determination of tissue growth rate [8]. Undoubtedly, temperature dramatically affects the mechanical, optical, and structural properties of important biomolecules, similar to proteins that can undergo denaturation, when exposed to other conditions than those that they have been produced, such as temperature variations (a few degrees above 37 °C) and change in pH, among others. Thus, for the full understanding of the dynamics of biosystems, the simultaneous monitoring of their temperature is fundamental to elucidate the origin of the observed behavior. Therefore, the interest in understanding the thermal behavior of the biosystem is a crucial point for early detection and treatment of diseases [9–13]. Usually one of the earliest indications of any disease (such as inflammation, cancer, or heart problems) is the arising of thermal singularities [3]. In fact, cancer cells have higher temperatures compared to healthy cells due to increased metabolic activity.

High-resolution thermal sensing is also required in cancer therapy processes, such as hyperthermia which is based on elevated tumor temperature induced by an external source to cytotoxic levels (43–45 °C). The increase in temperature can be done in a controlled manner in order to minimize thermally induced damage to the tissues around the treated region [9, 10].

Nanoscale thermometry therefore seeks a new standard in the use of both thermometric materials and properties. Due to the particularity of each system, different thermal sensors are required for different applications. In biomedicine, the NThs must be nontoxic (biocompatible), soluble in water, and very stable under the irradiation of light, in order not to release toxic components to the biosystem.

Examples of such miniaturization contain:

- Luminescent thermometers (LNThs) based on temperature dependence with emission intensity and/or lifetime of dyes, semiconductors such as quantum dots (QDs), and rare-earth ion-doped nanoparticles (RE-NPs)
- Infrared (IR) nanoscale thermometers from metallic NPs based on black-body radiation
- Mapping by thermal microscopes based on doped NPs

In principle, any luminescence NP (LNPs), whose luminescent property (in terms of intensity, spectral position, spectral form, or decay time) is heavily temperature dependent, may be considered as a potential LNTh [2, 3, 7]. However, the real-world application of LNThs as an intracellular thermal sensor requires several additional features. These LNPs should be readily dispersible in aqueous media such that they allow easy incorporation into the cell and, in addition, must have relatively high fluorescence quantum efficiency (η), so as to provide high-quality image contrasts. Also, they should show long thermal stability without degradation during optical excitation and have minimal interaction with the medium investigated. It is also necessary for them to be excited by photons in the IR (in the spectral range from 700 to 1400 nm). The use of radiation in the IR (where the "biological transparency window" or simply "biological window" is) allows a greater depth of penetration and, therefore, the possibility of real three-dimensional thermal images.

8.2 Luminescent Thermometers

The limitations of the contact thermometer to work on submicron scales are well known, and such difficulties led to the development of thermometry techniques without the need for contact with the medium, such as infrared thermography, thermoreflectance, optical interferometry, and luminescence.

Infrared thermography – because all bodies are at a temperature above absolute zero, they emit thermal radiation. Infrared thermography systems capture this radiation and convert it into an image representing the temperature distribution on the surface of the observed object. These methods have been applied in several areas (medical, industrial, civil construction, electronics, etc.).

Thermoreflectance – explores the temperature dependence with respect to the reflection coefficient of a material (where the refractive index is temperature dependent), thus obtaining a temperature profile through the analysis of the images formed by the reflection; however, the spatial resolution is limited by the optical imaging system of the equipment.

Optical interferometry – is based on measuring the variation of the radiation provided by the superposition of two optical beams. It provides local temperature information as well as performs precise measurements of thermal expansion or deformation of a given surface.

Optical nanothermometry – Among the methods of contactless thermometry, the determination of temperature from the thermal dependence of luminescence has attracted much interest, since it is a very precise technique. It is based on temperature-induced changes in the optical properties of luminescent materials. This is undoubtedly a promising technique for nanoscale thermal sensing. Among the many existing luminescent thermal sensors, we can mention organic dyes, polymers, metallic NPs, quantum dots, and nanomaterials doped with Ln^{3+} ions. In these cases, the temperature evaluation is made by the spectroscopic measurements (intensity, band shape, spectral position, polarization, lifetime, bandwidth, etc.) of the parameters that characterize the luminescence sensor.

(i) Luminescence Intensity-Based Thermometry

In this case, the thermal parameter (to be used for the sensing) is obtained by analyzing the luminescence intensity. When variations in temperature occur, a portion of the total number of photons emitted per second also changes, causing the emission spectrum to become less (or more) intense. These changes in luminescence intensity are usually caused by the thermal activation of the luminescence quenching process and/or due to the increased probability of non-radiative decays.

(ii) Thermometry Based on the Luminescence Band Shape

In this case we seek to relate the relative intensities of different spectral lines that compose the luminescence spectrum. In many cases the temperature induces variations in the shape of the luminescent band, and in general this occurs when the electronic states from which the emissions are being generated are thermally coupled. This method is also known as ratiometric and it is a self-referencing one. In this case, the intensity measurements are not compromised by the well-known disadvantages of experiments based on the intensity of only one transition.

(iii) Luminescence Spectral Position Thermometry

This method is based on the analysis of the spectral position of the emission lines, which are determined by the energy gap between the two electronic levels involved in this process. Moreover, the variation of the energy gap width depends on both the temperature and other parameters that are part of the material emission process, for example, the refractive index and the interatomic distances (density). Thus, the emission lines of any emitting material are expected to be temperature dependent. In fact, this method exploits the shift of a given material due to variations in temperature.

(iv) Thermometry Based on Polarization of Luminescence

In an anisotropic media, the emitted radiation is usually non-isotropically polarized, and consequently the form and intensity of emitted radiation are strongly dependent on its polarization. This fact allows us to define a parameter that we will call "anisotropy of polarization," which is the ratio of the intensity of the emitted luminescence in the two orthogonal states of the polarization. In summary, this approach is based on the influence of temperature on the "polarization anisotropy."

(v) Luminescence Bandwidth-Based Thermometry

The width of several emission lines, which are part of a luminescence spectrum, is determined by the properties of the material (e.g., the degree of disorder of the atoms) as well as its temperature. It is already known that many luminescent materials undergo an increase in the local temperature, due to the increase in the density of phonons in the material, thus contributing to the nonhomogeneous spectral widening of the luminescence spectrum. Variations in the width of the luminescence spectrum are exploited by this method, in order to perform a thermal reading of the medium.

(vi) *Luminescence Lifetime-Based Thermometry*

The lifetime, τ, of the luminescence is defined as the time in which the emitted luminescence intensity is reduced to $1/e$ of the initial value. This is a good indication of the total probability of decay of the emitted intensity (in fact, this probability is defined as the inverse of the lifetime of the the the emitting energy level). The decay probability from the electronic levels depends on an enormous number of factors, and many of them have a certain dependence on the temperature (such as energy transfer processes assisted by phonons and multiphonon decays). The dependence on temperature makes it possible to extract (perform) the thermal reading from the determination of the lifetime.

Figure 8.1 shows the schematic representative of the luminescence spectra at low and high temperature for the abovementioned induced effects. In summary, luminescence-based thermometry provides many options for monitoring the

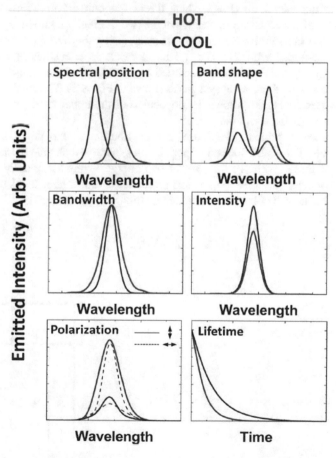

Fig. 8.1 Possible temperature-induced effects on the luminescence properties. The red lines represent higher temperature. The schematic representation is adopted from Ref. [3]

temperature of a given material from the emission spectra analyses. It is important to note that the thermal sensitivity varies from system to system. Those ones that show remarkable changes in the luminescence properties when submitted to small variations in the temperature are classified as excellent thermal sensors, since they are endowed with great thermal sensitivity.

8.3 Theoretical Backgrounds

The populations of thermally coupled energy levels could be described in terms of the Boltzmann population distribution. Thus, the emission of two levels, thermally coupled, of a trivalent rare-earth ion (RE^{3+}) can provide information about the temperature at which the system is in thermal equilibrium. This technique is called fluorescence intensity ratio (FIR), where the temperature information is obtained from the ratio of the emission intensities of the two thermally coupled levels, which was introduced in 1990 by Berthou e Jorgensen [14] using an optical fiber fluorinated glass co-doped with Yb^{3+}/Er^{3+}. The fluorescence intensities of two closely spaced energy levels (1 and 2) are recorded as a function of the temperature to be analyzed in a simple three-level system (as shown in Fig. 8.2). The small energy gap between two excited states allows the population of the upper level from the lower one through a thermal excitation.

The increase in temperature causes a redistribution of the population of the emitting levels. This redistribution of population generates a change in the emission intensities of these levels. In the case of two energy levels sufficiently close and thermally coupled, the ions' population promoted to the higher level by thermal energy can be described by the Boltzmann distribution law [15, 16]:

$$N_2 = N_1 e^{\left(\frac{-\Delta E_{21}}{K_B T}\right)} \tag{8.1}$$

Fig. 8.2 Simplified diagram for three energy level thermometry system

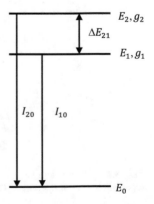

where N_1 (or N_2) is the population of level 1 (or 2), (look at Fig. 8.2); ΔE_{21} is the energy gap between the thermalized levels 1 and 2, which can be experimentally determined from the absorption spectrum; K_B is the Boltzmann's constant; and T is the absolute temperature. The population ratio can be described as

$$\frac{N_2}{N_1} = e^{\left(\frac{-\Delta E_{21}}{K_B T}\right)} \tag{8.2}$$

The thermalized emission intensity I_{20} (or I_{10}) originated from radiative decay from level 2 (or 1) to the ground state 0 (or other down level) is proportional to the ion population N_2 (or N_1), the spontaneous radiative emission rate A_{20} (or A_{10}), the branching ratio associated with the corresponding transition β_{20} (or β_{10}), the average photon energy $h\nu_{20}$ (or $h\nu_{10}$), and the degeneracy g_2 (or g_1) of the level. Thus, the thermalized emission intensity, I_{20}, could be written as

$$I_{20} = N_2 h\nu_{20}\beta_{20}A_{20}g_2 \tag{8.3}$$

Other way, I_{20} is proportional to the integrated area under the luminescence curve of this level $A_{E_{20}}$, $I_{20} \propto A_{E_{20}}$. Calculating the ratio of the integrated areas of the two transitions, i.e., the FIR of the two transitions corresponding to the thermalized energy levels, levels 2 and 1, we have

$$\text{FIR} = R = \frac{A_{E_{20}}}{A_{E_{10}}} = \frac{N_2 h\nu_{20}\beta_{20}A_{20}g_2}{N_1 h\nu_{10}\beta_{10}A_{10}g_1} \tag{8.4}$$

Since $\frac{N_2}{N_1} = e^{\left(\frac{-\Delta E_{21}}{K_B T}\right)}$, the fluorescence intensity ratio between two thermalized emissions can be associated with the temperature of the optical sensor through the following relation [17, 18]:

$$R = \frac{\nu_{20}\beta_{20}A_{20}g_2}{\nu_{10}\beta_{10}A_{10}g_1} e^{\left(\frac{-\Delta E_{21}}{K_B T}\right)} \tag{8.5}$$

or in a simpler form,

$$R = Ce^{\left(\frac{-\Delta E_{21}}{K_B T}\right)} \tag{8.6}$$

where $C = \frac{\nu_{20}\beta_{20}A_{20}g_2}{\nu_{10}\beta_{10}A_{10}g_1}$. Thus, calculating the ratio of the fluorescence intensities of the two thermalized emission bands, it is possible to determine the absolute temperature at which the system is in thermal equilibrium. As can be easily observed, calculating $\ln(R)$ in Eq. (8.6), a linear equation could be found as a function of T^{-1}, i.e., $\ln(R) = a + b\frac{1}{T}$ ($a = \ln(C)$ and $b = \frac{-\Delta E_{21}}{K_B}$). This linear behavior of $\text{Ln}(R)$ versus T^{-1} is of great interest in practical applications in sensors. The ratio of these intensities is independent of the source power intensity, since the intensity of each emission band is proportional to the population of each level. Such characteristic of FIR technique is promising for practical applications. However, the parameter C depends on the properties of the host matrix and the involved electronic transitions. For the most varied reasons, the Eq. (8.6) does not fit to some systems with thermally coupled levels, or at least the values of the parameters obtained from the fitting procedure are quite different from the real ones. For example, Pereira et al. [19] investigated Yb/Tm-co-doped nanocrystals obtaining experimentally an energy gap of 65 cm^{-1} , while the real is close to 370 cm^{-1}. Different terminal $^3H_6$ ground-state stark levels for each transition were suggested as the reason; however, there are many. For any sensor system, an important observation for practical applications is that the calibration curve has to be obtained by direct measurement of fluorescence intensities, for example, at peak wavelengths without applying any spectral deconvolution procedure. This makes thermal measurements fast and simple. Another very important question is that, independent of the equation governing the temperature dependence of the optical parameter, a linear dependence provides an almost temperature-independent thermal sensitivity, making straightforward the temperature readout in the thermal sensor. Therefore, a direct linear relationship of the optical parameter to the temperature is desired [4, 5, 13].

Thermal sensitivity is an important parameter in a temperature sensor, which is a scale of how much sensitive it is at a given temperature variation. The quality of an optical sensor is given by its thermal sensitivity S or the response speed of a sensor to the changes of temperature. This parameter is defined as the rate of the change of the intensity ratio R with temperature, and it allows comparison between different optical temperature sensors ($S = \frac{dR}{dT}$). Therefore, S for thermally coupled system is calculated by

$$S = R\left(\frac{\Delta E_{21}}{K_B T^2}\right) \tag{8.7}$$

The thermometry performance is mostly measured by the relative sensitivity S_R [20]. The S_R indicates the relative change of R per degree of temperature change and is defined by

$$S_R = \frac{1}{R}\left|\frac{dR}{dT}\right| = \frac{S}{R} = \frac{\Delta E_{21}}{K_B T^2} \tag{8.8}$$

This parameter is usually expressed in units of percent change per Kelvin, %K^{-1}, and has been commonly used as a figure of merit to compare different thermometers [2, 10, 21–23]. This means that different pairs of energy levels can be used to customize the active material to have high sensitivity in specific temperature ranges. Depending on the working temperature range, the most suitable material based on its phonon energy is chosen, as well the energy gap separation between the thermalized levels of a RE^{3+}.

8.4 Rare-Earth Ion-Doped Glasses for Thermal Sensing

The upconversion (UC) processes of rare-earth-doped materials could be used for multiple applications. One of the most important is the luminescent temperature sensing [24, 25], mainly in a technological point of view. Optical thermometry emerged as an outstanding substitute for the well-known traditional electric temperature sensors, thanks to its unique profiles such as contactless requirement method that offers an electromagnetic passive fast response and high temperature sensitivity. The optical sensors work based on the temperature-dependent variation of some spectroscopic properties of the RE^{3+} ions embedded in the matrix, as discussed earlier. However, two important parameters which show significant temperature-induced changes are the emission intensities and excited state lifetime values [26, 27]. Two thermally coupled emitting levels of the RE^{3+} ion could be examined by the FIR technique. The optimized energy separation between the emitting levels is usually indicated to be larger than 200 cm^{-1} and shorter than 2000 cm^{-1} to avoid strong overlapping of the two emissions and to allow the upper level to have a minimum population of optically active ions in the temperature range of interest. Some trivalent rare earths are known as suitable candidate whose emitting levels could provide the optical sensing, such as Er^{3+}[28, 29], Nd^{3+} [30], Sm^{3+} [31], Pr^{3+} [27, 32], and Eu^{3+} [33]. Another important factor for the optical sensors based on luminescent ions is enough radiative probabilities of the emitting levels to provide relatively large emission intensities, which is highly related to the incorporating matrix. Thermal sensing is indeed a hot field of research in both bulk and nano-sized luminescent systems, where RE^{3+} ions can be used as optical thermometers [34].

Among the important bulk system known up to date, tellurite glasses are promising candidates as host matrix for RE^{3+} ion-based optical temperature sensors. This type of glass family composed principally of TeO_2 glass former units, which need a modifying compound to facilitate the glass formation. For this reason, other oxides such as ZnO, Na_2O, PbO, Al_2O_3, AlF_3, BaF_3, etc., are usually incorporated in the final chemical composition [35]. The main advantages of tellurite glasses are their wide optical transparency window (from 350 to 5000 nm), good chemical and thermal stabilities, high linear and nonlinear refractive index, good rare-earth solubility, and their relatively low-energy phonons [35–38]. Here, we would emphasize some recent results on the applications of tellurite glasses doped with RE^{3+} ions as optical sensors.

8.4.1 Er^{3+}-Doped Glasses

Erbium ion has two thermalized levels, $^2H_{11/2}$ and $^4S_{3/2}$, from which two emissions are generated by radiative decays to the $^4I_{15/2}$ ground state (in the range of 500–570 nm). The emission spectra can be achieved either by direct excitation with the blue laser or by UC pumping at 800 or 980 nm, even with cheap commercial laser diodes. Er^{3+} tellurite glasses and fluorotellurite-based glasses have been studied a lot for their efficient UC emission and near-infrared band, which benefit from the advantages of these matrices in comparison to other oxides glasses. Tellurite and fluorotellurite glasses show low phonon energy, low melting point, high refractive index, and high chemical, mechanical, and thermal stabilities [38, 39].

The UC emission of the Er^{3+} ion-doped glasses is not possible without the involved energy transfer mechanisms between the Er^{3+} ions and/or excited state absorption (ESA). The main energy transfer takes place by populating the $^4I_{11/2}$ intermediate excited level (see Fig. 8.3). However, the dynamic of the UC emission intensity depends on the lifetime of the $^4I_{11/2}$ state, whereas for an ESA process, no delay is expected in the intermediate state and then the dynamic is independent of this level. Such energy transfer mechanism is already proven by many authors. For example, the comparison between the lifetime values of a directly and indirectly excited $^4S_{3/2}$ state was studied by Leon-Luis et al. [40]. The glass samples highly doped with rare-earth ions showed a slower luminescence decay curve originated from the $^4S_{3/2}$ state to $^4I_{15/2}$ ground state (green emission band) when excited at 522 nm ($^4I_{15/2} \rightarrow {}^2H_{11/2}$) than the case of indirect excitation at 800 nm ($^4I_{15/2} \rightarrow {}^4I_{9/2}$), as shown in Fig. 8.4. The green upconverted emission is generated mainly through

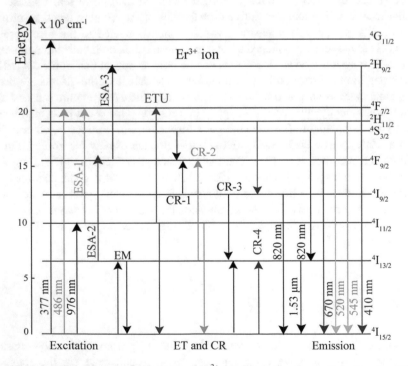

Fig. 8.3 Schematic energy level diagrams of Er^{3+} ion illustrating the excited state absorptions (ESA) and energy transfer mechanisms via energy migration (EM), energy transfer upconversion (ETU), and cross-relaxation (CR). (Figure was adapted from [36])

Fig. 8.4 Luminescence decay curves of the Er^{3+} ion-doped tellurite glasses after a 522 (direct excitation) or 800 nm laser pulse (upconversion) monitoring the green emissions for the TG10 and/or TG25 glasses ($85TeO_2$–10 PbF_2–(5–x) AlF_3–x ErF_3). Taken from [40]. The decay process is faster in heavily doped samples or via direct excitation at 522 nm

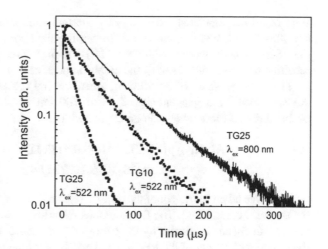

an energy transfer upconversion (ETU) mechanism, as $^4I_{11/2}, ^4I_{11/2} \rightarrow ^4I_{15/2}, ^4F_{7/2}$, where two neighboring Er^{3+} ions in the $^4I_{11/2}$ metastable level sum up to $^4F_{7/2}$ excited state, thanks to an energy transferred from an de-excitation to ground state. The $^4F_{7/2}$ state is depopulated to $^2H_{11/2}$ and $^4S_{3/2}$ emitting levels via a multiphonon decay process. This ETU process is much probable in high concentrated Er^{3+}-doped glasses, and this is why the UC emission in light-doping samples is hardly observed.

In the last two decades, a lot of interesting research on optical temperature sensing based on Er^{3+} ion-doped phosphors, liquids, and solid state media were developed [41–47]. Some examples of thermometry based on Er^{3+}-doped glassy materials are given in this chapter. Li et al. [48] studied temperature-dependent investigation of upconversion emission in Er^{3+}-doped PLZT transparent ceramic. Under 980 nm laser excitation, they observed three typical emission bands at around 540, 564, and 682 nm, whose intensity varied by increasing the temperature. It was reported that the intensity of the 564 nm peak at 10 K is about five times higher than that at 300 K, and the intensity of the 540 nm peak increases initially and then becomes saturated at temperatures above 260 K. However, the intensities of the 564 and 682 nm emission bands reduce when the temperature increases. They discussed the results by developing a rate equation theory, and a good agreement with the theoretically calculated data and experimental observations was obtained.

Li et al. [48] also discussed that the luminescence decay lifetime of the $^4S_{3/2}$ level (green emission) is more influenced by temperature than $^4F_{9/2}$ level (red-emitting state), and the maximum intensities of the upconversion emissions at 564 and 682 nm were obtained only at 10 K, where non-radiative transition rates are minimized. On the other hand, the upconversion emission band at 540 nm showed a maximum intensity at room temperature. Such behavior of thermal dependence of two green upconversion emission is due to the thermalization following the Boltzmann thermal distribution [49]. This thermalization is attributed to the small energy difference of two emitting states, $^4S_{3/2}$ and $^2H_{11/2}$ levels, from which $^2H_{11/2}$

state could be populated from $^4S_{3/2}$ by thermal excitation, resulting in a thermal equilibrium between the two levels. Increasing the temperature the populations of the $^4S_{3/2}$ state decreases, while the $^2H_{11/2}$ higher state gains more populations, resulting in the enhancement of the emission peak at 540 nm [48].

Two emitting states of Er^{3+} ions in green spectral region, $^2H_{11/2}$ and $^4S_{3/2}$ levels, are separated by a gap energy of about 700 cm^{-1}. The spontaneous emission probabilities of these levels can be expressed as

$$A\left(^2H_{11/2}\right) \propto 0.7158\,\Omega_2 + 0.4138\,\Omega_4 + 0.0927\,\Omega_6$$
$$A\left(^4S_{3/2}\right) \propto 0.2225\,\Omega_6$$

where it was taken into account the double-reduced matrix elements given by Weber for the LaF$_3$ crystal [34]. The Ω_2 parameter describes the hypersensitive transitions and is influenced drastically by the short-range crystal field modifications, such as the covalent character of the RE^{3+}-ligand bond and/or the local structural changes in the vicinity of the RE^{3+} ion. On the other hand, Ω_4 and Ω_6 are long-range parameters that can be related to the bulk properties, such as rigidity of the network [50, 51]. Leon-Luis et al. [40] reported a fluorotellurite glass with a very large Ω_2 parameter, resulting in the radiative rates of around 22,510 s^{-1} and 4455 s^{-1} for the $^2H_{11/2}$ and $^4S_{3/2}$ levels, respectively, with calculated lifetimes of around 45 μs and 225 μs. They discussed that theoretically no emission from the $^2H_{11/2}$ level would be expected since the one-phonon de-excitation probability to the $^4S_{3/2}$ level is quite high (relaxation time in the order of 1 ns) as the maximum energy phonon found in the fluorotellurite glasses is around 750 cm^{-1}, similar to other tellurite glasses [40]. However, thermalization process between these two levels results in the thermal population of the upper level ($^2H_{11/2}$) following the Boltzmann distribution function, as discussed above.

The thermal sensing could be evaluated by analyzing the ratio of the luminescence from each thermally coupled state based on a simple three-level system (see Fig. 8.2). The temperature dependence of the green upconverted emission from the two thermally coupled $^2H_{11/2}$ and $^4S_{3/2}$ levels of the Er^{3+} ion in a fluorotellurite glass under a cw laser diode excitation at 800 nm is shown in Fig. 8.5. Various concentrations of Er^{3+} ions were examined by Leon-Luis et al. [40]. Using the FIR technique, a maximum sensitivity of 54×10^{-4} K^{-1} at 540 K was obtained for the glass having a low Er^{3+} ion concentration. Manzani et al. [52] studied the thermal sensing from upconversion emission of Er^{3+} ions in a glass system having the following composition: $(97-x)[70TeO_2–15GeO_2–5K_2O–10Bi_2O_3]:xEr_2O_3/3Yb_2O_3$ (mol%), where $x = 0.1, 0.5, 1.0$, and 1.5 mol%. They found a relatively very large maximum sensitivity of 8.9×10^{-4} K^{-1}.

Ytterbium ions have been generally used to improve the excitation efficiency of acceptor/activator RE^{3+} ions by means of energy transfer since the absorption cross section of Yb^{3+} ions is broad and matches exactly with emission wavelengths of commercial and of cost-benefit lasers (920–980 nm). In a RE^{3+}/Yb^{3+}-co-doped system, the excitation takes place via the addition of photons by energy transfer

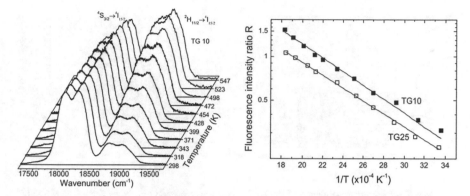

Fig. 8.5 (Left) Normalized temperature evolution of the Er^{3+} green upconverted emission for 85TeO$_2$–10 PbF$_2$–(5−x) AlF$_3$–x ErF$_3$ glasses (x = 1, TG10). (Right) Linear fit on the green fluorescence intensity ratio R vs. T^{-1} for the TG10 and TG25 (x = 2,5) glasses from room temperature to 540 K range. (Figures are adopted from [40])

(APTE, from "Addition de Photons par Transferts d'Energie") from the Yb^{3+} sensitizer ions [53, 54]. The mentioned APTE process is about 100 times more efficient than those cooperative luminescence process for the same Yb^{3+}–Yb^{3+} distances in highly doped Yb^{3+}-doped samples [55]. In fact, high Yb^{3+} ion concentration is used to generate strong green emission from Er^{3+} ions in relatively low Er^{3+} concentrated (up to 1.5 mol%) systems. The large absorption cross section of Yb^{3+} at 980 nm excitation wavelength and a fast and relatively easy energy transfer to neighboring, for example, Er^{3+}, provide sufficient population for $^4I_{11/2}$ level to get excited by a secondary excitation photon and generate a frequency UC (FUC) emission. dos Santos et al. [56] investigated optical temperature sensing using FUC emission in Er^{3+}/Yb^{3+} chalcogenide glass under excitation at 1.06 μm, in the temperature range of 20–225 °C, and obtained a resolution of approximately 0.5 °C using excitation power of a few tens of milliwatts. In the work [28], dos Santos et al. did a comparison of optical FUC thermal sensors with Er^{3+} and Er^{3+}/Yb^{3+} systems (chalcogenides, fluoroindate, germanoniobate, and fluoride) under excitation at 1.54 and 1.06 μm. Manzani et al. [52] concluded that high green emission intensity and thermal sensitivity for the probe could be optimized by controlling the concentration of RE^{3+} ions. In their work, the proposed thermometer is able to work linearly in the range of 5–50 °C and 50–200 °C, with a suitable accuracy, and precisions of ±0.5 and ±1.1 °C, respectively. The upconversion emission spectra were recorded under a 980 nm diode laser excitation (150 mW) for the 0.1 mol% (TG01), 0.5 mol% (TG05), 1.0 mol% (TG10), and 1.5 mol% (TG15) glass samples (not shown here). The intensity of both the green and red emissions initially increased by increasing Er$_2$O$_3$ concentration from 0.1 to 0.5 mol%, while they quenched for concentrations from 0.5 to 1.5 mol%. The decreasing of Er^{3+}-Er^{3+} and Er^{3+}-Yb^{3+} distances improve the efficiency of ion-ion interactions and drastically reduce the green FUC emission [57].

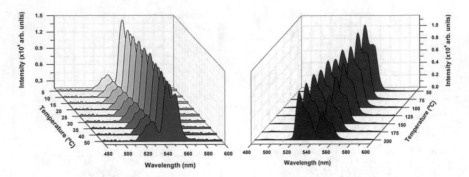

Fig. 8.6 Temperature dependence of the Er^{3+} green upconversion emissions for a $0.5Er^{3+}/3Yb^{3+}$ co-doped tellurite glass at (left) low temperature and (right) high temperature ranges. All the spectra were normalized to the maximum intensity of each spectrum. (Taken from [52])

Figure 8.6 shows the temperature dependence FUC emission of TG05 glass sample, at an optimized Er^{3+} concentration (0.5 mol%) by Manzani et al. [52]. The integrated intensity ratio between two observed green FUC emission shows a linear dependence to temperature ($R^2 > 0.99$). In this matter, they observed a relatively large value for the sensitivity of the FUC emission for TG05 sample. The largest value for absolute sensitivity is 9.1×10^{-3} K^{-1} at 423 K, whereas the maximum sensitivity is 8.9×10^{-3} K^{-1} at 473 K. Their results are shown in Table 8.1 and compared to the ones reported in literature for various glassy host.

The maximum relative thermal sensitivity (S_R) for different glasses and glass ceramics doped with Er^{3+} ions or co-doped with Er^{3+}/Yb^{3+} ions is listed in Table 8.1. As could be read on this table, the maximum sensitivity varies drastically from host to host. However, at a very fast glance on the table, it could be concluded that tellurite glasses in general show higher sensitivity than other glassy or ceramic hosts. For example, tellurite glasses show more thermal sensitivity ($S_R^{max} \sim 8.9 \times 10^{-3}$ K^{-1} for a Er^{3+}/Yb^{3+}-doped glass at 200 K [52]) than ceramics, although they do not reach as high as the sensitivity values of chalcogenide (10.2×10^{-3} K^{-1}) and PKAZLF glasses (7.9×10^{-3} K^{-1}) [59]. Moreover, it should be also noted that the choice of the excitation wavelength could also change the thermal sensitivity. For example, the maximum thermal sensibility of Er^{3+}-doped Te-Pb-Al glasses is higher (7.9×10^{-3} K^{-1}) when excited at 488 nm than when excited at 800 nm (5.4×10^{-3} K^{-1}).

Although a lot of research has been devoted to study the effect of temperature on the FUC emissions of Er^{3+} ions in green spectral region, and the effect of Er^{3+} or Yb^{3+} ion concentration is investigated to optimize the maximum thermal sensitivity, a large lack of discussions on the effect of the glass composition and modifiers on this parameter is clear.

Table 8.1 The maximum sensitivity values, excitation, and emission wavelengths of Er^{3+} doped in some glasses, ceramics, and glass ceramics

Ions	Host [Ref]	λ_{exc} (nm)	T_{max} (K)	Temperature range (K)	S_R (max, 10^{-3} K^{-1})	S_{Rm}(% K^{-1})
Er^{3+}	Tellurite glass [58]	379	596	313–713	8.5	–
	Te-Pb-Al glass [18]	488	541	293–773	7.9	
	Te-Pb-Al glass [40]	800	540	300–550	5.4	
	Fluorotellurite glass [40]	800	267	27–267	5.4	0.35
	PKAZLF glass [59]	488	630	298–773	7.9	–
	In-Zn-Sr-Ba [60] Fluoride glass [60]	406	425	125–425	2.8	–
	Sr-Ba-Nb-B glass [61]	532	600	300–700	1.7	
	ZBLALiP glass [25]	805	495	150–850	2.3	
	Chalcogenide glass [28]	1540		293–493	10.2	
	Fluorozirconate glass [25]	976	27	−123–577	0.6	0.58
	Fluoroindate glass [29]	980	152	−150–152	2.8	0.55
	Fluoroindate glass	1480	450	298–450	5.2	–
	Oxyfluoride glass [29]	980	240	−23–177	2.7	0.41
	Lead-silica glass [62]	980	317	23–377	2.6	0.24
	PLZT ceramic powder [63]	980	337	310–383	4.0	
	Ba(Zr,Ca)TiO$_3$ ceramic [64]	980	443	200–443	4.4	
	α-NaYF$_4$ glass ceramic [18]	488	540	300–720	2.4	
Er^{3+} /Yb^{3+}	TeO$_2$-WO$_3$ glass [65]	980	417	27–417	2.6	0.20
	Te-Ge-Bi glass [52]	980	200	5–200	8.9	0.53
	Ge-Te glass [29]	976	220	20–220	3.6	0.39
	Fluorophosphate glass [66]	980	6	−196–227	1.5	0.71
	Chalcogenide glass [28]	1060	220	20–220	5.2	0.38
	Silicate glass [67]	978	23	23–450	3.3	0.63
	Na$_{0.5}$Bi$_{0.5}$TiO$_3$ ceramics [68]	980	493	173–553	3.5	
	NaBiTiO$_3$ ceramics [69]	980	400	163–613	3.1	

8.4.2   Nd³⁺-Doped Glasses

One of the key factors of thermalized energy levels to show large emission intensity for sensing applications is the radiative probabilities of those states. In this sense, Er^{3+} ion is the most widely used one for this purpose, although other lanthanide ions could also provide comparatively and sufficient radiative properties as, for example, Nd^{3+}, Pr^{3+}, Ho^{3+}, and Sm^{3+} ions in glasses [20, 31, 32, 70] and other phosphors [71–73]. Contrary to the case of Er^{3+} ions, there are not much articles involved on the optical sensors based on Nd^{3+} ions. This could be due to the fact that emission of Nd^{3+} ions is located in the NIR region, whose detection is relatively difficult.

The emergence of the new generations of NIR detectors, such as CCD cameras and extended photomultipliers, allows the detection of the signals of new NIR optical materials as the temperature sensors. In fact, recently optical sensors based on Nd^{3+}-doped systems have attracted a great attention thanks to their multifunctionality and promising potential as in medical applications such as bio-imaging and bio-labeling, cellular and deep-tissue temperature probes, or anticancer therapies [3, 74–78]. One of the chief keys in this scenario is the position of the absorption and emission bands of the Nd^{3+} ion within the near-infrared biological window (NIR-BW) of the human tissues, which covers 650–950 nm (the first BW), 1000–1400 nm (second BW), and 1500–1700 nm (third BW), where the absorption of the light by human tissue is minimized and only scattering is taken into account as the main parameter for the light extinction. However, even if rare, interesting research has been published in the last years on the thermometry based on Nd^{3+} ion-doped phosphors [17, 30, 61, 79, 80].

Marcin Sobczyk [81] studied the effect of temperature on the optical response of Nd^{3+}-doped telluride glasses xNd_2O_3–$(7-x)La_2O_3$–$3Na_2O$–$25ZnO$–$65TeO_2$, where $x = 0, 0.1, 0.5$ and 7 mol%. The glasses were prepared by the melt quench technique, and the optical properties were studied in a temperature range of 298–700 K. The decay time of the $^4F_{3/2}$ level in this host was calculated using the Judd-Ofelt theory to be around 164 μs, while the experimentally measured lifetime was of 162 μs. The experimental lifetime, as expected, decreased from 162 to 5.6 μs by increasing the Nd_2O_3 concentration from 0.5 to 7.0 mol%. Based on these data, the fluorescence quantum efficiency diminishes from 100% (for 0.5 mol% Nd_2O_3–6.5 mol%La_2O_3) to 3.4% (for 7.0 mol% Nd_2O_3–0 mol%La_2O_3). The fluorescence intensity ratio (FIR) was used for the $^4F_{3/2} \rightarrow {}^4I_{9/2}$ and $^4F_{5/2} \rightarrow {}^4I_{9/2}$ transitions in the determined range of temperature. At 638 K, the sensitivity reached its maximum value of about $0.15\%K^{-1}$. In addition, two-photon excited infrared-to-visible upconversion emissions at 603.0 ($^4G_{5/2} \rightarrow {}^4I_{9/2}$) and 635.3 nm ($^4G_{5/2} \rightarrow {}^4I_{11/2}$) were observed at high temperatures using an 804 nm laser, and the temperature-stimulated upconversion excitation processes were also analyzed in details.

Lalla et al. [30] reported on the temperature dependence of the infrared luminescence of a fluorotellurite glass doped with 0.01 and 2.5 mol% of Nd^{3+} ions (Fig. 8.7). They proposed this glass as a high-temperature sensing probe. They studied the variations of the emission intensities of the ($^4S_{3/2}$, $^4F_{7/2}$), ($^2H_{9/2}$, $^4F_{5/2}$), $^4F_{3/2} \rightarrow {}^4I_{9/2}$ transitions using the FIR technique in the temperature range from 300 to 650 K. A strong temperature dependence of the calibration on the Nd^{3+} ion concentration was observed, and the best response was to the low-doped concentrated sample. The maximum value for the thermal sensitivity was $17 \times 10^{-4}\ K^{-1}$ at 640 K, which is claimed to be one of the largest values found in the literature for a Nd^{3+}-doped glassy optical temperature sensor. The differences in the temperature rates and sensitivities with the variation of Nd^{3+} concentration are discussed in terms of various interaction

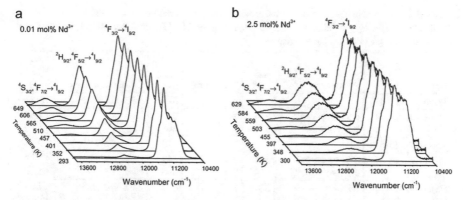

Fig. 8.7 Temperature-dependent NIR emissions of fluorotellurite glasses doped with (**a**) 0.01 mol% and (**b**) 2.5 mol% of Nd^{3+} ions. All the spectra were normalized to the maximum intensity of the transition at around 11,200 cm^{-1}. (Adopted from [30])

mechanisms such as the non-radiative (e.g., multiphonon relaxation) and energy transfer processes. Taken into account that the multiphonon relaxation probability does not depend upon dopant concentration and knowing that energy transfer probability could not affect the emission band profile, they conclude that the only significant process involved is the reabsorption of Nd^{3+} ions. Such radiative energy transfer processes occur due to the strong overlapping of absorption and emission of Nd^{3+} ions.

The thermometry parameters of some Nd^{3+}-doped glasses are listed in Table 8.2. Similar to what is concluded for the Er^{3+} ions, the maximum relative thermal sensitivity of Nd^{3+}-doped glasses depends on the choice of the glass host, concentrations of the Nd^{3+} ions, and excitation wavelength. By and large, tellurite glasses $(2.4\%K^{-1})$ show better response than ZBLAN $(1.08\%K^{-1})$ and fluorotellurite $(1.1–1.7\%K)$ glasses and a comparable thermal sensitivity to phosphate glasses $(2.5\%K^{-1})$. The authors have already studied the excitation wavelength dependence of some phosphate glasses. It could be concluded that the Nd^{3+} phosphate glasses show better sensing response under an 800 nm excitation wavelength than an 850 nm one. Looking at the thermal sensitivity of samples with different concentrations, one could conclude that for samples having higher concentration of Nd^{3+} ions, a smaller increase in intensity occurs with increasing temperature and vice versa. This causes changes in the FIR with the Nd^{3+} concentration, affecting its thermal sensitivity. In summary, the relative thermal sensitivity is higher for the sample with lower Nd^{3+} concentrations.

Effect of high temperature on luminescence lifetimes of $^4F_{3/2}$ level of Nd^{3+} ions in glass was studied by Sobczyk [81]. The lifetime measurements of this excited state were recorded in the 298–673 K temperature range, applying a direct excitation of the emitting $^4F_{3/2}$ level. Both of the examined glasses showed a single exponential luminescence decay profile. Effects of temperature on decay times of glasses containing 0.5 and 2.0 mol% of Nd_2O_3 are shown in Fig. 8.8. By increasing the

Table 8.2 The maximum relative thermal sensitivity values, excitation, and transition ($^4F_{7/2}$,$^4F_{3/2}$ → $^4I_{9/2}$[1] and $^4F_{7/2}$,$^4F_{5/2}$ + $^2H_{9/2}$ → $^4I_{9/2}$[2], $^4F_{5/2}$,$^4F_{3/2}$ → $^4I_{9/2}$[3]) emission wavelengths of Nd^{3+} doped in some glasses and glass ceramics

Host	Sample	λ_{exc} (nm)	Transitions	T_{max} (K)	Equation, ln(R)=	S_{Rmax} (% K^{-1})
Theoretical [30]			3	670	5.93–1050/kT	2.13
Fluorotellurite glass [30]	0.01 mol% Nd^{3+}	514	3	649		1.43
	2.5 mol% Nd^{3+}	514	3	666		1.10
Theoretical [30]			2	670		2.54
Fluorotellurite glass [30]	0.01 mol% Nd^{3+}	514	2	649		1.77
	2.5 mol% Nd^{3+}	514	2	666		1.22
La_2O_3-Na_2O-ZnO-TeO_2 [81]	1 mol%	584	3	638	3.82–963.6/kT	1.5
Silica glass [21]				700		2.7
Tellurite[a]	NR1:Nd^{3+} 1 wt.%		1	300	3.3–2178.1/T	2.416
Phosphate[a]	Q100:Nd^{3+} 9 wt.%	800	1	300	−3.4–1419/T	1.574
	Q98:Nd^{3+} 6 wt.%	800	1	300	4.2–2212.8/T	2.455
	Q98:Nd^{3+} 1 wt. %	800	1	300	4.7–2299.5/T	2.551
ZBLAN[a]	ZBLAN:Nd^{3+} 1 wt. %	800	1	300	1.24–1750.7/T	1.942
Fluorogermate[b]	A1:Nd^{3+} 1 wt.%	800	1	300	4.2–2210/T	2.452
Phosphate[a]	Q100:Nd^{3+} 9 wt.%	850	2	300	−0.22–813.8/T	0.903
	Q98:Nd^{3+} 6 wt.%	850	2	300	−0.22–1011.7/T	1.122
	Q98:Nd^{3+} 1 wt.%	850	2	300	0.58–1164.7/T	1.292
ZBLAN[a]	ZBLAN:Nd^{3+} 1 wt.%	850	2	300	−0.52–974.3/T	1.081
Fluorogermanate[b]	A1:Nd^{3+} 1 wt.%	850	2	300	0.99–1229.6/T	1.364
	A2:Nd^{3+} 0.75 wt.%	850	2	300	0.81–1146/T	1.271
P-K-Ba-Al glass [82]	69	532	3	300	–	0.0153
PbF_2 glass ceramic [82]	Nd^{3+}/Yb^{3+}	980	1, 2, 3	–	–	–

[a]The data presented by authors are not published yet
[b]70% $PbGeO_3$-15%$PbF_2$15%CdF_2

temperature from 298 to 568 K, the excited state lifetimes of the studied glasses are temperature independent. However, in the 568–673 K temperature range, decay time shows interesting changes. Initially, the lifetime of the $^4F_{3/2}$ level slowly increases by increasing the temperature and then drops until 626 K where it restarts to grow sharply. This dependence is of great importance since the minimum of the lifetime value took placed in the temperature range of the glass transition of the studied glasses. The author (Sobczyk) suggested that this observation could be used as a technique to determine the phase transitions in glasses by decay time measurement.

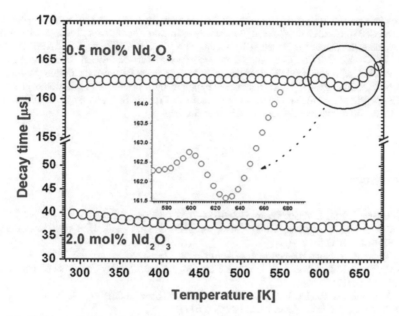

Fig. 8.8 Temperature-dependent lifetime of the $^4F_{3/2}$ level in $xNd_2O_3-(7-x)La_2O_3-3Na_2O-25ZnO-65TeO_2$ glasses. (Taken from [81])

8.5   Summary

In this chapter, we have revisited some recent and important researches carried out on the optical sensing of rare-earth ion-doped glasses, giving a special emphasis to tellurite glasses doped with Er^{3+} or Nd^{3+} ions. Er^{3+} ion-doped tellurite glasses show important thermalized levels in green spectral region which is achievable either by a Stokes or an anti-Stokes excitation approach. The upconversion emission of Er^{3+} ions has recently attracted a large attention. When co-doped with Yb^{3+} ions, a stronger upconversion emission could be generated, which is also suitable for thermal measurements based on luminescence intensity ratio method. A table was dedicated to the thermometry data of Er^{3+}-doped glasses and some ceramics. It is concluded that Er^{3+}-doped tellurite glasses show relatively higher sensitivity to temperature than other glass or ceramic compositions, up to the coverage of this book. On the other hand, the thermalized near-infrared emissions of Nd^{3+} ions could be a suitable case for the biological applications, where existence of more than one pair of thermalized levels gives more opportunities to analyze the optical sensitivity of such system. The maximum sensitivity obtained for Nd^{3+} ion-doped glasses, in general, is lower than those obtained for Er^{3+} ions. The effect of the concentration of both, Er^{3+} and Nd^{3+} ions, on the maximum thermal sensitivity is discussed in detail. However, the effect of glass host composition on the thermometry parameters is not discussed much in the literature and requires further attention. Tellurite glasses have shown significant optical sensing characteristics and are promising materials to develop optical sensor at a wide range of temperature.

Acknowledgments We would like to thank the Brazilian funding agencies for the support of this research: CNPq (Conselho Nacional de Desenvolvimento Científico e Tecnológico), FINEP (Financiadora de Estudos e Projetos) by means of CT-INFRA projects (INFRAPESQ-11 and INFRAPESQ-12), CAPES (Coordenação de Aperfeiçoamento de Pessoal de Ensino Superior) Grant PNPD-CAPES, and FAPEAL (Fundação de Amparo à Pesquisa do Estado de Alagoas), Grant 60030-000384/2017. W. Q. Santos is supported by a postdoctoral fellowship grant from CAPES (PNPD program). The authors are thankful to Prof. R. El-Mallawany, the editor of this book, for his kind invitation to contribute to this chapter.

References

1. X. Huang, J. Lin, J. Mater. Chem. C **3**, 7652–7657 (2015)
2. C.D.S. Brites, P.P. Lima, N.J.O. Silva, A. Mill, V.S. Amaral, D. Carlos, Nanoscale **4**, 4799–4829 (2012)
3. D. Jaque, F. Vetrone, Nanoscale **4**, 4301–4326 (2012)
4. A.F. Pereira, J.F. Silva, A.S. Gouveia-Neto, C. Jacinto, Sensors Actuators B Chem. **238**, 525–531 (2017)
5. C. Ximendes, U. Rocha, T.O. Sales, N. Fernández, F. Sanz-Rodríguez, I.R. Martín, C. Jacinto, D. Jaque, Adv. Funct. Mater. **27**, 1702249 (2017)
6. E.C. Ximendes, W.Q. Santos, U. Rocha, U.K. Kagola, F. Sanz-Rodríguez, N. Fernández, A.D.S. Gouveia-Neto, D. Bravo, A.M. Domingo, B. Del Rosal, C.D.S. Brites, L.D. Carlos, D. Jaque, C. Jacinto, Nano Lett. **16**, 1695–1703 (2016)
7. L.D. Carlos, F. Palacio, *Thermometry at the Nanoscale: Techniques and Selected Applications* (Royal Society of Chemistry, Cambridge, UK, 2015)
8. J.F. Gillooly, J.H. Brown, G.B. West, V.M. Savage, E.L. Charnov, Science (80-.) **293**, 2248–2251 (2001)
9. U. Rocha, K. Upendra Kumar, C. Jacinto, J. Ramiro, A.J. Caamarro, J. Garcia Sole, D. Jaque, Appl. Phys. Lett. **104**, 2012–2017 (2014)
10. E. Carrasco, B. Del Rosal, F. Sanz-Rodríguez, Á.J. De La Fuente, P.H. Gonzalez, U. Rocha, K.U. Kumar, C. Jacinto, J.G. Solé, D. Jaque, Adv. Funct. Mater. **25**, 615–626 (2015)
11. D. Jaque, C. Jacinto, J. Lumin. **169**, 394–399 (2016)
12. E.C. Ximendes, U. Rocha, C. Jacinto, K.U. Kumar, D. Bravo, F.J. López, E.M. Rodríguez, J. García-Soléb, D. Jaque, Nanoscale **8**, 3057–3066 (2016)
13. E.C. Ximendes, U. Rocha, K.U. Kumar, C. Jacinto, D. Jaque, Appl. Phys. Lett. **108**, 253103 (2016)
14. H. Berthou, C.K. Jörgensen, Opt. Lett. **15**, 1100–1102 (1990)
15. K.U. Kumar, W.Q. Santos, W.F. Silva, C. Jacinto, J. Nanosci. Nanotechnol. **13**, 6841–6845 (2013)
16. E. Saïdi, B. Samson, L. Aigouy, S. Volz, P. Löw, C. Bergaud, M. Mortier, Nanotechnology **20**, 115703/1–115703/8 (2009)
17. C. Pérez-Rodríguez, L.L. Martín, S.F. León-Luis, I.R. Martín, Sensors Actuators B Chem. **195**, 324–331 (2014)
18. S.F. León-Luis, U.R. Rodríguez-Mendoza, P. Haro-González, I.R. Martín, V. Lavín, Sensors Actuators B Chem. **174**, 176–186 (2012)
19. A.F. Pereira, K.U. Kumar, W.F. Silva, W.Q. Santos, D. Jaque, C. Jacinto, Sensors Actuators B Chem. **213**, 65–71 (2015)
20. V.K. Rai, Appl. Phys. B Lasers Opt. **88**, 297–303 (2007)
21. S.A. Wade, S.F. Collins, G.W. Baxter, J. Appl. Phys. **94**, 4743 (2007)

22. A. Benayas, B. Del Rosal, A. Pérez-Delgado, K. Santacruz-Gómez, D. Jaque, G.A. Hirata, F. Vetrone, Adv. Opt. Mater. **3**, 687–694 (2015)
23. E.N. Cerõn, D.H. Ortgies, B. Del Rosal, F. Ren, A. Benayas, F. Vetrone, D. Ma, F. Sanz-Rodríguez, J.G. Solé, D. Jaque, E.M. Rodríguez, Adv. Mater. **27**, 4781–4787 (2015)
24. A.K. Kewell, G.T. Reed, F. Namavar, Sensors Actuators **65**, 160–164 (1998)
25. Z.P. Cai, H.Y. Xu, Sensors Actuators A Phys. **108**, 187–192 (2003)
26. B.A. Weinstein, Rev. Sci. Instrum. **57**, 910–913 (1986)
27. V.K. Rai, S.B. Rai, Appl. Phys. B Lasers Opt. **87**, 323–325 (2007)
28. P.V. Dos Santos, M.T. de Araujo, A.S. Gouveia-Neto, J.A. Medeiros Neto, A.S.B. Sombra, IEEE J. Quantum Electron. **35**, 395–399 (1999)
29. M. Kochanowicz, D. Dorosz, J. Zmojda, J. Dorosz, P. Miluski, J. Lumin. **151**, 155–160 (2014)
30. E.A. Lalla, S.F. León-Luis, V. Monteseguro, C. Pérez-Rodríguez, J.M. Cáceres, J. Lumin. **166**, 209–214 (2015)
31. V.K. Rai, IEEE Sensors J. **7**, 1110–1111 (2007)
32. V.K. Rai, D.K. Rai, S.B. Rai, Sensors Actuators A Phys. **128**, 14–17 (2006)
33. H. Kusama, O.J. Sovers, T. Yoshioka, Jpn. J. Appl. Phys. **15**, 2349–2358 (1976)
34. F. Vetrone, R. Naccache, A. Zamarrón, A. Juarranz de la Fuente, F. Sanz-Rodríguez, L. Martinez Maestro, E. Martín Rodriguez, D. Jaque, J. García Solé, J.A. Capobianco, ACS Nano **4**, 3254–3258 (2010)
35. R.A.H. El-Mallawany, *Tellurite Glasses Handbook: Physical Properties and Data*, 1st edn. (CRC Press, Boca Raton, 2002.), 2.nd Ed. (2011)
36. M.R. Dousti, R.J. Amjad, M.R. Sahar, Z.M. Zabidi, A.N. Alias, A.S.S. De Camargo, J. Non-Cryst. Solids **429**, 70–78 (2015)
37. I. Jlassi, H. Elhouichet, M. Ferid, C. Barthou, J. Lumin. **130**, 2394–2401 (2010)
38. R. El-Mallawany, J. Appl. Phys. **72**, 1774 (1992)
39. H. Zhan, A. Zhang, J. He, Z. Zhou, J. Si, A. Lin, Appl. Opt. **52**, 9–11 (2013)
40. S.F. León-Luis, U.R. Rodríguez-Mendoza, E. Lalla, V. Lavín, Sensors Actuators B Chem. **158**, 208–213 (2011)
41. S.F. León-Luis, U.R. Rodríguez-Mendoza, P. Haro-González, I.R. Martín, V. Lavín, Chem. Sensors Actuators B **174**, 176–186 (2012)
42. G. Jiang, S. Zhou, X. Wei, Y. Chen, C. Duan, M. Yin, B. Yang, W. Cao, RSC Adv. **6**, 11795–11801 (2016)
43. X. Huang, B. Li, C. Peng, G. Song, Y. Peng, Z. Xiao, X. Liu, J. Yang, L. Yu, J. Hu, Nanoscale **8**, 1040–1048 (2016)
44. A. Pandey, V.K. Rai, V. Kumar, V. Kumar, H.C. Swart, Sensors Actuators B Chem. **209**, 352–358 (2015)
45. O.I.A. Savchuk, J.J. Carvajal, C. Cascales, M. Aguiló, F. Díaz, Appl. Mater. Interfaces **8**, 7266–7273 (2016)
46. A. Skripka, A. Benayas, R. Marin, P. Canton, E. Hemmera, F. Vetrone, Nanoscale **9**, 3079–3085 (2017)
47. X. Wang, Q. Liu, Y. Bu, C.-S. Liu, T. Liu, X. Yan, RSC Adv. **5**, 86219–86236 (2015)
48. X. Li, Y. Yu, Z. Zheng, Ceram. Int. **42**, 490–494 (2015)
49. V.K. Bogdanov, D.J. Booth, W.E.K. Gibbs, J. Non-Cryst. Solids **311**, 48–53 (2002)
50. L. E. K. A. Gschneidner Jr. (ed.), *Handbook on the Physics and Chemistry of Rare Earths* (Elsevier Science Publisher, Cambridge, UK, 1998), pp. 101–264
51. S. Tanabe, T. Ohyagi, S. Todoroki, T. Hanada, N. Soga, J. Appl. Phys. **73**, 8451–8454 (1993)
52. D. Manzani, J.F.d.S. Petruci, K. Nigoghossian, A.A. Cardoso, S.J.L. Ribeiro, Sci. Rep. **7**, 41596 (2017)
53. S.F. León-Luis, U.R. Rodríguez-Mendoza, I.R. Martín, E. Lalla, V. Lavín, Sensors Actuators B Chem. **176**, 1167–1175 (2013)

54. F. Auzel, Energy Transfer and Migration of Excitation in Solids and Confined Structures. In: B. di Bartolo (ed.), *Spectroscopy and dynamics of collective excitations in solids* (Plenum Press, New York; London, 1997), pp. 1–559
55. F. Auzel, P. Goldner, G.F. De Sa, J. Non-Cryst. Solids **265**, 185–189 (2000)
56. P.V. dos Santos, M.T. De Araujo, A.S. Gouveia-Neto, J.A. Medeiros Neto, A.S.B. Sombra, Appl. Phys. Lett. **73** (1998). https://doi.org/10.1063/1.121861
57. K. Annapoorani, N. Suriya Murthy, T.R. Ravindran, K. Marimuthu, J. Lumin. **171**, 19–23 (2016)
58. G.Z. Sui, X.P. Li, L.H. Cheng, J.S. Zhang, J.S. Sun, H.Y. Zhong, Y. Tian, S.B. Fu, B.J. Chen, Appl. Phys. B Lasers Opt. **110**(4), 471–476 (2013)
59. N. Vijaya, P. Babu, V. Venkatramu, C.K. Jayasankar, S.F. León-Luis, U.R. Rodríguez-Mendoza, I.R. Martín, V. Lavín, Sensors Actuators B Chem. **186**, 156–164 (2013)
60. P. Haro-González, S.F. León-Luis, S. González-Pérez, I.R. Martín, Mater. Res. Bull. **46**, 1051–1054 (2011)
61. P. Haro-González, I.R. Martín, L.L. Martín, S.F. León-Luis, C. Pérez-Rodríguez, V. Lavín, Opt. Mater. (Amst). **33**, 742–745 (2011)
62. W.A. Pisarski, J. Pisarska, R. Lisiecki, W. Ryba-Romanowski, Opt. Mater. (Amst). **59**, 87–90 (2016)
63. A.S.S. De Camargo, J.F. Possatto, L.A.D.O. Nunes, E.R. Botero, E.R.M. Andreeta, D. Garcia, J.A. Eiras, Solid State Commun. **137**, 1–5 (2006)
64. P. Du, L.H. Luo, W.P. Li, Q. Yue, H. Chen, Appl. Phys. Lett. **104**, 152902 (2014)
65. A. Pandey, S. Som, V. Kumar, V. Kumar, K. Kumar, V. Kumar, H.C. Swart, Sensors Actuators B Chem. **202**, 1305–1312 (2014)
66. B. Lai, L. Feng, J. Wang, Q. Su, Opt. Mater. (Amst). **32**, 1154–1160 (2010)
67. C. Li, B. Dong, S. Li, C. Song, Chem. Phys. Lett. **443**, 426–429 (2007)
68. P. Du, J.S. Yu, Ceram. Int. **41**, 6710–6714 (2015)
69. P. Du, L.H. Luo, W. Li, Q. Yue, J. Appl. Phys. **116**, 14102 (2014)
70. R.K. Kumar Rai, Sensors Actuators B Chem. **210**, 581–588 (2015)
71. W. Xu, H. Zhao, Y. Li, L. Zheng, Z. Zhang, W. Cao, Sensors Actuators B Chem. **188**, 1096–1100 (2013)
72. O.A. Savchuk, J.J. Carvajal, M.C. Pujol, E.W. Barrera, J. Massons, M. Aguilo, F. Diaz, J. Phys. Chem. C **119**, 18546–18558 (2015)
73. S. Ćulubrk, V. Lojpur, S.P. Ahrenkiel, J.M. Nedeljković, M.D. Dramićanin, J. Lumin. **170**, 395–400 (2016)
74. U. Rocha, C. Jacinto da Silva, W.F. Silva, I. Guedes, A. Benayas, L.M. Maestro, M.A. Elias, E. Bovero, F.C.J.M. van Veggel, J.A.G. Solé, D. Jaque, Subtissue thermal sensing based on meodymium-doped LaF nanoparticles. ACS Nano **7**(2), 1188–1199 (2013)
75. U. Rocha, K.U. Kumar, C. Jacinto, I. Villa, F. Sanz-Rodríguez, M.d.C.I. de la Cruz, A. Juarranz, E. Carrasco, F.C.J.M. van Veggel, E. Bovero, J.G. Solé, D. Jaque, Neodymium-doped LaF nanoparticles for fluorescence bioimaging in the second biological window. Small **10**(6), 1141–1154 (2014)
76. I. Villa, A. Vedda, I.X. Cantarelli, M. Pedroni, F. Piccinelli, M. Bettinelli, A. Speghini, M. Quintanilla, F. Vetrone, U. Rocha, C. Jacinto, E. Carrasco, F.S. Rodríguez, Á. Juarranz, B. del Rosal, D.H. Ortgies, P.H. Gonzalez, J.G. Solé, D. Jaque, 1.3 μm emitting SrF2:Nd3+ nanoparticles for high contrast in vivo imaging in the second biological window. Nano Res. **8** (2), 649–665 (2015)
77. E. Carrasco, B. del Rosal, F. Sanz-Rodríguez, Á.J. de la Fuente, P.H. Gonzalez, U. Rocha, K.U. Kumar, C. Jacinto, J.G. Solé, D. Jaque, Intratumoral thermal reading during photo-thermal therapy by multifunctional fluorescent nanoparticles. Adv. Funct. Mater. **25**(4), 615–626 (2015)
78. B. del Rosal, A. Pérez-Delgado, M. Misiak, A. Bednarkiewicz, A.S. Vanetsev, Y. Orlovskii, D. J. Jovanović, M.D. Dramićanin, U. Rocha, K.U. Kumar, C. Jacinto, E. Navarro, E.M. Rodríguez, M. Pedroni, A. Speghini, G.A. Hirata, I.R. Martín, D. Jaque, Neodymium-doped

nanoparticles for infrared fluorescence bioimaging: The role of the host. J. Appl. Phys. **118**(14), 143104 (2015)
79. I.E. Kolesnikov, E.V. Golyeva, A.A. Kalinichev, M.A. Kurochkin, E. Lähderant, M.D. Mikhailov, Sensors Actuators B Chem. **243**, 338–345 (2017)
80. S. Balabhadra, M.L. Debasu, C.D.S. Brites, L.A.O. Nunes, O.L. Malta, J. Rocha, M. Bettinellie, L.D. Carlos, Nanoscale **7**, 17261–17267 (2015)
81. M. Sobczyk, J. Quant. Spectrosc. Radiat. Transf. **119**, 128–136 (2013)
82. W. Xu, H. Zhao, Z.G. Zhang, W.W. Cao, Sensors Actuators B Chem. **178**, 520–524 (2013)

Chapter 9
NIR Emission Properties of RE³⁺ Ions in Multicomponent Tellurite Glasses

M. S. Sajna, V. P. Prakashan, M. S. Sanu, Gejo George, Cyriac Joseph, P. R. Biju, and N. V. Unnikrishnan

Abstract The unique physical and optical properties of tellurium-based glasses such as low melting point, low phonon energy, higher linear and nonlinear refractive index, and wide transparency in the visible to IR region make them excellent candidates for telecommunication applications. This chapter reviews trivalent rare-earth (RE³⁺)-doped tellurium-based glasses developed over a wide range of compositions formed from the viscous melts of more than two glass formers. This chapter includes some important tellurium-based glasses as potential host materials for RE³⁺ ions having near-infrared (NIR) emissions. The influences of the composition on the spectral as well as laser parameters of certain rare-earth transitions investigated by several researchers are also detailed so as to apply them in a wide variety of practical applications. It also covers some basic theories necessary to explain the spectroscopic features of interest, the required experimental evidences, and the representative data related to the topic from the previous reports. The recent developments in the intensification in NIR luminescence of lanthanide-embedded tellurite-based hosts due to the co-doping of the metal nanoparticles are also addressed.

9.1 Introduction

Rare-earth-doped glasses have wide applications in the fields of sensors, lasers, optical amplifiers, optical switches, etc. Various glass compositions embedded with lanthanide ions have been prepared by quenching of their respective oxide melts. The luminescence properties of lanthanide ions depend on the matrix in which they are incorporated. Unlike crystals, glasses are inherently disordered medium, and hence the environment of the rare-earth ions in the glass host is not uniform, which causes a site-to-site variation in the energy levels of the rare-earth ion, and their radiative and non-radiative transition probabilities also changes. In glassy hosts, lanthanide ions are known to exhibit laser action [1]. Oxide glasses possessing

M. S. Sajna · V. P. Prakashan · M. S. Sanu · G. George · C. Joseph · P. R. Biju
N. V. Unnikrishnan (✉)
School of Pure & Applied Physics, Mahatma Gandhi University, Kottayam, India

high chemical durability and stability are required for the doping of lanthanides for the fabrication of optoelectronic devices. The glass hosts like silicates, borates, phosphates, tellurites, germanates, chalcogenides, fluorozirconates, sulfides, antimony, fluoroindates, and tungstates are differed mainly by their phonon energy, preparation methods, physicochemical characteristics, and transmission windows. The choice of host glass matrix is very significant for rare-earth doping in a working medium as it is intended to be utilized in application point of view for designing optical fibers or lasers. In general, the suitability relies on the phonon energy of the glass system intended to be used as an optical host matrix. The efficiency of the energy level could be considerably enhanced by using low phonon energy systems by reducing non-radiative losses, favorable to achieve increase in population inversion that is a necessary criterion to increase the efficiency of the emission levels. Now a days, exploring new host systems with better optical properties is a wide area of extensive investigation by many scientists.

Good corrosion resistance, chemical durability, and mechanical strength in addition to the ease of preparation due to low melting temperature of the precursors are the major advantages of tellurium-based glasses. The significant attraction of the glasses on the spectroscopic features is their excellent IR transmission in the visible as well as near-infrared region up to 4.5 μm, slow corrosion rate, high rare-earth ion solubility, higher refractive index, as well as optical quality [2, 3].

In this review article, particular attention has been devoted to tellurium-based glasses. Comparing with the common oxide glasses like silicates, borates, phosphates, and germanates, the tellurite glasses have lower phonon energies [4]. By monitoring the optical absorption and emission spectra in the ultraviolet-visible-near-infrared (UV-VIS-NIR) region, the spontaneous emission probabilities, intensity parameters, radiative lifetimes, and branching ratios have been obtained. It is also possible to compute the peak stimulated emission cross sections for the selected NIR transitions for ensuring the utility of these glasses for practical applications. The broad near infrared luminescence properties of the glasses are attributed to their being promising nominees for tunable lasers and optical broadband fiber amplifiers.

9.2 Significance of Near-Infrared (NIR)Emissions and Their Applications

The wavelength ranging from 0.75 to 1000 μm is termed as infrared (IR) radiation, and while taking into account the detector limitations, it is divided into three different regions: 0.750–3 μm near-infrared (NIR), 3–30 μm mid-infrared (M-IR), and 30–1000 μm far-infrared (F-IR), respectively. For biomedical imaging applications, tunable infrared emissions are used [5]. There are extensive uses for the IR spectroscopy for the element identification applications. Since the performance of an optical component or system depends on basic material, it is crucial to optimize them in order to yield a maximum utility. The transmission, refractive index, gradient index, and dispersion are all to be taken into account to realize much more in-depth information for them to be used extensively in a variety of applications ranging from the thermal imaging to design of plano-optics.

Optical window is the low attenuation wavelength range of the fiber-optic cable used in telecommunication systems. For fiber-optic communication purposes, LEDs were used in the beginning as the light source mainly operated at the wavelength of 780 or 850 nm, and this region is referred to be the first transmission window. For high-bandwidth transmissions over long distances, these LEDs could not be utilized and were replaced by lasers. The wavelength regions referred to as the second and third optical transmission windows in which the lasers operated at 1310 and 1550 nm are commonly used. Praseodymium and erbium ions have near-infrared emissions in the 1.3 and 1.5 μm wavelengths and hence could be used in telecommunication fields. Enormous interest has been drawn to increase the capacity of data transmission in optical communication systems and investigations are ongoing worldwide to develop a much more efficient optical amplifier to be operated in the low-loss transmission window in the near-infrared region. The second telecommunication window comprises of the short-wavelength edge (~1200 nm) and E (1360–1460 nm)-bands. Femtosecond laser writing technology is implemented in such a way that it is possible to make a refractive index variation within a transparent glass for the fabrication of 3D geometry photonic circuit. Rare-earth ions in low phonon glass matrices providing emissions in the NIR region are very significant to fabricate, examine, and explore their utility and potential in the mentioned context.

9.3   Multicomponent Tellurite Glass Hosts for the NIR Emission

For doping lanthanide ions, the glassy matrices were chosen after a systematic search in order to evolve apt glass materials having improved properties. It is necessary to incorporate various glass-forming oxides while choosing tellurium as the base or main glass host since under normal atmospheric conditions, tellurium cannot form glass by itself and is a conditional glass former. Hence it is a prerequisite that modifiers like other glass formers, transition metal oxides, alkali, and/or alkaline earth elements should be added in stoichiometric amount to yield tellurium in glassy form, and for the mentioned purpose, adding other elements is justified by several authors [6, 7]. Adding network formers like boric oxide (B_2O_3) and phosphorus pentoxide (P_2O_5) can form ionic-covalent bonds with oxygen atoms which constitute the basic glass structure along with TeO_2. Adding B_2O_3, which is a characteristic network former and also a flux material, can enhance the population accumulation in the $^4I_{13/2}$ level and 980 nm pumping efficiency even though the phonon energy of the lattice of the glass matrix increases [8]. An appreciably good heat stability is offered by the network former P_2O_5. The alkali metal potassium

(K) serves the role of a network modifier which will increase the local symmetry around the lanthanide ions incorporated in the matrix; also via non-bridging oxygen (NBO) atoms, they attach themselves to the structure and help in the formation of a stable glass. The formation of a more connected network in the tellurium-based glassy systems was offered with the introduction of Zn in the network with more non-bridging oxygens (NBOs) in $(Te-O)^-$ bonds as well as Te-O-Te linkages [9]. The electronegativity of fluorine is higher than oxygen, and hence with the modifier ZnF_2 addition or the fluorine ion substitution for the oxide ions in the glass matrix can lower the viscosity and improve the glass formation range and also increase the stability of the glass to a great extent [10, 11].

Incorporating rare earths into various oxide glasses is important to develop various optical devices such as waveguide amplifiers, IR to visible upconverters, infrared lasers, etc. Glass compositions which are stable against devitrification are selected for doping lanthanides in order to yield near-infrared emission. Fluoride and halide glass matrices or their crystals are extensively examined since they have lower phonon energy among the glass systems, but their chemical and mechanical stabilities are relatively poor compared to that of oxides [12, 13].

One other crucial parameter in the processing for making glass or fiber drawing is their thermal stability and is found from the differential scanning calorimetry analysis. The ΔT value is a suitable measure for the thermal stability against crystallization, and it represents the interval during which the nucleation takes place.

$$\Delta T = T_x - T_g \tag{9.1}$$

where T_x is the onset of crystallization and T_g is the glass transition temperature. The difference should be at least >100 °C for the glass composition for using in fiber drawing purposes. The tellurite glasses have higher thermal stabilities, and they exhibit high thermal resistance against crystallization as compared to the binary glasses and hence are suitable for various fiber device applications. Certain multicomponent tellurite glasses were developed in this context and their thermal characterizations were already reported [2, 14] (Table 9.1).

Among the oxide glasses, the peak stimulated emission cross sections of certain NIR transitions of RE ions in tellurite glasses rank first. Opting tellurite as the glass host was motivated by taking into consideration of their high mechanical strength,

Table 9.1 Some tellurite-based glass compositions and their thermal characteristics

Sample name	Composition (mol %)	T_g (°C)	ΔT (°C)
TP10	$90TeO_2$-$10P_2O_5$	348	124
TP20	$80TeO_2$-$20P_2O_5$	380	⋯
TBSNP8	$70TeO_2$-$3.5BaO$-$10.5SrO$-$8Nb_2O_5$-$8P_2O_5$	404	176
TBSNWP8	$66TeO_2$-$3.5BaO$-$10.5SrO$-$8Nb_2O_5$-$4WO_3$-$8P_2O_5$	419	199
TBSNWTP8	$64TeO_2$-$3.5BaO$-$10.5SrO$-$8Nb_2O5$-$4WO_3$-$2Ta_2O_5$-$8P_2O_5$	433	224
TBSNWP16	$58TeO_2$-$3.5BaO$-$10.5SrO$-$8Nb_2O_5$-$4WO_3$-$16P_2O_5$	440	224
TZLB	$78TeO_2$-$5ZnO$-$12Li_2O$-$5Bi_2O_3$	275	⋯

Reused with permission from Ref. [15]

Table 9.2 Comparison of phonon energy ($h\omega$) of glasses [23]

Glass	Phonon energy $h\omega$ (cm^{-1})
Fluoride	500–600
Tellurite	600–850
Germanate	975–800
Phosphate	1350–1200
Borate	1340–1480

low melting point, good corrosion resistance, low phonon energies, and high chemical durability. The high polarizability of the tellurium ions causes higher refractive index of the tellurite-based glasses which makes them apt for the opto-electronic device applications [16, 17]. A local field correction at the rare-earth ion site is induced by the high refractive index which prompts an enhancement in the radiative transition rates of the RE^{3+} ions in the matrix [2, 18].

Another important requirement for a fiber amplifier glass host is its phonon energy, and it should be minimum for maximum amplifier performance. It has also been evidently proved that the intensity of fluorescence of the rare-earth ion is enhanced to several folds while doped in a glass host of lower phonon energy [19, 20]. Comparison of phonon energy for various host glasses are given in Table 9.2. Due to the low phonon energy, tellurite glasses have relatively low probability of non-radiative transitions and lower energy transfer between active dopants compared with various oxide glasses like silicate, borate, and phosphate glasses, which also makes tellurium-based glasses as attractive hosts for lanthanide incorporation [21]. A detailed analysis of various glass compositions for laser writing applications especially multicomponent tellurite glasses is described by Toney et al. [22].

9.4 Methods Employed for Fabrication of the TeO$_2$-Based Multicomponent Glasses

Common preparation techniques employed for the preparation of tellurium-based glasses are melt quenching method, non-hydrolytic sol-gel method, and chemical vapor deposition. Conventional melt quenching has been usually adopted to prepare Ln^{3+}-doped multicomponent tellurite glasses for NIR emission applications. From the oxide precursors of the hosts, glass and the dopant lanthanides of the respective glasses of interest were weighed in appropriate quantities in molar ratio to yield a glass of fixed weight. They were then well mixed using an agate mortar, transferred to a platinum crucible, and melted in a high-temperature electric furnace. The molten liquid is suddenly quenched to a brass mold preheated in an annealing furnace at a temperature below the glass transition temperature. The quenched glassy material in the brass mold is again kept in an annealing furnace for several hours and then gradually cooled down to room temperature in order to reduce any thermal stress. The final glass samples were then sliced in required dimensions and polished for further use. The optical quality tellurite glasses are prepared in inert platinum or gold crucible to obtain refractive index homogeneity required for high-power laser applications, but there may be present traces of dissolved (ionic) platinum. It was reported that the optical performances of the tellurite glasses have negligible effect on these platinum ion inclusions compared to phosphate glasses particularly in UV-visible absorption range [24]. Even though, in order to avoid any possible effect of these impurities or OH content on the optical absorption properties and any contamination during the melt quenching or processing, high-purity precursor oxides were used and in some cases, melting in dry atmosphere is also employed [25–27].

9.5 Spectral Parameters Quantifying NIR Emission Properties

The understanding of the spectroscopic properties of rare-earth activators is of key significance for laser action, upconversion, phosphors, energy transfer, optical fibers, and so on. There is a lot of spectral parameters for the optical assessment of NIR emission. These properties comprise absorption-emission transitions, peak wavelengths and linewidths, branching ratios, transition probabilities, cross sections, lifetimes, quantum efficiencies and fluorescence quenching mechanisms, etc. Based on experimentally measured absorption spectrum and utilizing the Judd-Ofelt and McCumber theoretical approaches, the quantities like the radiative transition rates, fluorescence lifetime, branching ratios and stimulated emission cross sections can be calculated for the transitions of interest.

9.5.1 Absorption and Emission Cross Sections

The ability to absorb or emit light is quantified by the term cross section. The emission cross section is a very significant parameter to quantify the emission characteristics of a lanthanide-doped host system and is generally calculated by evaluating the absorption cross section from the measured absorption spectrum and the emission cross section from the emission spectrum. In addition, in the case of certain rare earths like erbium and holmium, a theoretical approach named as McCumber analysis has been adopted to find the emission cross section from absorption cross section itself by making use of the absorption spectrum. The stimulated emission cross section of a transition is calculated from the emission spectra by the Fuchtbauer-Ladenburg equation [28].

$$\sigma_e = \frac{\lambda_p^4 A[(S\,L)J, (S'\,L')J']}{8\pi c n^2 \Delta\lambda_{\text{eff}}} \qquad (9.2)$$

where λ_p denotes peak wavelength of the emission band, $\Delta\lambda_{\text{eff}}$ is the effective linewidth, c is the velocity of light, n is the refractive index and A is the transition probability [29]. The value of stimulated emission cross section indicates the rate of energy extraction from the lasing material. For the present glasses which have high stimulated emission cross sections, it is entailed that they have high potential laser applications.

The figure of merit (FOM) for bandwidth is defined by the effective width of the emission peak and the stimulated emission cross section and is given as

$$\Delta G = \Delta\lambda_{\text{eff}} \times \sigma_e(\lambda_p) \qquad (9.3)$$

9.5.2   McCumber Spectral Evaluation

Dean E. McCumber at Bell Laboratories in the 1960s, following the prior hypothetical examinations of Albert Einstein and utilizing the thermodynamic principles, worked out and put forward a theory [30] – now named as McCumber theory. This theory is applicable to both the absorption and emission properties of the laser gain media, specifically to the solid-state media, for example, transition metal-doped or rare-earth-doped gain media.

The ability to absorb and emit light was quantified by the factor cross section. From the experimentally measured absorption coefficient and the RE^{3+} concentration (N) in the glass, the absorption cross section (σ_a) was calculated using the following equation [31]:

$$\sigma_a = \frac{2.303}{N \times t} E(\bar{\nu}) \qquad (9.4)$$

where t is the thickness of the glass and E is the optical absorbance in wave number ν.

The stimulated emission cross section σ_e^{M} of the transition of RE^{3+} ions is calculated by using absorption cross section σ_a by the following relation [30]:

$$\sigma_e^{\text{M}} = \sigma_a \exp\left[(\varepsilon - h\nu)/k_B T\right] \qquad (9.5)$$

where ν is the frequency of radiation, h is the Planck's constant, k_B is the Boltzmann's constant, and ε represents the free energy needed to excite one RE^{3+} ion from the ground state to excited state at room temperature which is found by the method proposed by Miniscalco et al. [32].

9.5.3 Gain Spectra Characteristics

To evaluate the gain characteristics, the gain coefficient has been computed. On the basis of absorption (σ_a) and emission (σ_e) cross sections, one can calculate the wavelength dependence of gain coefficient ($G(\lambda)$) as a function of population inversion between the upper and ground levels to determine the gain property quantitatively. It is assumed that there is a simplified two level system for which the RE^{3+} ions are distributed between the ground state and the upper state. The net gain coefficient ($G(\lambda)$) at wavelength (λ) is evaluated by the following equation [33]:

$$G(\lambda) = N[P\sigma_e(\lambda) - (1 - P)\sigma_a(\lambda)] \tag{9.6}$$

where P is the population inversion parameter defined as the ratio of number of active ions in the excited state RE^{3+} lasing level to the total number of active ions in the glass. It is dependent on the pump energy density, and its value falls into the region 0 to 1; N is the concentration of RE^{3+} ions. The gain coefficient spectrum for a lasing transition of the glass has been computed as a function of wavelength using the above equation for different P values.

9.5.4 Lifetime Measurements

Measuring the lifetime of transition levels is important for quantifying the emission. A profound knowledge about the mechanisms involved in the excitation process can be elucidated by examining the fluorescence excitation and de-excitation owing to intra-4f electronic transition. The curve is illustrated by integral solutions to proper rate equations which accounts for the possible excitation and relaxation mechanism. In most of the cases at lower concentrations of RE^{3+} ion dopants, the interaction between the optically active RE ions is insignificant, and hence the fluorescence decay curves can be fitted to the single exponential function. The experimental lifetime (τ_{exp}) was estimated by fitting the following function to the single exponential equation.

$$y = y_0 + Ae^{-t/\tau} \tag{9.7}$$

where "t" denotes the time after excitation. However, at high RE ion concentrations, the interaction among the ions becomes so prominent such that the energy transfer occurs from an excited ion to a non-excited ion. Hence the decay curve deviates from the single exponential curve to a double exponential behavior and follows the equation

$$y = y_0 + A_1 e^{-t/\tau_1} + A_2 e^{-t/\tau_2} \tag{9.8}$$

where y is the intensity of luminescence at time t. τ_1 and τ_2 are the short and long lifetimes corresponding to the each de-excitation mechanisms. A_1 and A_2 are the intensity coefficients used as the fitting parameters. The average lifetime was calculated by the equation

$$\tau_{exp} = \frac{A_1 \tau_1^2 + A_2 \tau_2^2}{A_1 \tau_1 + A_2 \tau_2} \tag{9.9}$$

Figure of merit (FOM) for amplifier gain (G) is an imperative parameter so as to evaluate the performance of the amplifier while designing devices. It is specified as the product of lifetime and stimulated emission cross section and is given by [34].

$$G = \tau_{exp} \times \sigma_e (\lambda_p) \tag{9.10}$$

The non-radiative decay rate can be expressed as [35].

$$W_{NR} = \frac{1}{\tau_{exp}} - \frac{1}{\tau_R} \tag{9.11}$$

The luminescence quantum efficiency (η) is defined as the ratio of the number of photons emitted to the number of photons absorbed. For an emission level in the case of RE^{3+} ions, it is equal to the ratio of measured experimental lifetime to the predicted radiative lifetime obtained from the JO theory

$$\eta = \frac{\tau_{exp}}{\tau_R} \times 100\% \tag{9.12}$$

9.6 Lanthanide Ions Having NIR Emission and Their Electronic Energy Level Structure

The main rare-earth ions emitting near-infrared emissions are Tm^{3+}, Pr^{3+}, Er^{3+}, Nd^{3+}, Ho^{3+}, Ce^{3+}, and Yb^{3+}. The origin of the highly efficient selective emissions in rare-earth ions in different host matrices is arising from their particular energy level structure. In the technological point of view, there are a lot of applications for these emissions especially in the NIR region.

9.6.1 Thulium

For a broadband amplifier at 1.47 μm, Tm^{3+}-doped glasses were widely investigated in the past decades. For the designing and development of new lasers, Tm^{3+}-doped materials playing key roles are considered because of their 1.8 μm NIR emissions [36]. The $^3H_4$ level can be directly excited by a 790–800 nm pumping. The fluorescence emissions of Tm^{3+} doped glasses excited by 798 nm wavelength are

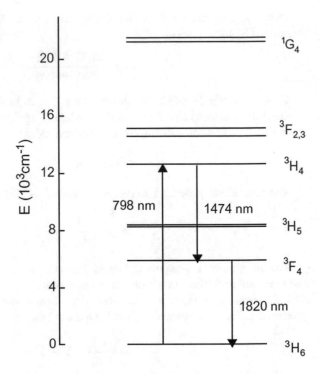

Fig. 9.1 Energy level diagram of Tm^{3+}. (Reused with permission from Ref. [38])

at around 1.4 μm and 1.8 μm which are due to $^3H_4 \rightarrow {}^3F_4$ and $^3F_4 \rightarrow {}^3H_6$ transitions, respectively, and the partial energy level diagram is shown in Fig. 9.1. By developing thulium-doped fiber amplifier (TDFA), it is possible to extend the bandwidth of transmission than the available range of the erbium-doped fiber amplifier (EDFA). Choosing tellurite as the host for the EDFA is justified by the ultra-broadband gain arising from the inherent properties of the tellurite host [37].

Followed by the investigations of two-module amplifier comprising a tellurite-glass-based EDFA and fluoride TDFA which are reported by Yamada et al. [39], and investigations by Mira Naftaly et al. [40] analyzed the effect by changing the host from fluorozirconate (ZBLAN)-based glass to tellurite-based TDFA with the same doping concentration of Tm^{3+} ions. There occurs a significant broadening in the FWHM from 76 to 114 nm, and this broadening can be attributed to the presence of multiple sites in tellurite glass. In addition the emission peak of the Tm^{3+} is red shifted, and it can possibly associate to the higher refractive index of the tellurium-based glass. It can also be related to the nephelauxetic effect due to which the absorption and emission lines tend to shift to lower energy side in high-refractive-index hosts and/or in covalently bound hosts. It is also obtained from the studies that the high refractive index of tellurite glass results in the increase in the emission cross section of the order of two times and shorter lifetime compared with the ZBLAN glass.

Fig. 9.2 Energy level diagram of Pr^{3+}. (Reused with permission from Ref. [45])

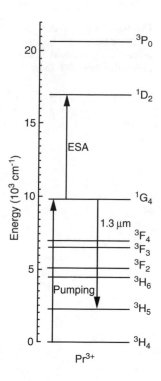

Pr³⁺

9.6.2   Praseodymium

Trivalent praseodymium (Pr^{3+}) doped in various glass/crystal lattices is well studied, and their energy levels almost up to 47,000 cm⁻¹ are well established [41–43]. Followed by the development and successful establishment of the optical fiber amplifier doped with erbium (EDFA) which amplifies signals in the 1.53 μm in the third telecommunication window, praseodymium-doped fiber amplifiers (PDFA) were also developed and investigated and successfully operated in the 1.3 μm wavelength in the second telecommunication window. Taking into considerations of the low phonon energy, initially it was developed in the fluoride glass systems [44]. Partial energy level diagrams of Pr^{3+} in a glass system are shown in Fig. 9.2.

9.6.3   Holmium

Ho^{3+} ion electronic transitions comprise in the visible as well as in the infrared regions for which the $^5I_7$ level has relatively long lifetime and larger peak stimulated emission cross sections [46, 47]. The infrared laser emission range of the Ho^{3+} ion is in the 1.2–4.9 μm regions [48, 49]. Holmium-doped laser operating at the wavelength of 1.9 μm is an eye-safe potential laser emission even at room temperature

Fig. 9.3 The partial energy level scheme of Ho^{3+} ions in tellurite glass. (Reused with permission from Ref. [54])

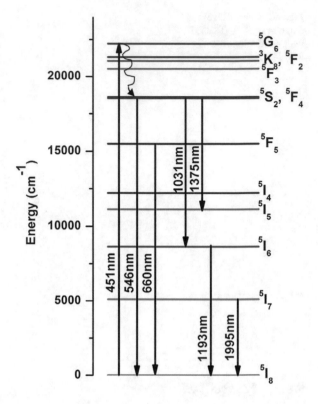

(RT) with a low threshold action outside the retinal hazard region [50] and is available commercially with output power 100 watt. For Ho^{3+} ions, the most appropriate transition to generate light with wavelength close to 2.0 μm region is the $^5I_7 \rightarrow {}^5I_8$ laser transition. At this wavelength, the emission band is broader, and the stimulated emission cross section is much larger and has longer fluorescence lifetime and is a favorable candidate for Q-switched laser [51]. Due to the rich multiple energy levels, Ho^{3+} exhibits near-infrared emissions at 1.20, 1.38, 1.46, and 1.65 μm $(^5I_6 \rightarrow {}^5I_8)$, $((^5F_4, {}^5S_2) \rightarrow {}^5I_5)$, $(^5F_5 \rightarrow {}^5I_6)$, and $(^5I_5 \rightarrow {}^5I_7)$; and also, the lasing action at 2.08 μm $(^5I_7 \rightarrow {}^5I_8)$ along with other visible emissions [52]. Several researchers have investigated the absorption and emission spectra of Ho^{3+} ions in various glass hosts [53]. The energy level structure and the emission channels of Ho^{3+} ions are shown in Fig. 9.3.

Seshadri et al. reported the multicomponent tellurite glass of the composition $(78 - x)TeO_2 + 4.5Bi_2O_3 + 5.5ZnO + 10.5Li_2O + 1.5Nb_2O_5 + xHo_2O_3$ and the cross-section value calculated for $^5I_6 \rightarrow {}^5I_8$ transition by McCumber spectral evaluation is 13.1×10^{-21} cm^2, and is shown in Fig. 9.4. This value is greater than the fluoro-tellurite and lithium-barium-bismuth-lead glasses [55, 56]. For the same glass, the $^5I_7 \rightarrow {}^5I_8$ transition has the cross-section value of 14.5×10^{-21} cm^2 and is greater than the silicate, germanate, gallate-bismuth-germanium-lead, and lead phosphate glasses [57]. The obtained lifetime, cross section, and figure of merit for

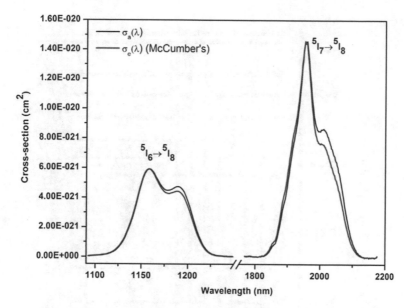

Fig. 9.4 The cross section of $^5I_6 \leftrightarrow {}^5I_8$ and $^5I_7 \leftrightarrow {}^5I_8$ transitions for Ho^{3+}-doped tellurite glass. (Reused with permission from Ref. [54])

amplifier gain values of the relevant NIR transitions of the above mentioned tellurite glass were higher than those for the other reported glasses [56, 58–60] and hence are promising host materials for achieving high gain coefficients.

9.6.4 Neodymium

Snitzer in 1961 demonstrated the lasing of Nd^{3+}-doped glass, and from then onwards glasses doped with transition metal ions and rare-earth ions were increasingly used to develop solid-state materials [61]. Nd^{3+}-doped tellurite single-mode fiber laser has been demonstrated in the past decades itself [62]. The emission spectra have three prominent peaks in the near-infrared region at 874, 1057, and 1331 nm corresponding to the $^4F_{3/2} \rightarrow {}^4I_{9/2, \, 11/2}$, and $_{13/2}$ transition levels, respectively. It is interesting to extend the spectroscopic investigations to the Nd^{3+} ions in glassy systems since it is a significant dopant for NIR laser applications at around 1.06 μm. The energy level scheme for the Nd^{3+} ions in tellurite glasses is depicted in Fig. 9.5. It was reported by Jing et al. that the TeO$_2$ is beneficial to the maximum emission linewidth and it increases the emission cross section. It is also reported that the optical gain properties of these Nd^{3+}-doped glasses can have a low threshold for laser operation of $^4F_{3/2} \rightarrow {}^4I_{11/2}$ transition [63].

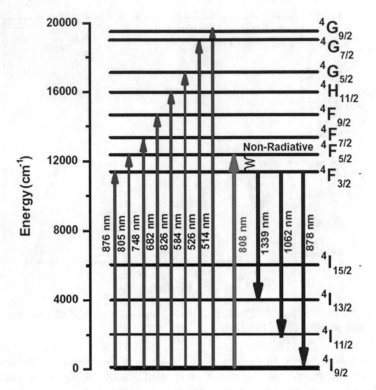

Fig. 9.5 Energy level scheme of Nd^{3+} ions in tellurite glass. (Reused with permission from Ref. [64])

9.6.5 Erbium

Out of the trivalent rare-earth ions, Er^{3+} has due importance because of its popularity and high efficiency. Development of Er^{3+}-doped waveguides, fiber lasers and amplifiers has drawn wide attention because of the commercialization of 800 and 980 nm laser diodes [65, 66]. The Er^{3+}-doped glasses yield the $^4I_{13/2} \rightarrow ^4 I_{15/2}$ transition with either 805 or 980 nm excitation. From the Er^{3+} centers in glass hosts, strong NIR emission at 1.53 μm is observed at room temperature. Tellurite single-mode fiber laser with Er^{3+} doping has been demonstrated initially in the 1990's [67]. Tellurite glasses doped with erbium have been developed with various chemical and optical properties apt for use in optical applications [68, 69]. Figure 9.6 represents the NIR emission of Er^{3+}-doped niobic tellurite glass [70]. The host dependency of significant spectral parameters of the $^4I_{13/2} \rightarrow ^4I_{15/2}$ transition of Er^{3+} ions is illustrated in Table 9.3.

Rolli et al. investigated some other tellurite glasses of the composition $75TeO_2:12ZnO: 10 Na_2O:2PbO/2GeO_2:1Er_2O_3$ and reported that the cross section at 1.5 μm is about 7.85×10^{-21} cm^2 due to the higher refractive index of the tellurite host (~above 2) [75]. The effective bandwidth obtained for the mentioned glass is

Fig. 9.6 Typical NIR emission of Er^{3+}-doped tellurite glass. (Reused with permission from Ref. [70])

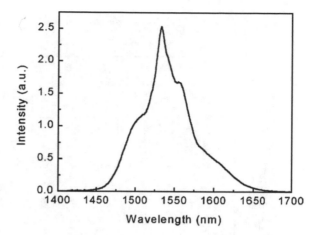

Table 9.3 Significant spectral parameters of 1.53 μm emission of Er^{3+} ions in various glass hosts

Glass code	$\Delta\lambda_{\mathrm{eff}}$ (nm)	σ_e ($\times 10^{-21}$ cm^2)	ΔG ($\times 10^{-28}$ cm^3)	Reference
TPBKZFEr05	63	9.67	609.21	–
PBSEr(38/62)	65	5.7	370.5	[72]
LTBE0.5Y	80	5.9	472	[73]
TZF16-0.5	67.7	8.2	555	[74]
TZF35-0.5	66.7	7.1	474	[74]
TPEr	58.7	6.9	405.03	[28]

Reused with permission from Ref. [71]

63 nm and is similar to other reported tellurite glasses [73] and is very much greater than other hosts like phosphates or silicates [18, 76]. The absorption as well as the emission cross section of the $^4I_{13/2} \leftrightarrow {}^4I_{15/2}$ transition of Er^{3+} ions in a multicomponent tellurite glass is shown in Fig. 9.7. Since Er^{3+}-doped tellurite glasses have very large emission cross sections, these glasses are promising candidates for broadband amplification in the third telecommunication window.

Er^{3+}-doped tellurite glasses with broad and intense 1.53 μm infrared emissions are valuable in developing optical amplifiers and lasers.

Erbium-doped tellurite glasses can have long fluorescence lifetimes compared with other erbium-doped host glasses at certain pump wavelengths because of their low phonon energies [78]. The stimulated emission cross section, lifetimes of the $^4I_{13/2}$ level of Er^{3+} are significant parameters for broadband optical amplifiers or lasers at the 1.53 μm, and in this context the respective parameters are discussed in the report of Sajna et al. [71]. Corresponding to the $^4I_{13/2} \rightarrow {}^4I_{15/2}$ NIR emission transition for the TPBFEr05 glass, the value of figure of merit for amplifier gain (G) is found to be 34.72×10^{-24} cm^2 s. This value indicates that the current glass system is potentially valuable as a laser material at 1.53 μm. The radiative lifetime of the $^4I_{13/2}$ level of Er^{3+} ion in TPBFEr05 glass was predicted as 3.24 ms using Judd-Ofelt analysis. Using the measured lifetime (τ_{exp}) and radiative lifetime (τ_R), the

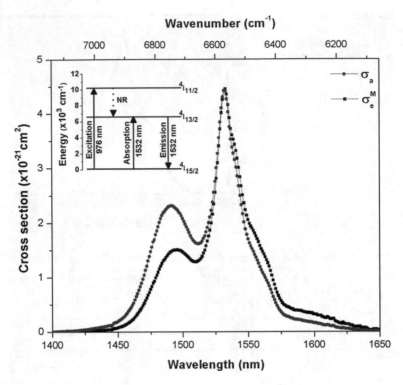

Fig. 9.7 The absorption and emission cross sections of the Er^{3+}-doped tellurium-based glass. (Reused with permission from Ref. [77])

quantum efficiency was estimated for the $^4I_{13/2}$ metastable state of the Er^{3+} ions. While calculating the quantum efficiency of TPBFEr05 glass, the value is found to be beyond 100%. This discrepancy is mainly due to the fact that self-absorption introduces a delay in the excited luminescence by successive reabsorption and reemission processes. The reabsorption of early emitted photons lengthens the emission process inside the glass depending on the absorption cross section (σ_a) and thickness (L). From Auzel's approach, the lifetime corrected for the self-absorption effect is obtained by the equation [74].

$$\tau_{\exp} = \tau_{PL}(1 + \sigma_a NL) \tag{9.13}$$

where τ_{PL} is the non-affected value of lifetime by self-absorption and N is the ion concentration. The corrected lifetime was found as 2.56 ms for $^4I_{13/2}$ level of Er^{3+} in TPBFEr05 glass. The comparative study on the spectral properties of the NIR emission recommends that the glass material under examination might be a promising candidate for Er^{3+}-doped fiber amplifiers.

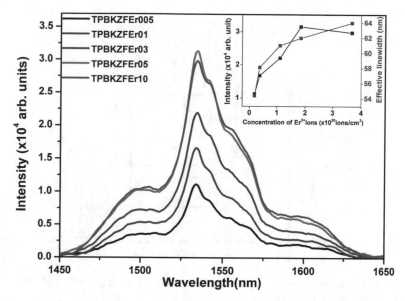

Fig. 9.8 NIR emission spectra of a tellurite glass with variation of the Er^{3+} doping. (Reused with permission from Ref. [71])

9.7 Co-Doping

While adding the rare-earth ions to improve the luminescence properties, difficulty arises due to concentration quenching which limits the possibilities of higher doping [79, 80], and an illustration of this is given as Fig. 9.8.

To overcome this difficulty, alternative approaches such as introducing metallic nanoparticles (NPs), adding quench reduction agents, co-doping semiconductors, or other RE with a resonant energy level have been developed [81].

9.7.1 Rare Earths

To investigate the possibility of evolving efficient amplifiers in the NIR region, the co-doping of rare earths is also promoted widely. Trivalent Yb can serve the role of a sensitizer to efficiently enhance 980 nm absorption and energy transfer to a number of rare-earth ions (Ho^{3+}, Pr^{3+}, Tm^{3+}, Er^{3+}) emitting infrared emissions, and the same is depicted in Fig. 9.9 [82].

Tellurite glass of the composition 75TeO$_2$ + 20ZnO + 5Na$_2$O doped with Pr^{3+} is co-doped with Yb^{3+} ion to sensitize the Pr: 1.3 μm emission by the Yb → Pr energy transfer [83]. In this, at higher temperature, there is a rapid energy migration between Yb ions which promotes Yb^{3+}: $^2F_{5/2}$ → Pr^{3+}: $^1G_4$ transfer.

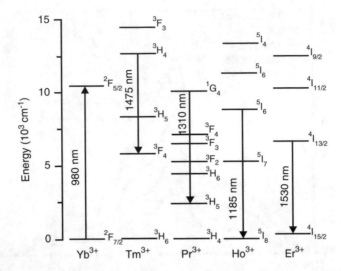

Fig. 9.9 The excitation and emission pathways in the energy levels of the Yb^{3+}, Tm^{3+}, Pr^{3+}, Ho^{3+}, and Er^{3+} ions. (Reused with permission from Ref. [82])

G. Lakshminarayana et al. presents the NIR emission spectrum of $50TeO_2$–$20ZnO$–$10B_2O_3$–$10Li_2O$–$9.0Na_2O$–$1.0Er_2O_3$-doped glass in comparison with the same glass co-doped with the $2.0Yb_2O_3$. From this spectrum, it is evident that the broad emission centered at 1532 nm ($^4I_{13/2}$-$^4I_{15/2}$ transition) under 980 nm excitation is enhanced along with the change in the full-width at half-maximum (FWHM) 53–58 nm [84].

9.7.2 Metal Nanoparticles

Co-doping with metal nanoparticles along with the lanthanide ions into glassy frameworks opens up possibilities for the manufacture of new materials attributable to their enhanced luminescence properties induced by the surface plasmon resonance (SPR) mechanism. By the excitation radiation, the metal nanoparticles resonantly excited and due to which the electromagnetic field of the order of 10^2 in the vicinity of the particles enhanced remarkably [85]. This prompts increased charge densities accumulated principally about to the sharp edges of nanostructures. Henceforth the nanostructures assume crucial part in the SPR phenomenon by transferring the excited radiation into localized electric field. By controlling the parameters like the size and shape of the co-doped nanoparticles, the SPR effects can be regulated. In this context, the rods or ellipsoid-like anisotropic nanostructures are referred to as light-harvesting nano-optical antennas and have many advantages [86]. As the wavelength of excitation or emission of a rare-earth ion is close to the SPR wavelength of the embedded metallic NPs, the respective transition probability is enhanced considerably due to the increase in local surface charge densities of the nanoparticles. Therefore in host matrices particularly in the chalcogenide and inorganic glasses by the in situ formation, the incorporation of gold and silver nanoparticles can significantly favour the optical properties for various applications. Hence particular scientific attention has been attracted for the investigations on metal−/rare-earth-incorporated tellurite glasses. In several reports, Er^{3+}/Au co-doped systems are proved to be potential candidates for optoelectronic applications due to the strong SPR effects by gold nanoparticles [87, 88].

9.8   Upconversion Properties

Several studies are ongoing in order to develop infrared-laser-pumped upconversion solid-state lasers. Halide glasses have good upconversion characteristics, even though its utility in practical purposes is limited due to their hygroscopic nature. The other glass hosts for Ln^{3+} ions like chloride, iodide, bromide, fluoride, and chalcogenide glasses are widely studied for the upconversion emission [89–93]. Tellurite glasses have proven its role as better oxide glass hosts for upconversion emission due to their higher refractive index along with the relatively lower phonon energy. Hai Lin et al. discussed the upconversion mechanism relating with the excitation intensity [70]. In this, the emission intensity of the upconversion (I_{UP}) is a function of the excitation intensity in the infrared region (I_{IR}).

$$I_{UP} \propto I_{IR}{}^{m} \qquad (9.14)$$

where "m" represents the number of infrared photons absorbed per visible photon emitted. The number of photons contributed to the upconversion emission in the visible region is by the slope of the straight line obtained while plotting $\log I_{UP}$ versus $\log I_{IR}{}^{m}$.

Co-doping of rare earths of metallic nanoparticles can enhance the upconversion emissions also. Yb^{3+} is also a best known lanthanide element which can act as a sensitizer for the visible upconversion emissions with the 1 μm laser pumping in Ho^{3+}, Er^{3+}, or Tm^{3+} co-doped glasses and crystals.

9.9   Conclusions

In this chapter, we center our focus on the tellurite glasses doped with the rare-earth ions having near-infrared emission and their optical properties for various practical applications. Tellurite glasses with better thermal stability and broad near-infrared emissions have drawn considerable interest in tunable laser sources and broadband optical amplification at low-loss telecommunication windows. This work reviews the researches in the development of promising Ho^{3+}-doped tellurite glasses for optical amplifiers operated at 1.2 and 2.0 μm wavelengths. The laser transition of trivalent erbium ($^4I_{13/2} \rightarrow {}^4I_{15/2}$, at ≈1.54 μm) in tellurite glasses widely utilized in the field of optical communications and in the low-loss optical waveguides is detailed. The strategies of co-doping of other rare earths/metal nanoparticles for the enhancement of the characteristic emissions primarily in the NIR region of optically active rare-earth ions are discussed.

References

1. M.R. Sahar, K. Sulhad, J. Non-Cryst. Solids **354**, 1179 (2008)
2. S. Wang, E.M. Vogel, E. Snitzer, Opt. Mater. **3**, 187 (1994)
3. M.R. Sahar, A.K. Jehbu, M.M. Karim, J. Non-Cryst. Solids **213-214**, 164 (1997)
4. M. J. Weber, P. Mazzoldi (eds.), *From Galilers 'Occhialino' to Optoelectronics* (World Scientific, Singapore, 1993), p. 332
5. L.A. Sordillo, Y. Pu, S. Pratavieira, Y. Budansky, R.R. Alfano, J. Biomed. Opt. **19**(5), 056004 (2014)
6. G.D. Khattak, M.A. Salim, J. Electron Spectrosc. Relat. Phenon **123**, 47 (2002)
7. M.S. Sajna, S. Gopi, V.P. Prakashan, M.S. Sanu, C.C. Joseph, P.R. Biju, N.V. Unnikrishnan, Opt. Mater. **70**, 31 (2017)
8. S. Hocde, S. Jiang, X. Peng, N. Peyghambarian, T. Luo, M. Morrell, Opt. Mater. **25**, 149 (2004)
9. S. Marjanovic, J. Toulouse, H. Jain, C. Sandmann, V. Dierolf, A.R. Kortan, N. Kopylov, R.G. Ahrens, J. Non-Cryst. Solids **322**, 311 (2003)
10. V. Nazabal, S. Todoroki, A. Nukui, T. Matsumoto, S. Suehara, T. Hondo, T. Araki, S. Inoue, C. Rivero, T. Cardinal, J. Non-Cryst. Solids **325**, 85 (2003)
11. M.M. El-Desoky, M.M. Abo-Nafs.M, J. Mater. Sci. Mater. Electron. **15**, 425 (2004)
12. S. Kirkpatrick, L. Shaw, S. Bowman, S. Searles, B. Feldman, J. Ganem, Opt. Express **1**(4), 76 (1997)
13. Y. Zheng, B. Chen, H. Zhong, J. Sun, L. Cheng, X. Li, J. Zhang, Y. Tian, W. Lu, J. Wan, T. Yu, L. Huang, H. Yu, H. Lin, J. Am. Ceram. Soc. **94**(6), 1766 (2011)
14. V. Dorofeev, A.N. Moiseev, M.F. Churbanov, G.E. Snopatin, A.V. Chilyasov, I.A. Kraev, A.S. Lobanov, T.V. Kotereva, L.A. Ketkova, A.A. Pushkin, V.V. Gerasimenko, V.G. Plotnichenko, A.F. Kosolapov, E.M. Dianov, Opt. Mater. **33**, 1911 (2011)
15. G. Senthil Murugan, A. Suzuki, Y. Ohishi, Appl. Phys. Lett. **86**, 221109 (2005)
16. V. Dimitrov, S. Sakka, J. Appl. Phys. **79**, 1736 (1996)
17. P. Hagenmuller (ed.), *Inorganic Solid Fluorides, Chemistry and Physics* (Academic Press, New York, 1985)
18. L. Le Neindre, S. Jiang, B. Hwang, T. Luo, J. Watson, N. Peyghambarian, J. Non-Cryst. Solids **255**, 97 (1999)
19. K. Hirao, S.K. Kishimomo, K. Tanaka, S. Tanaka, N. Saga, J. NonCryst. Solids **151**, 1390 (1992)
20. R. Reisfeld, J. Hormadaly, J. Chem. Phys. **64**, 3207 (1976)
21. A. Mori, K. Kobayashi, M. Yamada, T. Kanamori, K. Oikawa, Y. Nishida, Y. Ohishi, Electron. Lett. **34**, 887 (1998)
22. T. Toney Fernandez, S.M. Eaton, G. Jose, R. Osellame, P. Laporta, J. Solis, Laser writing in tellurite glasses, in *Technological Advances in Tellurite Glasses*, Springer Series in Materials Science, ed. by V. Rivera, D. Manzani, vol. 254, (Springer, Cham, 2017)
23. V.V. Ravi Kanth Kumar, A.K. Bhatnagar, R. Jagannathan, J. Phys. D. Appl. Phys. **34**, 1563 (2001)
24. C.A. Click, R.K. Brow, P.R. Ehrmann, J.H. Campbell, J. Non-Cryst. Solids **319**(1-2), 95 (2003)

25. X. Feng, J. Shi, M. Segura, N.M. White, P. Kannan, L. Calvez, X. Zhang, L. Brilland, W.H. Loh, Fibers **1**, 70 (2013)
26. A. Lin, A. Ryasnyanskiy, J. Toulouse, Opt. Lett. **36**(5), 740 (2011)
27. J. Massera, A. Haldeman, J. Jackson, C. Rivero-Baleine, L. Petit, K. Richardson, J. Am. Ceram. Soc **94**(1), 130 (2011)
28. J. Song, J. Heo, Appl. Phys. **93**, 9441 (2003)
29. W.L. Barnes, R.I. Raming, E.J. Tarbox, P.R. Morkel, IEEE J. Quantum Electron. **27**, 1004 (1991)
30. D.E. McCumber, Phys. Rev. **136**(4A), A954 (1964)
31. H. Chen, Y.H. Liu, Y.F. Zhou, Z.H. Jiang, J. Alloys Compd. **397**, 286 (2005)
32. W.J. Miniscalco, R.S. Quimby, Opt. Lett. **16**, 258 (1991)
33. E. Desurvire, *Erbium Doped Fiber Amplifiers: Principles and Applications* (John Wiley, NewYork, 1994)
34. S.X. Shen, A. Jha, Opt. Mater. **25**, 321 (2004)
35. J. Coelho, J. Azevedo, G. Hungerford, N.S. Hussain, Opt. Mater. **33**, 1167 (2011)
36. A. Sennaroglu, I. Kabalci, A. Kurt, U. Demirbas, G. Ozen, J. Lumin. **116**, 79 (2006)
37. S. Shen, M. Naftaly, A. Jha, in *11th International Symposium on Non-Oxide Glasses*, Sheffield, 6–10 Sept 1998
38. R. Balda, L.M. Lacha, J. Fernández, J.M. Fernández-Navarro, Opt. Mater. **27**, 1771 (2005)
39. M. Yamada, A. Mori, K. Kobayashi, H. Ono, T. Kanamori, K. Oikawa, Y. Nishida, Y. Ohishi, IEEE Photon. Technol. Lett. **10**, 1244 (1998)
40. M. Naftaly, S. Shen, A. Jha, Appl. Opt. **39**, 4979 (2000)
41. J. Makovsky, W. Low, S. Yatsiv, Phys. Lett. **2**, 186 (1962)
42. W.T. Carnall, P.R. Fields, R. Sarup, J. Chem. Phys. **51**, 2587 (1969)
43. E. Loh, Phys. Rev. A **140**, 1463 (1965)
44. A. Bjarklev, *Optical Fiber Amplifiers: Design and System Applications* (Artech House, New York, 1993), p. 235
45. Y.G. Choi, J. Heo, J Non-Cryst. Solids **217**, 199 (1997)
46. L. Feng, J. Wang, Q. Tang, L.F. Liang, H.B. Liang, Q. Su, J. Lumin. **124**, 187 (2007)
47. C.H. Qi, X.R. Zhang, H.F. Hu, Acta Opt. Sinica **14**, 583 (1994)
48. D. Lande, S.S. Orloy, A. Akella, K. Inoue, Opt. Lett. **22**, 1722 (1997)
49. C.J. He, X.B. Chen, Y.G. Sun, L. Chen, Chinese J. Quant. Electron. **19**(2), 109 (2002)
50. A. Erbit, H.P. Jenssen, Appl. Opt. **19**, 1729 (1980)
51. S.D. Jackson, Laser Photonics Rev. **3**, 466 (2009)
52. J-b Chen, S-Q Man, N Zhang, in *Proceedings of the International Conference on Chemical, Material and Food Engineering, Advances in Engineering Research*, vol. 346 (2015)
53. E. Rukmini, C.K. Jayasanker, Opt. Mater. **4**, 529 (1995)
54. M. Seshadri, L.C. Barbosa, M. Radha, J. Non-Cryst. Solids **406**, 62 (2014)
55. J. He, H. Zhan, Z. Zhou, A. Zhang, A. Lin, Opt. Commun. **320**, 68 (2014)
56. B. Zhou, D. Yang, H. Lin, E.Y.B. Pun, J. Non-Cryst. Solids **357**, 2468 (2011)
57. B. Peng, T. Izumitani, J. Opt. Mater. **4**, 797 (1995)
58. B.J. Chen, L.F. Shen, E.Y.B. Pun, H. Lin, Opt. Commun. **284**, 5705 (2011)
59. G. Gao, G. Wanga, C. Yu, J. Zhang, L. Hu, J. Lumin. **129**, 1042 (2009)
60. R. Reisfeld, Y. Kalisky, Chem. Phys. Lett. **75**(3), 443 (1980)
61. E. Sintzer, Phys. Rev. Lett. **7**, 444 (1961)
62. J.S. Wang, D.P. Machewirch, F. Wu, E.M. Vogel, E. Snitzer, Opt. Lett. **19**, 1448 (1994)
63. J. Wan, L. Cheng, J. Sun, H. Zhong, X. Li, W. Lu, Y. Tian, B. Wang, B. Chen, Physica B **405**, 1958 (2010)
64. M. Venkateswarlu, S.K. Mahamuda, K. Swapna, M.V.V.K.S. Prasad, A. Srinivasa Rao, A. Mohan Babu, S. Shakya, G. Vijaya Prakash, Opt. Mater. **39**, 8 (2015)
65. T.J. Whitley, C.A. Millar, R. Wyatt, M.C. Brierley, D. Szebesta, Electron. Lett. **27**, 1785 (1991)
66. D.L. Veasey, D.S. Funk, N.A. Sanford, J.S. Hayden, Appl. Phys. Lett. **74**, 789 (1999)
67. A. Mori, Y. Ohishi, S. Sudo, Electron. Lett. **33**, 863 (1997)

68. M.J.F. Digonnet, *Rare Earth Doped Fiber Lasers and Amplifiers* (Marcel Dekker, New York, 1993)
69. S. Sudo, *Optical Fiber Amplifiers—Materials, Devices, and Applications* (Artech House, Boston, 1997)
70. H. Lin, G. Meredith, S. Jiang, X. Peng, T. Luo, N. Peyghambarian, E. Yue-Bun Pun, J. Appl. Phys. **93**, 186 (2003)
71. M.S. Sajna, S. Thomas, C. Jayakrishnan, C. Joseph, P.R. Biju, N.V. Unnikrishnan, Spectrochim. Acta A **161**, 130 (2016)
72. B. Zhou, L. Tao, C.Y.Y. Chan, Y.H. Tsang, W. Jin, E.Y. Pun, Spectrochim. Acta Part A **111**, 49 (2013)
73. Y. Ding, S. Jiang, B. Hwang, T. Luo, N. Peyghambarian, Y. Miura, Proc. SPIE **3942**, 166 (2000)
74. J.G. Edwards, Nature **212**, 752 (1966)
75. R. Rolli, M. Montagna, S. Chaussedent, A. Monteil, V.K. Tikhomirov, M. Ferrari, Opt. Mater. **21**, 743 (2003)
76. M. Dejneca, B. Samson, MRS Bull. **24**(9), 39 (1999)
77. M.S. Sajna, S. Thomas, K.A. Ann Mary, C. Joseph, P.R. Biju, N.V. Unnikrishnan, J. Lumin. **159**, 55 (2015)
78. S. Dai, J. Wu, J. Zhang, G. Wang, Z. Jiang, Spectrochim. Acta Part A **62**, 431 (2005)
79. W.A. Pisarski, J. Pisarska, M. Maczka, W. Ryba-Romanowski, J. Mol. Struct. **207**, 792–793 (2006)
80. S. Da, C. Yu, G. Zhou, J. Zhang, G. Wang, L. Hu, J. Lumin. **117**, 39 (2006)
81. H. Zheng, D. Gao, Z. Fu, E. Wang, Y. Lei, Y. Tuan, J. Lumin. **131**, 423 (2011)
82. B. van Saders, L. Al-Baroudi, M.C. Tan, R.E. Riman, Opt. Mater. Express **3**, 567 (2013)
83. S. Tanabe, T. Kouda, T. Hanada, J. Non-Cryst. Solids **274**, 55 (2000)
84. G. Lakshminarayana, R. Vidya Sagar, S. Buddhudu, J. Lumin. **128**, 690 (2008)
85. M. Eichelbaum, K. Rademann, Adv. Funct. Mater. **19**, 2045 (2009)
86. P.P. Pompa, L. Martiradonna, A.D. Torre, F.D. Sala, L. Manna, M.D. Vittorio, F. Calabi, R. Cingolani, R. Rinaldi, Nat. Nanotechnol. **1**, 126 (2006)
87. A. Lin, S. Boo, D.S. Moon, H.J. Jeong, Y. Chung, W.T. Han, Opt. Express **15**, 8603 (2007)
88. M.R. Dousti, R.J. Amjad, Z.A.S. Mahraz, J. Mol. Struct. **1079**, 347 (2015)
89. H. Higuchi, M. Takahashi, Y. Kawamoto, K. Kadono, T. Ohtsuki, N. Peyghambarian, N. Kitamura, J. Appl. Phys. **83**, 19 (1998)
90. M. Shojiya, M. Takahashi, R. Kanno, Y. Kawamoto, K. Kadono, Appl. Phys. Lett. **67**, 245 (1995)
91. A. Gharavi, G.L. McPherson, Appl. Phys. Lett. **61**, 2635 (1992)
92. D.F. de Sousa, L.F.C. Zonetti, M.J.V. Bell, R. Ledullenger, A.C. Hernandes, L.A.O. Nunesl, J. Appl. Phys. **85**, 2502 (1999)
93. A.S. Oliveira, M.T. de Araujo, A.S. Gouveia-Neto, J.A. Medeiros Neto, A.S.B. Sombra, Y. Messaddeq, Appl. Phys. Lett. **72**, 753 (1998)

Chapter 10
Tellurite Glasses: Solar Cell, Laser, and Luminescent Displays Applications

Luciana R. P. Kassab, L. A. Gómez-Malagón, and M. J. Valenzuela Bell

Abstract Rare-earth-doped glasses can be exploited to control the solar spectrum in order to enhance the solar cell efficiency. We review recent results of the management of the solar spectrum on a solar cell using rare-earth ion-doped TeO_2-ZnO glasses, with and without metallic nanoparticles, as a cover slip. Transparent rare-earth-doped materials as glasses can absorb light at shorter wavelength and emit light at longer wavelengths, the well-known downconversion process; besides they have the advantage of easy preparation and high doping concentration of rare-earth ions. In this context tellurite glasses appear as potential candidates because of their wide transmission window (400–5000 nm), low phonon energy (800 cm^{-1}) when compared to silicate, and thermal and chemical stability. Few vitreous hosts have been investigated to be used as cover slip to enhance the performance of conventional solar cell; so the lack of studies using rare-earth-doped glasses on the top of standard solar cells has motivated the recent reports that are reviewed in this chapter. We discuss the role of the downconversion process to increase the solar cell efficiency. It is shown that the management of Tb^{3+} and Yb^{3+} ions concentration can be optimized to modify the solar spectrum and consequently increase the solar cell efficiency. It is demonstrated that plasmon-assisted efficiency enhancement could be obtained for commercial Si and GaP solar cells, respectively, covered with Eu^{3+}-doped TeO_2-ZnO glasses with silver nanoparticles. Tellurite glasses have also proven to be adequate hosts for rare-earth ions and for the nucleation of metallic nanoparticles (NPs). We review results of the modification introduced by different Nd_2O_3

L. R. P. Kassab (✉)
Faculdade de Tecnologia de São Paulo/ CEETEPS, São Paulo, SP, Brazil

L. A. Gómez-Malagón
Escola Politécnica de Pernambuco, Universidade de Pernambuco, Recife, PE, Brazil

M. J. Valenzuela Bell
Departamento de Física, Universidade Federal de Juiz de Fora, Juiz de Fora, MG, Brazil

© Springer International Publishing AG, part of Springer Nature 2018
R. El-Mallawany (ed.), *Tellurite Glass Smart Materials*,
https://doi.org/10.1007/978-3-319-76568-6_10

225

concentration on the laser operation of TeO_2-ZnO glasses. The control and improvement of the photoluminescence efficiency due to the nucleation of gold NPs in Yb^{3+}/Er^{3+}-doped TeO_2-PbO-GeO_2 glasses is also reviewed. It is shown that the nucleation of silver NPs in Tb^{3+}-doped TeO_2-ZnO-Na_2O-PbO glass contributes for the large enhancement in the blue-red spectrum.

10.1 Introduction

The demand for the development of new nanostructured materials with relevant properties to fulfill the different necessities of our society has been leading the researches to look for the production of new dielectric materials; in this context tellurite glasses appear as potential candidates because of their wide transmission window (400–5000 nm), high refractive index (≥ 2.0), low phonon energy (800 cm^{-1}) when compared to silicate glasses, high mechanical durability, large mechanical resistance, and high vitreous stability [1–4]. Rare-earth-doped tellurite glasses have demonstrated unique physical properties [5–14]. Low cutoff phonon energy is normally required to increase the upconversion efficiency as the multiphonon relaxation becomes less probable and high refractive index is interesting for ultrafast response devices based on the nonlinear response. They are also capable of incorporating rare-earth ions and metallic nanoparticles (NPs) with an extensive range of photonic applications.

This chapter presents some reports related to metal-dielectric nanocomposites based on tellurite glasses for photonic applications. Their emerging application in the field of plasmonics has demonstrated the possibility for the development of color displays, optical amplifiers, and sensors and biosensors based on the enhancement of the spectroscopic properties of rare-earth ions in the presence of metallic NPs [15–22]. Tellurite glasses have also demonstrated to be potential material for laser applications. For this reason in this chapter, we also focus our attention to this important application of Nd^{3+}-doped TeO_2-ZnO [23, 24].

Rare-earth-doped glasses can be exploited to control the solar spectrum in order to enhance the solar cell efficiency. Transparent rare-earth-doped materials as glasses can absorb light at shorter wavelength and emit light at longer wavelengths, by the well-known downconversion process; besides they have the advantage of easy preparation and high doping concentration of rare-earth ions. In this context tellurite glasses appear as potential candidates too. Few vitreous hosts have been investigated to be used as cover slip to enhance the performance of conventional solar cell; so the lack of studies using rare-earth-doped glasses on the top of standard solar cells has motivated the recent reports that are reviewed in this chapter [25, 26].

In this chapter we review some of our previous results obtained with samples based on different tellurite compositions. Firstly the control and improvement of the photoluminescence efficiency due to the nucleation of gold NPs in Yb^{3+}/Er^{3+}-doped TeO_2-PbO-GeO_2 glasses is reviewed [27]. Then it is shown that the nucleation of silver NPs in Tb^{3+}-doped TeO_2-ZnO-Na_2O-PbO glass contributes for the large enhancement in the blue-red spectrum [28]. The possibility to use tellurite glasses for laser applications is also reviewed, and recent results of the modification introduced by different Nd_2O_3 concentration on the laser operation of TeO_2-ZnO glasses [23, 24] are discussed. We discuss the role of the downconversion process to increase the solar cell efficiency. It is shown that the management of Tb^{3+} and Yb $^{3+}$ ions concentration can be optimized to modify the solar spectrum and consequently increase the solar cell efficiency [25]. It is demonstrated that plasmon-assisted efficiency enhancement could be obtained for commercial Si and GaP solar cells, respectively, covered with Eu^{3+}-doped TeO_2-ZnO glasses with silver nanoparticles [26].

Finally a summary of the results and further comments on the potential of tellurite glasses for different applications is discussed.

10.2   Experimental Details

10.2.1   Method Used for the Production of the Tellurite Glasses

In this chapter we review results of different tellurite compositions: TeO_2-ZnO, TeO_2-PbO-GeO_2, and TeO_2-ZnO-Na_2O-PbO [23–28]. The samples used were prepared by the conventional melt quenching technique. All reagents and doping species were oxide powders obtained commercially. The high-purity reagents (~99.999%) were melted in platinum crucibles, at different temperatures (750–1050 °C) depending on the composition, during 20–120 min, quenched in air, in a heated brass mold, annealed for 2 h at 270–350 °C to avoid internal stress, and then cooled to room temperature inside the furnace. Then, after the cooling the glass samples were cut and polished for the optical experiments. For the experiments with glasses having silver or gold NPs, the samples prepared with $AgNO_3$ or Au_2O_3 were submitted to additional heat treatment (HT), at the transition temperature, during different periods of time, to reduce the metallic ions and nucleate Ag or Au NPs, following the procedure already reported [15, 16, 19]; samples without metallic NPs were also produced to be used for comparison with those with silver or gold NPs.

10.2.2   Characterization Techniques

A transmission electron microscope (HR-TEM) operating at 200 kV was used to investigate the presence of gold or silver NPs embedded in the glass matrix and enabled the determination of the size and shape of the NPs.

Optical absorption spectra were measured at room temperature in the 400–1700 nm range using a commercial spectrophotometer in order to determine the absorption bands related to the transitions of the trivalent rare-earth ions as well as the absorption bands of the localized surface plasmon resonance (LSPR) associated to the metallic NPs [29].

Photoluminescence spectra were measured with different excitation sources, depending on the doping species and the goals of each experiment. The samples doped with Tb^{3+} were excited in two different ways: using a 30 W xenon lamp (pulses of ~3 μs; 80 Hz) at 377 nm to study the photoluminescence enhancement in the presence of metallic NPs and the third harmonic generation from a 1064 nm pulsed nanosecond Nd: YAG laser (QuantelUltra 50) and an OPA (Coherent Libra) tuned at 482 nm operating in a quasi-continuous (quasi-cw) mode as excitation lasers for samples singly doped with Tb^{3+} [28] and co-doped with Yb^{3+} and Tb^{3+} [26]. For the studies of the enhanced upconversion of the samples co-doped with Yb^{3+} and Er^{3+} [27], the excitation was performed with a CW diode laser operating at 980 nm, and the upconversion luminescence signals were dispersed by a monochromator fitted by a S-20 photomultiplier and computer. Cw laser operating at 473 nm together with the Ocean Optics spectrometer was used for the excitation of the samples prepared with Eu^{3+} that were used as cover slip to enhance the commercial solar cells performance [25].

For the case of the laser action study [23, 24] in TeO_2-ZnO glasses doped with Nd^{3+}, a titanium sapphire (Ti:Sa) laser emitting at 808 nm (300 mW) and chopped at 100 Hz was used. Then the light emitted by the sample was collected with an optical fiber detector, and the signal was analyzed with the aid of an optical spectrum analyzer (OSA). Exciting the sample with a pulsed optical parametric oscillator (OPO) system, emitting at 808 nm (7 mJ, 5 ns pulse), enabled the determination of the fluorescence lifetime.

The laser setup used to study the CW laser action in Nd^{3+}-doped TeO_2-ZnO glasses was consisted in a standard plane-concave laser resonator and was described in ref [23, 24]. A flat dichroic mirror highly reflective ($R > 99.5\%$) around 1064 nm and with high transmittance ($T > 95\%$) around 808 nm was used together with two different concave output mirrors with 100 mm radius curvature and transmissions of 0.8% and 4% around 1064 nm. The samples used for this experiment were carefully polished and stickled with silver paste on a copper sample holder without any particular cooling; a CW Ti: sapphire laser tuned at 806 nm and focused with a lens of 10 cm focal length pumped them through the dichroic input mirror.

The electrical characterization to determine the performance of the solar cell using the rare-earth ion-doped TeO_2-ZnO glasses placed on top of the solar cell [25, 26] was performed using a solar simulator (LCS-100 Newport), with a AM 1.5 filter and a sourcemeter (Keithley 2420) coupled to a personal computer, as reported in ref. [25]. The electrical parameters such as efficiency; fill factor (FF); short-circuit current, I_{sc}; and open-circuit voltage, V_{oc}, were obtained from the current-voltage (I-V) curves under 1000 W/m^2 irradiance of commercial silicon (BPW34 Vishay Semiconductors) and GaP (FGAP71 Thorlabs) semiconductor photodiodes, with energy gap of 1.1 eV and 2.26 eV, respectively.

10.3   Results and Discussion

10.3.1   Control and Improvement of the Photoluminescence Efficiency Due to the Nucleation of Gold NPs in Yb^{3+}/ Er^{3+}-Doped TeO_2-PbO-GeO$_2$

We review the effects of gold NPs in the infrared-to-visible frequency upconversion of Er^{3+}/Yb^{3+} co-doped TeO_2-PbO-GeO$_2$ glasses [27].

Glasses were prepared by adding 0.5 wt % of Er_2O_3, 2.0 wt% of Yb_2O_3, and 1.0 wt% of Au_2O_3 to the base glass composition 33.3 TeO_2–33.3 PbO-33.3 GeO$_2$ (in wt%), using the melt quenching method [27]. After the annealing performed to reduce the internal stress, the samples were heat treated at 350 °C during 24, 48, and 72 h (3 steps of 24 h) to thermally reduce Au^+ and Au^{3+} ions to Au^0 and consequently to nucleate and grow gold NPs, following the procedure used for different germanate and tellurite hosts successfully [17, 30–36]. The results obtained demonstrated that the enhanced local field contribution attributed to gold nanoparticles and the energy transfer processes between two different rare-earth ions can be used to control the efficiency of luminescent glasses.

Figure 10.1 shows the TEM image of the sample after heat treatment during 72 h where we observe gold NPs which average size around 10 nm. Figure 10.2 presents the upconversion emission spectra of all the samples prepared; the results for different heat treatment times are shown as well as the one of the sample without gold NPs. Intense emission bands at 527, 550, and 660 nm were observed corresponding to well-known Er^{3+} transitions $^2H_{11/2} \rightarrow {}^4I_{15/2}$, $^4S_{3/2} \rightarrow {}^4I_{15/2}$, and $^4F_{9/2} \rightarrow {}^4I_{15/2}$, respectively [27, 32, 37, 38]. We can notice enhancement of about 20% for the $^2H_{11/2} \rightarrow {}^4I_{15/2}$, $^4S_{3/2} \rightarrow {}^4I_{15/2}$ transitions by comparing the samples prepared without

Fig. 10.1 TEM image of Er^{3+}-Yb^{3+} co-doped TeO_2-PbO-GeO$_2$ glass containing gold NPs after heat treatment during 72 h [27]

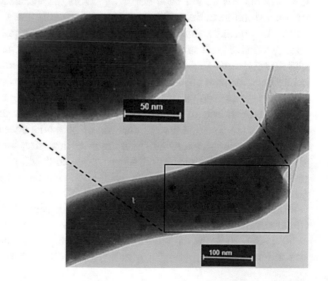

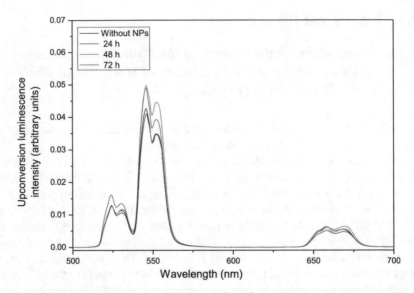

Fig. 10.2 Upconversion emission spectra of Er^{3+}/Yb^{3+} co-doped TeO_2-PbO-GeO_2 glasses containing gold NPs for different heat treatment times; the upconversion emission of the sample without gold NPs is shown as reference TEM image of Er^{3+}/Yb^{3+} co-doped TeO_2-PbO-GeO_2 glass containing gold NPs after heat treatment during 72 h [27]

gold NPs with the one heat treated during 72 h. These transitions have wavelength near the surface plasmon band, whereas the one at $^4F_{9/2} \rightarrow {}^4I_{15/2}$ does not. So this transition is less influenced by the presence of gold NPs than the other observed ones. These results corroborated the studies presented in ref [32, 37] that demonstrated that the proximity of the infrared-to-visible frequency upconversion luminescence with the plasmon absorption band normally favors the intensity enhancement caused by the intensified local field effect.

So the present results showed that the influence of gold NPs and the efficiency energy transfer mechanism between Yb^{3+} and E^{+3} ions can change the upconversion visible spectrum. The role of the gold NPs could be demonstrated as the luminescence improvement observed, due to the intensified local field around the Er^{3+} ions, took place even using a low concentration of the doping species.

The dependence of the upconversion intensities with the laser intensity was studied and showed that the introduction of gold NPs does not affect the dependence of the upconversion intensities with the laser intensity; for all the samples studied, a slope ~2.0 was observed for $^4S_{3/2} \rightarrow {}^4I_{15/2}$ and $^4F_{9/2} \rightarrow {}^4I_{15/2}$ transitions in the presence and in the absence of gold NPs, demonstrating that two photons are participating in the upconversion emission [27, 32].

Figure 10.3 presents the energy level diagram of Er^{3+} and Yb^{3+} ions and illustrates the possible upconversion pathways. The energy transfer pathways indicated in Fig. 10.3 (the dashed lines 1, 2 and 3) represent the dominant processes. We observe that the $^2H_{11/2} \rightarrow {}^4I_{15/2}$ and $^4S_{3/2} \rightarrow {}^4I_{15/2}$ transitions originate the green

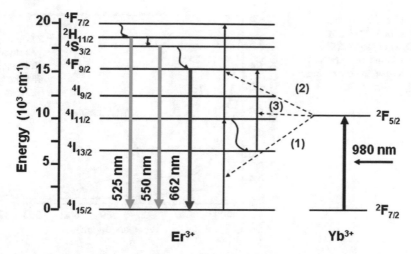

Fig. 10.3 Energy level diagram of Er^{3+} and Yb^{3+} ions illustrating possible upconversion pathways for Er^{3+}-Yb^{3+} co-doped TeO_2-PbO-GeO_2 glasses with gold NPs. The solid straight lines with upward and downward arrows indicate optical transitions; dotted lines and wavy arrows denote ET processes and nonradiative relaxation, respectively [27]

emissions at 527 nm and 550 nm and the red emission at 660 nm is due to the $^4F_{9/2} \rightarrow {}^4I_{15/2}$ transition.

This work showed that the introduction of gold NPs enabled the control and improvement of the photoluminescence efficiency of Yb^{3+}/Er^{3+} co-doped TeO_2 -PbO-GeO_2 glasses and the use of Yb^{3+} lower concentrations, due to the local field growth around Er^{3+} ions located in the vicinity of the metallic NPs. The present procedure can be used for different hosts in order to achieve luminescent materials of high performance.

10.3.2   Large Enhancement in the Blue-Red Spectrum Due to the Nucleation of Silver NPs in Tb^{3+}-Doped TeO_2 -ZnO-Na_2O-PbO

We review the luminescence properties of Tb^{3+}-doped TeO_2-ZnO-Na_2O-PbO glasses containing silver NPs [28]. As the Tb^{3+} ions located in the vicinity of the NPs were in the presence of an intensified local field, the luminescence efficiency increased. The whole spectrum was intensified by the appropriate heat treatment of the samples that were prepared using the melt quenching technique using as doping species Tb_4O_7 (5 wt %) and Ag_2O (10.0 wt %). Different heat treatment times at 270 °C were used to reduce the Ag^+ ions to Ag^0 and to nucleate silver NPs.

Figure 10.4 shows the absorption band due to the surface plasmon resonance of silver NPs [28, 29], and whose amplitude increases with the annealing time because of the growth of the NPs concentration. Figure 10.5 shows a TEM micrograph that

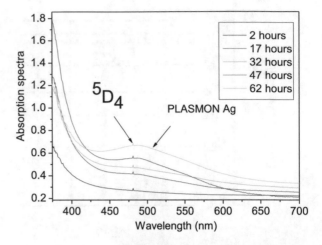

Fig. 10.4 Absorption spectra of Tb^{3+}-doped TeO$_2$-ZnO-Na$_2$O-PbO samples containing NPs for various heat treatment times [28]

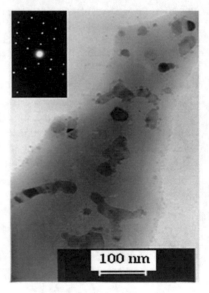

Fig. 10.5 Transmission electron microscope image of the sample annealed during 62 h. The inset shows the electron diffraction pattern of the silver NPs [28]

corroborates the presence of silver NPs and aggregates with dimensions varying from 2 to 150 nm. The inset of the figure shows the diffraction patterns characteristic of silver crystals.

Figure 10.6a presents the emission bands at 485 nm, 550 nm, 585 nm, and 623 nm, related to Tb^{3+} transitions, detected at 377 nm. We observe increase of the whole spectrum of the Tb^{3+} luminescence for samples heat treated for 62 h. The highest enhancement is observed for the luminescence at 550 nm (~200). A simplified energy level scheme of Tb^{3+} ion with indication of the luminescence transitions observed is presented in Fig. 10.6b.

The emission at 550 nm is more affected than the one at 485 nm and can be attributed to the fact that electric dipole transitions are more sensitive to the local

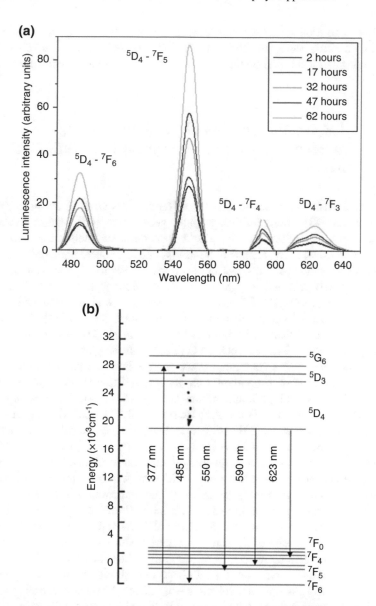

Fig. 10.6 (a) Emission spectra of Tb^{3+}-doped TeO$_2$-ZnO-Na$_2$O-PbO samples containing NPs for different annealing times (excitation wavelength: 377 nm). (b) Simplified energy level scheme of Tb^{3+} ion with indication of the luminescence transitions observed. The dashed line indicates nonradiative decay to level ^{5}D$_3$ followed by cross-relaxation among excited ions and neighbors in the ground state [28]

field increase due to the NPs than magnetic dipole transitions. The Tb^{3+} ions located in the vicinity of the NPs are in the presence of an intensified local field, and consequently the luminescence efficiency increases. We recall that tellurite glasses co-doped with Tb^{3+} and Eu^{3+} also showed enhancement ~100% as reported before

[39]. The large luminescence enhancement was obtained due to the simultaneous contribution of the energy transfer from Tb^{3+} ions to Eu^{3+} ions and the contribution of the intensified local field on the Eu^{3+} ions located near silver nanostructures.

10.3.3   Modification Introduced by Different Nd_2O_3 Concentration on the Laser Operation of TeO_2-ZnO Glasses

Nd^{3+} laser emission has been reported in different glass hosts as fluorides [40–42], chalcogenides [43], aluminosilicates [44], germanates [45], and tellurite [46–50]. Among oxi-tellurites, the TeO_2-ZnO glass deserves attention as it combines good mechanical stability, chemical durability, and high linear and nonlinear refractive indices, together with low phonon energies (~750 cm^{-1}), a wide transmission window (0.4–6.0 μm) and a high rare-earth solubility [51, 52]. The large linear refractive index (1.97) [53] of TeO_2-ZnO glasses enables large stimulated emission cross sections, normally larger than the one of phosphate glasses [54]. We review in this section the modification introduced by different Nd_2O_3 concentration on the laser operation of Nd^{3+}-doped TeO_2-ZnO (TZO) tellurite glass.

We demonstrated that continuous-wave laser action can be achieved with this TZO bulk tellurite glass by pumping the sample inside a standard plane-concave mirror laser cavity with different output couplers [23, 24]. The results that are reviewed in this section together with those previously reported with higher concentration of Nd_2O_3 (1.0 wt%) [23] could determine the adequate Nd_2O_3 concentration for laser action. So we review a CW laser action in a bulk Nd^{3+}-doped TZO glass, at 1062 nm, prepared using the melt quenching technique and with 0.5 wt% of Nd_2O_3. In this case acceptable laser threshold of 73 mW was obtained as well as slope efficiency and output mirror transmission. When compared to our previous report [23] in which the sample was prepared with 1.0 wt% of Nd_2O_3, the laser action is modest as it was obtained a low laser threshold of 8 mW and a laser slope efficiency of 21%. However the results of the sample prepared with lower Nd_2O_3 concentration deserve attention and are reviewed in this section as they complement those obtained by Bell et al. [23] and also demonstrate the best concentration for low threshold pump power [24] in TZO glass. As the objective was to test the best concentration for laser action in order to complement the results of Bell et al., three samples were prepared with the following concentrations of Nd_2O_3 (in wt%): 0.5, 2.0 and 3.0. Figure 10.7 presents the near-infrared (NIR) luminescence spectra of these samples. They were obtained with excitation at 806 nm and consist of three broadband emissions centered around 882, 1062 and 1335 nm associated, respectively, to the three usual Nd^{3+} emission transitions, $^4F_{3/2} \rightarrow {}^4I_{9/2}$, $^4F_{3/2} \rightarrow {}^4I_{11/2}$, and $^4F_{3/2} \rightarrow {}^4I_{13/2}$.

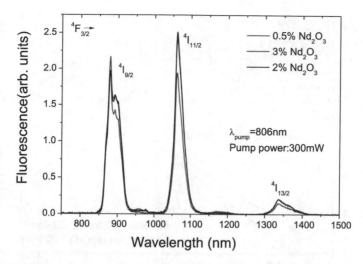

Fig. 10.7 NIR Fluorescence spectra of the TZO:Nd samples. They were obtained with excitation at 806 nm within the $^4I_{9/2} \rightarrow {}^4F_{5/2} + {}^2H_{9/2}$ absorption band. It consists of three broadband emissions peaking around 882 nm, 1062 nm, and 1335 nm. They are assigned to the three usual Nd^{3+} emission transitions, $^4F_{3/2} \rightarrow {}^4I_{9/2}$, $^4F_{3/2} \rightarrow {}^4I_{11/2}$, and $^4F_{3/2} \rightarrow {}^4I_{13/2}$ [24]

The three samples were tested for laser emission, using the experimental procedure presented in experimental details; however true CW laser action could only be obtained with the lowest concentration: 0.5 wt% of Nd_2O_3. Figure 10.8 exhibits the results of the laser output versus the absorbed pump power obtained at 806 nm. Threshold pump powers of 53 mW and 73 mW (as extrapolated with the straight lines reported in the figure) and slope efficiencies of 6.6% and 8.2% were obtained for the output coupler transmissions of 0.8% and 4%, respectively.

Table 10.1 reviews the comparison for TZO samples prepared with the different Nd_2O_3 concentrations [24]. We observe that for lower concentration (0.5 wt%) laser action is possible because of the reasonable quality achieved, attested by low internal losses, together with a fairly long emission lifetime of about 158 μs, emission quantum efficiency ($\eta_{Quantum}$) of 0.8 and a high stimulated emission cross section (σ_{em}) of 4.2×10^{-20} cm^2 for a emission bandwidth ($\Delta\lambda$) of 24 nm.

So the modification introduced by different Nd_2O_3 concentration on the laser operation of TZO glasses is reviewed in this section and shows that it is possible to determine the adequate concentration range for laser action. The present review also demonstrates the adequate method for glasses preparation for solid-state laser applications. We recall that the laser slope efficiency (21%) of the TZO sample doped with 1 wt% of Nd_2O_3 is the highest obtained considering the other tellurite glasses already reported for laser action [46–50]. A higher slope efficiency was reported for a 60 cm long Nd-doped tellurite glass fiber ($76.9\%TeO_2$–$6.0\%Na_2O$–$15.5\%ZnO$–$1.5\%Bi_2O_3$–$0.1\%Nd_2O_3$) for which it was reported a laser slope efficiency of 46% for a lasing threshold of 27 mW [55].

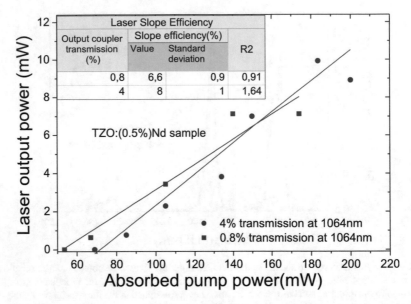

Fig. 10.8 Laser output power versus absorbed pump power curves obtained after pumping the TZO glass sample around 806 nm, for the output coupler transmissions of 0.8%, and 4% for the TZO:0.5%Nd sample. Threshold pump powers (P_{th}) of 53 mW and 73 mW (as extrapolated with the straight lines reported in the figure) and slope efficiencies of 6.6% and 8.2% were obtained for the output coupler transmissions of 0.8% and 4%, respectively [24]

Table 10.1 Laser cavity parameters of TZO: Nd^{3+} tellurite glasses

Parameter	TZO:x%Nd glasses			
	Bell et al. [25]	This work		
Nd $_2$O$_3$ (%wt)	1	0.5	2	3
λ_{laser} (nm)	1062	1062	N/A	N/A
λ_{pump} (nm)	806	806	806	806
η_{slope} (%)	21	8	N/A	N/A
P_{th} (mW)	8	73	N/A	N/A
T (%)	2.7	4	N/A	N/A
τ_f, τ_R (µs)	210,217	158,198	124,182	90,162
$\eta_{Quantum}$ (%)	95	80	68	56
$\Delta\lambda$ (nm)	29	24	24	24
σ_{em} (10^{-20} cm^2)	3.1	4.2	4.6	5.1

λ_{laser}, λ_{pump}, η_{slope}, T, P_{th}, σ_{em}, τ_f, τ_R, $\eta_{Quantum}$ and $\Delta\lambda$ mean emission wavelength, pumping wavelength, slope efficiency, output coupled transmission, threshold, stimulated emission cross section, fluorescence lifetime, radiative lifetime, quantum conversion efficiency and emission bandwidth [24]

10.3.4 Management of the Solar Spectrum Incident on a Solar Cell Using TeO₂-ZnO Glasses Doped with Tb³⁺ and Yb³⁺ Ions as a Cover Slip

Glasses doped with Tb^{3+} and Yb^{3+} ions have been largely studied in the literature, and their optical properties demonstrated that they are potential candidates for solar cell applications due to their capability of transferring energy from the UV/VIS region to the NIR region [56–63]. However only few reports [64] showed the possibility to use them as cover slips to improve commercial solar cell efficiency. The absence of such researches motivated us to study the possibility to use tellurite glasses as cover slips for commercial solar cells, whose results were reported recently [25] and are presented in this chapter. So we review the downconversion process in tellurite glasses doped with Tb^{3+} and Yb^{3+} ions and also the modification of the solar spectrum using these materials as cover slip to increase the efficiency of commercially available silicon and GaP solar cells [25].

The samples were prepared by the melt quenching technique, as reported in ref [25], using the TeO₂-ZnO (TZO) composition and the following doping species: Tb₄O₇ (1 and 2 wt%) and Yb₂O₃ (5 and 7 wt%). Figure 10.9 presents the emission in the VIS-IR region under UV excitation (355 nm). The visible emission in the region of 500–700 nm corresponds to the electronic transitions of $^5D_4 \rightarrow \ ^7F_j$ ($j = 6, 5, 4, 3$) of Tb^{3+} ions. For the sample prepared with 1% of Tb^{3+} ion, under excitation at 355 nm, the emission at $^5D_4 \rightarrow \ ^7F_5$ (~ 548 nm) increases for the co-doped sample prepared with 5 wt% of Yb^{3+}ion, whereas for co-doping of 7 wt% of Yb^{3+} ion, it decreases. Also with the increase of Yb^{3+} enhancement of the VIS emission in the co-doped TZO, samples can be observed under UV excitation. This is probably due to the following cross-relaxation mechanism: the upper lying states (Tb) + $^3F_6$ (Yb) $\rightarrow \ ^5D_4$ (Tb) + $^2F_{5/2}$ (Yb). Concerning the IR region, the emission increases for the sample co-doped with 5 wt% of Yb^{3+} ion, and decreases for the one co-doped with 7 wt%, as can be seen in Fig. 10.3b, demonstrating that the quenching processes is present for high Yb^{3+} ion concentration.

Using the measurements of the solar radiation transmittance emitted by a solar simulator (LCS-100 Newport) through the samples using a mini-spectrometer (HR4000 Ocean Optics) and a calibrated reference solar cell (91,150 V Newport), it was possible to obtain the modification of the solar cell.

The solar spectrum has components in the UV-VIS-NIR region; then its interaction with Tb^{3+}- and Yb^{3+}-doped TZO glasses modifies the solar spectrum. This change was studied using the transmitted irradiance spectra shown in Fig. 10.10 in which we can observe the downshift in the range of 400–800 nm and the UV radiation reduction. These effects were attributed to downconversion mechanism and to the absorption of the tellurite glass, respectively. The expected increase of the radiation in the 900–1100 nm region was not observed, indicating that the interplay between the self-absorption of the Yb^{3+} ion and the energy transfer process between the Tb^{3+} and Yb^{3+} ions must be improved.

Fig. 10.9 Emission spectra of tellurite glasses under 355 nm laser excitation. VIS spectra for samples doped with 1%and 2% (**a**) of Tb^{3+} ions co-doped with 5% and 7% of Yb^{3+} ions. (**b**) NIR spectrum for samples doped with 1% and 2% of $Tb^{3+}$and co-doped with 5% and 7% of Yb^{3+} ions [25]

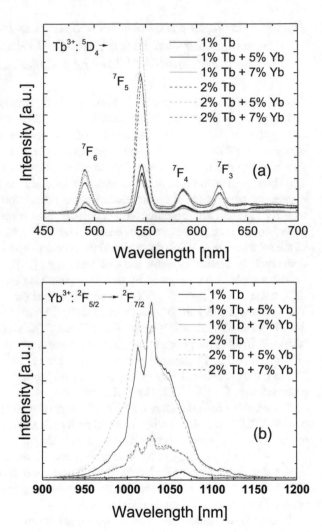

We review the results of voltage-current for silicon and GaP solar cells covered with TZO glasses doped with Tb^{3+} and Yb^{3+} ions that were obtained using the experimental setup presented in Fig. 10.11 whose results are shown in Figs. 10.12 and 10.13, respectively. In Figs. 10.12 and 10.13 the undoped TZO glass results are presented to be used as reference. Using the voltage-current results, the following electrical parameters, shown in Table 10.2, were obtained [25]: short-circuit current (I_{sc}), the open-circuit voltage (V_{oc}), the filling factor (FF $= V_{mp}I_{mp}/V_{oc}I_{sc}$), and the efficiency (eff $= V_{oc}I_{sc}$FF/AI), where A is the cell area and I is the incident irradiance (1000 W/m^2).

Efficiency enhancement of about 7% was observed when TZO glass singly doped with 1 wt% of Tb^{3+}ion is used as cover slip for the silicon solar cell (Eg = 1.1 eV); with the addition of Yb^{3+} ions, the silicon solar cell efficiency decreases and assumes the

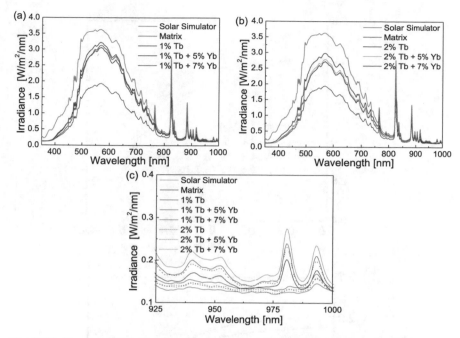

Fig. 10.10 Transmitted irradiance spectra for (**a**) samples doped with 1% of Tb^{3+} and (**b**) 2% of Tb^{3+}ions. IR spectra for all samples (**c**). The concentration of Yb^{3+}ions was 5% and 7% [25]

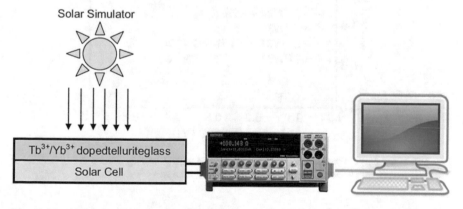

Fig. 10.11 Experimental setup for the electrical characterization of Tb^{3+}/Yb^{3+}-doped TZO glasses [25]

lowest value for the highest Yb^{3+} concentration (1 wt% of Tb^{3+} and 7 wt% of Yb^{3+}). As far as we are concerned, this is the highest enhancement reported for silicon solar cells using Tb^{3+}/Yb^{3+}-doped glasses as cover slips. Recently it was reported lower enhancement of 0.34% for phosphate glasses co-doped with 1.0 wt% of Tb^{3+} and 0.5 wt% of Yb^{3+}ions [65].

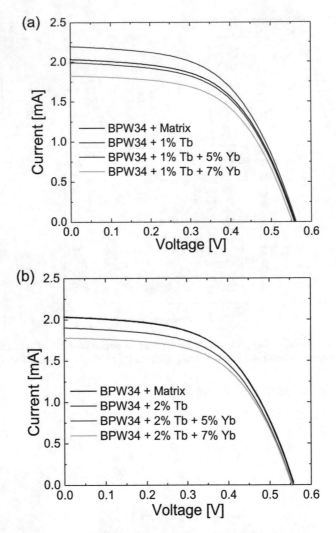

Fig. 10.12 V-I curves for the BPW34 solar cell covered with tellurite glass doped with (**a**) 1% of Tb^{3+} and (**b**) 2% of Tb^{3+} ions. The concentration of Yb^{3+} ions was 5% and 7% [25]

For the case of the GaP solar cell covered with the undoped tellurite glass, efficiency enhancement of ~1.1% was observed when covered with the same tellurite glass but co-doped with 1% of Tb^{3+} and 5% of Yb^{3+}. For the case of the samples doped with 2 wt% of Tb^{3+}, the efficiency increases with the Yb^{3+} concentration, up to 5 wt% of Yb^{3+}, and decreases for the sample co-doped with 7 wt% of Yb^{3+}.

So the results reviewed in the present section show that the management of Tb^{3+} and Yb^{3+} ions concentration can be optimized to modify the solar spectrum and

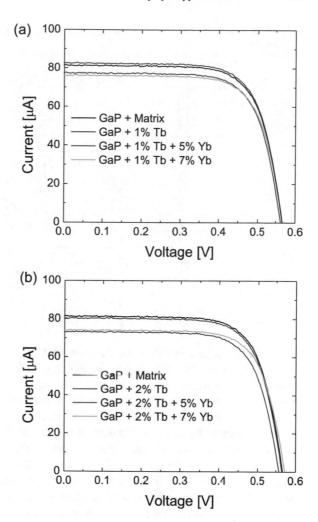

Fig. 10.13 V-I curves for the GaP solar cell covered with tellurite glass doped with (**a**) 1% of Tb^{3+} and (**b**) 2% of Tb^{3+} ions. The concentration of Yb^{3+} ions was 5% and 7% [25]

increase the solar cell performance. Efficiency enhancement depends on the rare-earth ions concentrations, and the results were attributed to the modification of the incident radiation spectral profile in the IR region.

10.3.5   The Use of the Localized Surface Plasmon to Enhance the Performance of Solar Cell with Eu^{3+}-Doped TeO$_2$-ZnO Glasses with Silver Nanoparticles

Several publications have been reporting the spectroscopic characterization of rare-earth-doped glasses for photovoltaic applications and also how the luminescence increase can be obtained using metallic nanostructures [66–68]. However, few

Table 10.2 Electrical parameters (I_{sc} and V_{oc}), filling factor (FF), and efficiencies of the studied samples

Solar cell	Matrix doping [%] Tb	Yb	I_{sc} (10^{-5}) [A]	V_{oc} [V]	FF	Efficiency [%]
BPW34	0	0	203	0.55778	0.5550	6.9818
BPW34	1	0	220	0.56136	0.5443	7.4692
BPW34	1	5	196	0.5542	0.5643	6.8110
BPW34	1	7	183	0.55076	0.5647	6.3237
BPW34	2	0	197	0.55419	0.5737	6.9597
BPW34	2	5	190	0.55072	0.5607	6.5184
BPW34	2	7	197	0.55077	0.5187	6.2527
GaP	0	0	8.16	0.56481	0.7194	0.6906
GaP	1	0	7.76	0.56126	0.7170	0.6504
GaP	1	5	8.25	0.56474	0.7193	0.6985
GaP	1	7	7.61	0.56487	0.7235	0.6480
GaP	2	0	7.29	0.55429	0.7215	0.6077
GaP	2	5	8.03	0.56481	0.7194	0.6794
GaP	2	7	7.41	0.57179	0.7194	0.6349

Uncertainty in efficiency calculation is ± 0.00001 [25]

works have been reported showing the improved efficiency of solar cells under the modified spectrum [69–73]. More recently the SPR has also been explored to enhance the thin-film solar cell efficiency with metallic nanoparticles incorporated into or on the solar cell [74]. These works show that the SPR is an interesting mechanism to improve the solar cell efficiency and is discussed in this section.

So this section focus on the review of the use of TZO glasses to enhance the performance of commercial solar cell making use of the localized surface plasmon of silver NPs that were responsible for a considerable enhancement of the Eu^{3+} luminescence as reported in ref [26]. It was shown that for excitation at 473 nm the emissions related to Eu^{3+} transitions were enhanced by almost 100% due to the enhanced local field around the silver NPs (because of the mismatch between the dielectric function of the metallic NPs and the glass) and also to the energy transfer from silver NPs to the Eu^{3+} ions [17, 20, 29, 64].

So we review the electrical characterization of commercial solar cells covered by Eu^{3+}-doped TZO glasses, with and without silver NPs used as cover slips [26]. It was used the same procedure presented in the last section for the electrical characterization of commercial silicon and gallium phosphide cells. Current-voltage (I-V) characteristics of the modified solar cells were measured for 1000 W/m^2 irradiance using commercial silicon (BPW34 Vishay Semiconductors) and GaP (FGAP71 Thorlabs) semiconductor photocells, with energy gaps of 1.1 eV and 2.26 eV, respectively. For the electrical characterization, the glasses were placed on top of the solar cells without any glue or special arrangement, following the same procedure used in the last section. TZO samples were prepared with the addition of Eu_2O_3 (1.0 wt. %) and $AgNO_3$ (2.0 wt. %). Undoped sample (without Eu_2O_3 and $AgNO_3$) was also prepared to be used as reference. Using the I-V results presented in

Fig. 10.14 Electrical characterization of silicon solar cell covered with undoped TeO_2-ZnO glass, Eu_2O_3-doped TeO_2-ZnO glass with and without Ag nanoparticles [26]

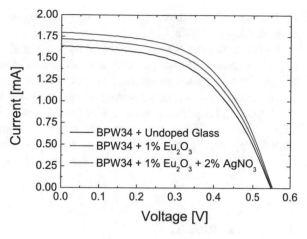

Fig. 10.15 Electrical characterization of GaP solar cell covered with undoped TeO_2-ZnO glass, Eu_2O_3-doped TeO_2-ZnO glass with and without Ag nanoparticle [26]

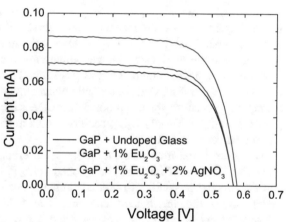

Figs. 10.14 and 10.15, the short-current circuit (I_{sc}), the open-circuit voltage (V_{oc}), and the current (I_{mp}) and voltage (V_{mp}) at the maximum incident optical power were obtained and allowed the calculation of the efficiency (η), the filling factor (FF), and the efficiency enhancement (EE).

For silicon and GaP solar cells, efficiency enhancement of 7.0% and 5.4% was observed when 1 wt% of Eu_2O_3 was added to the undoped sample. Efficiency measurements of silicon and GaP solar cells covered with glasses doped with 1 wt % of Eu_2O_3 and 2 wt% of $AgNO_3$ revealed an enhancement of 14.0% and 34.5% in comparison with the undoped sample, respectively.

So this review demonstrated another possible use of TZO glasses for solar cell applications. In conclusion, it was possible to corroborate that the efficiency enhancement of silicon and GaP solar cells covered with TZO glasses can be obtained adding Eu^{3+} ions into the host. Also it was demonstrated that further improvement is possible with the nucleation of Ag NPs that enabled the increase of the local field around the Eu^{3+} ion and the energy transfer from the Ag nanoparticles to the Eu^{3+} ions.

Similar experiments were performed by Song et al. [75] using Tb^{3+}- and Eu^{3+}-doped phosphate glasses. However in this case the glasses were not prepared with metallic NPs; also to ensure good optical contact, a matching oil with refractive index around 1.5 was inserted between the glass and the solar cell. So when the Eu^{3+}/Tb^{3+}-doped phosphate glasses were placed on the top of amorphous and multicrystalline silicon solar cells, the efficiency enhanced by 3.6%/3.3% and 4.6%/6.3%, respectively. In our case as the experiments were performed without the matching oil, we intend to use it in the future in order to optimize our results.

10.4   Conclusions

In this chapter we reviewed recent studies related to different applications of tellurite glasses prepared using the melt quenching technique. The large potential of rare-earth-doped tellurite glasses with and without metallic NPs was demonstrated.

Different tellurite compositions (TeO_2-ZnO, TeO_2-ZnO-Na_2O-PbO and TeO_2-PbO-GeO_2) were discussed for several applications using for all the cases the melt quenching procedure for the production of the samples. For the case of samples prepared with $AgNO_3$ or Au_2O_3, additional heat treatment was performed to enable the nucleation of metallic NPs.

The control and improvement of the photoluminescence efficiency due to the nucleation of gold NPs in Yb^{3+}/Er^{3+}-doped TeO_2-PbO-GeO_2 glasses was reviewed and showed the possibility to produce luminescent materials of high performance making use of the simultaneous enhanced local field contribution of gold NPs and the energy transfer processes between two different rare-earth ions.

This chapter also reviewed the nucleation of silver NPs in Tb^{3+}-doped TeO_2-ZnO-Na_2O-PbO glass that contributed for the large enhancement in the blue-red spectrum due to the presence of silver NPs and aggregates with dimensions varying from 2 to 150 nm. The whole spectrum was intensified by the appropriate heat treatment of the samples; also the growth of the surface plasmon resonance absorption band intensity could be observed indicating enhancement of the silver NPs concentration with the heat treatment.

We also reviewed results of the modification introduced by different Nd_2O_3 concentration on the laser operation of TeO_2-ZnO glasses. It was shown that it is possible to determine the adequate concentration range for laser action. Samples doped with 0.5 and 1.0 wt % of Nd_2O_3 demonstrated laser action. Threshold pump powers of 53 mW and slope efficiency of 6.6% were obtained, for TeO_2-ZnO samples prepared with 0.5 wt% of N_2O_3; for higher concentration of Nd_2O_3 (1.0 wt%), threshold pump power of 8 mW and slope efficiencies of 21% were obtained indicating then the best condition for laser action in Nd^{3+}-doped TeO_2-ZnO glass.

This chapter was also devoted to the review of the use of TeO_2-ZnO glasses for solar cell applications. It was shown that the management of Tb^{3+} and Yb^{3+} ions concentration can be optimized to modify the solar spectrum and increase the commercial silicon and GaP solar cells performance. Two different results were obtained and deserve attention. Efficiency enhancement of about 7% was observed when TeO_2-ZnO glass singly doped with 1 wt% of Tb^{3+} ion was used as cover slip for the silicon solar cell; on the other hand, for the case of the GaP solar cell covered with the undoped TeO_2-ZnO glass, efficiency enhancement of ~1.1% was observed when covered with TeO_2-ZnO glass co-doped with 1 wt% of Tb^{3+} and 5 wt% of Yb^{3+}.

It was possible to corroborate that the efficiency enhancement of commercial silicon and GaP solar cells covered with TeO_2-ZnO glasses can be obtained by adding Eu^{3+} ions or Eu^{3+} ions with silver NPs into the TeO_2-ZnO host. In the last case, it was the increase of the local field around the Eu^{3+} ions and the energy transfer from the silver NPs to the Eu^{3+} ions that enabled the large solar cell performance improvement.

In conclusion the works reviewed in this chapter demonstrated the possibility to use tellurite glasses in different applications. In particular it was shown the large potential for solar cell, laser action, and luminescent displays applications that certainly motivate new studies in the near future.

Acknowledgments We thank the financial support from the Conselho Nacional de Desenvolvimento Científico e Tecnológico (CNPq) through the National Institute of Photonics (INCT de Fotonica). The Nanotechnology National Laboratory (LNNano) from Centro Nacional de Pesquisa em Energia e Materiais (CNPEM) is acknowledged for the TEM images.

References

1. W.H. Dumbaugh, Phys. Chem. Glasses **19**, 121 (1978)
2. W.H. Dumbaugh, Phys. Chem. Glasses **27**, 119 (1986)
3. L.R.P. Kassab, C.B. de Araujo, Chapter 2. Linear and nonlinear optical properties of some tellurium oxide glasses, in *Technological Advances in Tellurite Glasses: Properties, Processing and Applications*, ed. by V. A. G. Rivera, D. Manzani, (Springer, Cham, 2017), pp. 15–39. ISBN: 978-3-319-53038
4. L.R.P. Kassab, D.M. da Silva, V.D. del Cacho, L. Bontempo, S.G. dos Santos Filho, M.I.A. Chavez, Chapter 1. Tellurite thin films produced by RF sputtering for optical waveguides and memory device applications, in *Technological Advances in Tellurite Glasses: Properties, Processing and Applications*, ed. by V. A. G. Rivera, D. Manzani, (Springer, Cham, 2017), pp. 241–257. ISBN: 978-3-319-53038
5. R. El-Mallawany, M. Sidkey, A. Khafagy, H. Afifi, Mater. Chem. Phys. **37**(2), 197 (1994)
6. M.M. El-Zaidia, A.A. Ammar, R. El-Mallawany, Phys. Status Solidi **91**(2), 637 (1985)
7. Z.C.K. Bouchaour, M. M Poulain, I. Bel-Hadji, R. Hager, El. Mallawany. J. Non-Cryst. Solids **351**(10), 818 (2005)
8. A. El-Adawy, R. El-Mallawany, J. Materials Sci. Lett. **15**(23), 2065 (1996)
9. R. El-Mallawany, A. Abd El-Moneim, Phys. Status Solidi **166**(2), 829 (1998)
10. N.S. Hussain, G. Hungerford, R. El-Mallawany, M.J.M. Gomes, M.A. Lopes, A. Nasar, J.D. Santos, S. Buddhudu, J. Nanosci. Nanotechnol. **9**(6), 3672 (2009)

11. R. El-Mallawany, Phys. Status Solidi **177**(2), 439 (2000)
12. Z. Hager, R. El-Mallawany, J. Mater. Sci. **45**(4), 897 (2010)
13. A. Mirgorodsky, M. Colas, M. Smirnov, T. Merle-Méjean, R. El-Mallawany, P. Thomas, J. Solid State Chem. **190**, 45 (2012)
14. M.M. Elkholy, R.A. El-Mallawany, Mater. Chem. Phys. **40**(3), 163 (1995)
15. C.B. de Araújo, L.R.P. Kassab, R.A. Kobayashi, L.P. Naranjo, P.A.S. Cruz, J. Appl. Phys. **99**, 123522 (2006)
16. L.P. Naranjo, C.B.de Araújo, O.L. Malta, P.A.S. Cruz, L.R.P. Kassab, Appl. Phys. Lett. **87**, 24194 (2005)
17. L.R.P. Kassab, D.S. da Silva, R. de Almeida, C.B. de Araújo, Appl. Phys. Lett. **94**, 101912 (2009)
18. L.R.P. Kassab, L.F. Freitas, T.A.A. Assumpção, D.M. da Silva, C.B. de Araújo, Appl. Phys. B Lasers Opt. **104**, 1029 (2011)
19. L.R.P. Kassab, C.B. de Araújo, R.A. Kobayashi, R.A. Pinto, D.M. da Silva, J. Appl. Phys. **102**, 103515 (2007)
20. R. de Almeida, D.M. da Silva, L.R.P. Kassab, C.B. de Araújo, Opt. Commun. **281**, 108 (2008)
21. V.P.P. de Campos, L.R.P. Kassab, T.A.A. de Assumpção, D.S. da Silva, C.B. de Araújo, J. Appl. Phys. **112**, 063519 (2012)
22. D.S. da Silva, T.A.A. de Assumpção, G.B.C. de Simone, L.R.P. Kassab, C.B. de Araújo, Appl. Phys. B Lasers Opt. **121**, 117 (2015)
23. M.J.V. Bell, V. Anjos, L.M. Moreira, F. Falci, L.R.P. Kassab, D.S. da Silva, J.L. Doualan, P. Camy, R. Moncorgé, J. Opt. Soc. Am. B. **31**, 1590 (2014)
24. L.M. Moreira, V. Anjos, M.J.V. Bell, C.A.R. Ramos, L.R.P. Kassab, D.J.L. Doualan, P. Camy, R. Moncorgé, Opt. Mater. **58**, 84 (2016)
25. L.A. Florencio, L.A.M. Gomez, B.C. Lima, A.S.L. Gomes, L.R.P. Kassab, Sol. Energy Mater. Sol. Cells **157**, 468 (2016)
26. B.C. Lima, L.A.M. Gomez, A.S.L. Gomes, J.A.M. Garcia, L.R.P. Kassab, J. Electron. Mater. J. Electron. Mater. **46**(2), 6750 (2017)
27. L.R.P. Kassab, M.E. Camilo, C.T. Amâncio, D.M. da Silva, J.R. Martinelli, Opt. Mater. **33**, 1948 (2011)
28. L.R.P. Kassab, R. de Almeida, D.M. da Silva, C.B. de Araújo, J. Appl. Phys. **104**, 093531 (2008)
29. P.N. Prasad, *NanoPhotonics* (Wiley, Hoboken, 2004)
30. M.E. Camilo, T.A.A. Assumpcao, D.M. da Silva, D.S. da Silva, L.R.P. Kassab, C.B. de Araujo, J. Appl. Phys. **113**, 153507 (2013)
31. C.B. de Araújo, L.R.P. Kassab, Chapter 5. Enhanced photoluminescence and planar wave-guides of rare-earth doped germanium oxide glasses with metallic nanoparticles, in *Glass Nanocomposites: Preparation, Properties, and Applications*, ed. by B. Karmakar, K. Rademann, A. L. Stepanov, (Elsevier, Oxford, 2016), pp. 132–144. ISBN: 978-0323-39309-6
32. D.M. Silva, L.R.P. Kassab, S.R. Luthi, C.B. Araujo, A.S.L. Gomes, M.J.V. Bell, Appl. Phys. Lett. **90**, 081913 (2007)
33. L.R.P. Kassab, D.M. Silva, J.A.M. Garcia, D.S. da Silva, C.B. de Aráujo, Opt. Mater. **60**, 25 (2016)
34. A.P. Silva, A.P. Carmo, V. Anjos, M.J.V. Bell, L.R.P. Kassab, R.A. Pinto, Opt. Mater. **34**, 239 (2011)
35. T.A.A. de Assumpção, L.R.P. Kassab, A.S.L. Gomes, C.B. de Araújo, N.U. Wetter, Appl. Phys. B. Lasers Opt. **103**, 165 (2011)
36. L. Bontempo, S.G.S. Filho, L.R.P. Kassab, Thin Solid Films **4**, 21 (2016)
37. L.R.P. Kassab, F.A. Bomfim, J.R. Martinelli, N.U. Wetter, J.J. Neto, C.B. de Araújo, Appl. Phys. B Lasers Opt. **94**, 239 (2009)
38. J. Jakutis, L. Gomes, C.T. Amancio, L.R.P. Kassab, J.R. Martinelli, N.U. Wetter, Opt. Mater. **33**, 107 (2010)

39. L.R.P. Kassab, R. de Almeida, D.M. da Silva, T.A.A. de Assumpção, C.B. de Araújo, J. Appl. Phys. **105**, 103505 (2009)
40. R.R. Petrin, M.L. Kliewer, J.T. Beasley, R.C. Powell, I.D. Aggarwal, R.C. Ginther, IEEE J. Quantum Electron. **27**, 1031 (1991)
41. J. Azkargorta, I. Iparraguirre, R. Balda, J. Fernández, E. Dénoue, J.L. Adam, IEEE J. Quantum Electron. **30**, 1862–1867 (1994)
42. J. Azkargorta, I. Iparraguirre, R. Balda, J. Fernández, Opt. Express. **16**(16), 11894 (2008)
43. T. Schweizer, D.W. Hewak, D.N. Payne, T. Jensen, G. Huber, Electron. Lett. **32**, 666 (1996)
44. D.F. de Sousa, L.A.O. Nunes, J.H. Rohling, M.L. Baesso, Appl. Phys. B Lasers Opt. **77**, 59 (2003)
45. J. Fernandez, I. Iparraguirre, R. Balda, J. Azkargorta, M. Voda, J.M. Fernandez-Navarro, Opt. Mater. **25**(2), 185 (2004)
46. J.C. Michel, D. Morin, F. Auzel, Rev. Phys. Appl. **13**, 859 (1978)
47. A. Miguel, J. Azkargorta, R. Morea, I. Iparraguirre, J. Gonzalo, J. Fermamdez, R. Balda, Opt. Express. **21**, 009298 (2013)
48. H. Kalaycioglu, H. Cankaya, G. Ozen, L. Ovecoglu, A. Sennaroglu, Opt. Commun. **281**, 6056 (2008)
49. I. Iparraguirre, J. Azkargorta, J.M. Fernández-Navarro, M. Al-Saleh, J. Fernández, R. Balda, J. Non-Cryst. Solids **353**(8–10), 990 (2007)
50. N. Lei, B. Xu, Z.H. Jiang, Opt. Commun. **127**(4–6), 263 (1996)
51. J.S. Wang, E.M. Vogel, E. Snitzer, Opt. Mater. **3**(3), 187 (1994)
52. A. Jha, S. Shen, M. Naftaly, Phys. Rev. B **62**(10), 6215 (2000)
53. H.A.A. Sidek, S. Rosmawati, Z.A. Talib, M.K. Halimah, W.M. Daud, Am. J. Appl. Sci. **6**(8), 1489 (2009)
54. M.J. Weber, J. Non-Cryst. Solids **123**(1–3), 208 (1990)
55. J.S. Wang, D.P. Machewi, F. Wu, F. Snitzer, E.M. Vogel, Opt. Lett. **19**, 1448 (1994)
56. I.A.A. Terra, L.J. Borrero-González, J.M. Carvalho, M.C. Terrile, M.C.F.C. Felinto, H.F. Brito, et al., J. Appl. Phys. **113**, 073105 (2013)
57. I.A.A. Terra, L.J. Borrero-González, T.R. Figueredo, J.M.P. Almeida, A.C. Hernandes, L.A.O. Nunes, et al., J. Lumin. **132**, 1678 (2012)
58. Q. Duan, F. Qin, Z. Zhang, W. Cao, Opt. Lett. **37**, 521 (2012)
59. P. Vergeer, T.J.H. Vlugt, M.H.F. Kox, M.I. den Hertog, J.P.J.M. van der Eerden, A. Meijerink, Phys. Rev. B **71**, 014119 (2005)
60. Y. Wang, L. Xie, H. Zhang, J. Appl. Phys. **105**, 023528 (2009)
61. Q.Y. Zhang, C.H. Yang, Z.H. Jiang, X.H. Ji, Appl. Phys. Lett. **90**, 061914 (2007)
62. L. Zhao, D. Wang, Y. Wang, Y. Tao, J. Am. Ceram. Soc. **97**, 3913 (2014)
63. X. Zhou, Y. Wang, G. Wang, L. Li, K. Zhou, Q. Li, J. Alloys Compd. **579**, 27 (2013)
64. L.R.P. Kassab, D.S. da Silva, C.B. de Araújo, J. Appl. Phys. **107**, 113506 (2010)
65. G. Li, C. Zhang, P. Song, P. Zhu, K. Zhu, J. He, J. Alloys Compd. **662**, 89 (2016)
66. Z.Q. Li, X.D. Li, Q.Q. Liu, X.H. Chen, Z. Sun, C. Liu, X.J. Ye, S.M. Huang, Nanotechnology **23**, 025402 (2012)
67. A.C. Atre, A. García-Etxarri, H. Alaeian, J.A. Dionne, J. Opt. **14**, 024008 (2012)
68. S. Derom, A. Berthelot, A. Pillonnet, O. Benamara, A.M. Jurdyc, C. Girard, G.C. des Francs, Nanotechnology **24**, 495704 (2013)
69. G. Chen, J. Seo, C. Yang, P.N. Prasad, Chem. Soc. Rev. **42**, 8304 (2013)
70. B.S. Richards, Sol. Energy Mater. Sol. Cells **90**, 2329 (2006)
71. J. Zhou, Y. Teng, S. Ye, G. Lin, J. Qiu, Opt. Mater. **34**, 901 (2012)
72. J. Merigeon, O. Maalej, B. Boulard, A. Stanculescu, L. Leontie, D. Mardare, M. Girtan, Opt. Mater. **48**, 243 (2015)
73. B. Han, Y. Yang, J. Wu, J. Wei, Z. Li, Y. Mai, Ceram. Int. **41**, 12267 (2015)
74. K.R. Catchpole, A. Polman, Opt. Express. **16**, 21793 (2008)
75. P. Song, C. Zhang, P. Zhu, IEEE J. Quantum. Electron. **51**, 4800105 (2015)

Chapter 11
Lanthanide-Doped Tellurite Glasses for Solar Energy Harvesting

Venkata Krishnaiah K, Venkatramu V, and Jayasankar C. K.

Abstract To meet an ever-increasing energy demand, it is of prime importance to utilize the solar energy effectively for improving the efficiency of silicon (Si)-based photovoltaic (PV) cells. So far, Si-based solar cells are ruling the global market due to their large availability and flexible price. On the other hand, lanthanide-doped materials exhibit the high photoluminescence quantum yield (PLQY) in the visible and near-infrared regions. Recently, these materials can be integrated with the Si solar cells to create an additional electron-hole pairs through optical conversion (upconversion and downconversion) processes. However, it is crucial to identify the suitable materials which convert light energy into electrical energy. Lanthanide-doped tellurite glasses have the advantages over other low phonon energy materials (i.e., fluoride glasses) that exhibit the properties including low phonon energy, wide transmission (ranging from visible to infrared region), and high refractive index. Low phonon energy of tellurite glasses favors in the enhancement of photoluminescence quantum yield for optical conversion. Moreover, high PLQY glasses could be employed on the top and bottom of PV cells to improve the photocurrent further. In addition, TiO_2-modified lanthanide-doped tellurite glasses may also enrich the photocatalytic activity in the visible region of electromagnetic spectrum.

Venkata Krishnaiah K (✉)
Laser Applications Research Group, Ton Duc Thang University, Ho Chi Minh City, Vietnam

Faculty of Applied Sciences, Ton Duc Thang University, Ho Chi Minh City, Vietnam

Department of Physics, RGM College of Engineering and Technology, Nandyal, India
e-mail: kvkrishaniah@tdt.edu.vn

Venkatramu V
Department of Physics, Yogi Vemana University, Kadapa, India

Jayasankar C. K.
Department of Physics, Sri Venkateswara University, Tirupati, India

© Springer International Publishing AG, part of Springer Nature 2018
R. El-Mallawany (ed.), *Tellurite Glass Smart Materials*,
https://doi.org/10.1007/978-3-319-76568-6_11

249

11.1   Introduction

Effective harvesting of light is of essential prominence for the development of highly proficient solar energy harvesting approaches such as photovoltaic (PV) and photo-electrochemical (PEC) water-splitting devices. During the past couple of decades, extensive studies on diverse materials and innovations in the design of a device have been accomplished for enhancing the conversion efficiencies of such devices. The Ln^{3+}-doped materials exhibit broad absorption bands that are more significant for solar energy harvesting applications. These materials having the property of spectral conversion become unique for developing the advanced photonic devices not limited to solid-state lasers [1–4], optical amplifiers [5], temperature sensors [6], laser cooling of solids [7, 8], radiation-balanced lasers [9], light converters [10], image production devices [11], light-emitting diodes [12], integrated circuits [13], and high-density data storage [14].

Among diverse oxide glass hosts that are available, tellurite glasses are attracted a significant interest because of their peculiar properties, viz., relatively low phonon energy and high refractive index compared to other oxide glasses such as borate, phosphate, and silicate glasses [15]. Low phonon energy glasses exhibit a high photoluminescence quantum yield (PLQY), low background losses, and low non-radiative relaxations. For example, fluoride glasses exhibit high PLQY for various metastable states of Ln^{3+} ions due to its low phonon energies ($\sim$500 cm^{-1}, 0.06 eV) [16]. However, they are not feasible for practical appliances due to their poor chemical and thermal stability and subject to corrosion. On the other hand, tellurite glasses have a lot of advantages including wide transmission region (0.35–6 μm, 3.54–0.2 eV), high density, high Ln^{3+} ion solubility, broad bandwidth ($\sim$80 nm, 15.49 eV), and high stimulated emission cross-section ($\sigma_e \geq 0.75$ pm^2) at 1.5 μm (0.82 eV) compared to silica-based erbium-doped fiber amplifiers (EDFAs), high mechanical and chemical stability, relatively low melting temperatures, slow corrosion rate, high linear and nonlinear refractive indices, high dielectric constant, and relatively low phonon energy around $\sim$700 cm^{-1} (0.08 eV). These features make them promising for wide range of photonic applications including fiber amplifiers [16, 17], waveguide devices [18], Raman amplifiers [19], high-power mid-infrared lasers [20], optical temperature sensors [21, 22], and supercontinuum sources [23]. However, there is still unexploited prospective of these glasses for energy harvesting and sustainable environment applications.

Transparent materials with high PLQY (e.g., Ln^{3+}-doped glasses) are highly essential to integrate with silicon photovoltaic (Si-PV) cells for enhancing their conversion efficiency by downconversion (DC) and upconversion (UC) processes. However, UC emission in borates, phosphates, and silicates are not very efficient due to their high energy of phonons, which always favors non-radiative relaxations from higher excited states of Ln^{3+} ions. This leads to restrict the favorable mechanisms such as excited state absorption (ESA) and energy transfer upconversion (ETU) for UC process. In this contest, low phonon energy ($\sim$700 cm^{-1}, 0.08 eV) tellurite glasses could be more suitable than the fluoride-based glasses ($\sim$500 cm^{-1},

0.06 eV) regarding the chemical stability. Hence, significant research has been described on Ln^{3+}-doped and Ln^{3+}/Yb^{3+} co-doped tellurite glasses for solar energy harvesting [24–26].

So far, researchers have developed different types of solar cells for harvesting the solar energy effectively. Moreover, continuous search on novel materials for developing Si solar cells [27], tandem or stacked solar cells [28], perovskite solar cells [29, 30], quantum dot solar cells (QDSCs) [31], dye-sensitized solar cells (DSSCs) [32], organic semiconducting polymers [33], and carbon nanotubes [34] is being investigated with an aim to enhance their conversion efficiency. However, carbon nanotubes and polymer-based solar cells are in the research stage and exhibit interesting perceptions for the next-generation technology. Among these cells, Si solar cells have occupied a major portion of the globe due to their large availability with low cost and relatively higher efficiency. About 90% of the solar cells are manufactured based on crystalline silicon (c-Si) due to its high efficiency compared to organic or amorphous materials [35]. However, they have restricted due to their limited efficiency [36]. So far, there is no other type of solar cells that can meet the efficiency of the Si solar cells. But the efforts are under way to replace them in near future with the emerging low-cost perovskite solar cells [37] as they enhanced the conversion efficiency significantly during the past couple of years.

A major limiting issue in the conversion efficiency of PV cells is their insensitivity to most of the solar spectrum. The energy dissemination of sunlight at the standard air mass (AM1.5) covers the range of 200–2500 nm (6.19–0.49 eV) as shown in Fig. 11.1. However, Si solar cell could absorb only the solar radiation with energies much larger than the bandgap of 1.12 eV which matches to the wavelengths shorter than 1100 nm. These solar cells can make use of the photons in the visible to NIR region up to 1100 nm (1.12 eV), because these photons with high energy surpass the bandgap energy of Si and generate the electron-hole pair. This leads to the production of electrical energy. However, the photons in the mid-infrared (MIR) region do not contribute as these photons do not exceed the Si bandgap. A significant enhancement in the efficiency of solar cell could be attained if part of this energy utilized by the Si solar cells, about 20% of the solar irradiation, has covered the region above 1100 nm (1.12 eV) [38]. The transmission losses of the radiation are minimized through converting the NIR (low energy) photons from the solar spectrum to UV/visible photons (high energy) upon the deployment of UC materials on the rear side of the solar cell [39, 40].

Therefore, there are two mechanisms of losses in Si PV cells to restrict its efficiency. Those are transmission and thermalization losses. These losses can be minimized by adopting the upconverting and downconverting (quantum cutting) materials [41], respectively. There are three approaches that are identified to boost the Si solar cells efficiency by utilizing the solar radiation effectively. Those are downconversion (conversion of one photon at high energy into two or more photons at low energy), downshifting (shifting of photons into the wavelength regions that can be exploited by the solar cell), and upconversion (conversion of two or more photons at low-energy into one photon at higher energy).

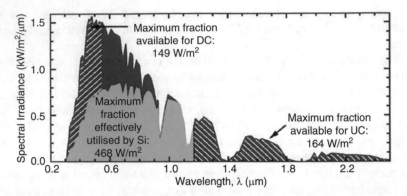

Fig. 11.1 Blue curve shows AM1.5 terrestrial solar spectrum. Part of the spectrum highlighted in the green is effectively absorbed by Si solar cells to convert into photocurrent. The blue shaded part above the green can be absorbed by Si solar cells but cannot be converted to photocurrent due to thermal losses of the excited carriers. The part shown in the blue comprises about 15% of terrestrial solar energy and can be utilized by DC and UC processes. (Reproduced from Ref. [40], copyright 2016, Elsevier)

In order to minimize the transmission losses, different designs have been proposed and developed including tandem solar cells [42], multiple exciton generation [43], hot carrier solar cells [44], and upconversion [45]. The thermalization losses become predominant because of the large energy difference between the bandgap of the semiconductor and the absorbed photon. Thus, if it can renovate the high-energy photons to photons of low energy with a quantum efficiency (QE) greater than 100%, then more number of photons can be absorbed by the Si PV device. This will increase the photocurrent and subsequently improve the energy conversion efficiency of the device [39]. This strategy is implemented in PV with luminescent downshifting (similar to DC but with maximum PLQY of unity). Significant enhancement of QE on thin-film PV modules is obtained [46]. Based on such results, DC could be even more efficient with PLQY of 200%. Such huge efficiency initiates from their broad absorption. The most significant property of silicon (Si) for photovoltaics (PVs) is its bandgap energy, 1.12 eV. This value is close to the ideal energy gap, and it matches with the solar spectrum (1.20 eV). Third-generation PVs are able to overcome the Shockley-Queisser theoretical efficiency (30%) limit for a single bandgap Si solar cell.

Solar cell efficiency can be improved by an effective utilization of solar spectrum by exploring the DC and UC Ln^{3+}-doped materials. The transmission loss mechanism can be resolved by using photon UC process in which two or more photons at low energy harvest one highly energetic photon. The Ln^{3+} ions are ideal ones for such light conversion due to their ladder-like metastable energy levels. Thermalization loss mechanism can be resolved by DC or QC process in which one high-energy photon (in the range of 300–500 nm, 4.13–2.47 eV) splits into two low-energy photons (close to bandgap energy of c-Si solar cell). In addition, the use of such materials in PV cell may fill the unexplored gaps for enhancing the solar cell efficiency by aiming with the enhancement in PLQY and conversion efficiency. These low phonon energy Ln^{3+}-doped tellurite glasses are found to be a suitable for solar energy harvesting compared to traditional glasses. On the other hand, these Ln^{3+}-doped tellurite glasses may also be useful for photocatalysis applications for the production of H_2 as they have the ability to transform the unutilized visible and NIR solar radiation into bandgap energy region close to TiO_2 ($E_g = 3.2$ eV).

11.2 Downconversion/Quantum Cutting

The thermalization of charge carriers is one of the key loss mechanisms, which are produced due to the absorption of photons at high energy by the solar cell. One of the ways is multiplication of charge carriers (e^-–h^+ pairs) to reduce these losses in a solar cell, i.e., the generation of significant number of e^-–h^+ pairs per each incident photon energy greater than twice the bandgap energy of the solar cell. To enhance more number of these pairs, a different mechanism, i.e., DC/QC is shown in Fig. 11.2, has been proposed by Trupke et al. in 2002 [47].

The DC or QC or photon cascade emission (PCE) is a process in which one highly energetic photon (UV or visible) divides into two or more photons at low energy (NIR or IR). This process can be enlightened by energy level diagrams of two Ln^{3+} ions (I and II), as shown in Fig. 11.3 [48]. Two-photon QC emission from the excited level of a sole Ln^{3+} ion is shown in Fig. 11.3a. However, contrastingly photon emission in the UV and IR (the solid lines of Fig. 11.3a) regions can also occur. This leads to prevention of visible QC emission from sole Ln^{3+} ion. Figure 11.3b–d presents the energy level structures for three ET DC processes in two dissimilar Ln^{3+} ions (I and II). The emission takes place from excited level in Type I ion and transfers to an activator ion, Type II. Figure 11.3b indicates two-photon emission in pair of Ln^{3+} ions from I to II by CR (denoted by "1" in circle) and ET from dissimilar ions I to II (denoted by "2" in circle) that leads to get emission from ion II. Figure 11.3c, d shows a CR mechanism followed by the emission of photons simultaneously from dissimilar ions I and II. If the two-step ET process is efficient, it is potential to achieve the QE close to 200%. The thermalization (IR region) and transmission (UV region) losses that can always exist in a single Ln^{3+} ion will be resolved with co-doping of Ln ions.

This QC process was theoretically proposed for the first time by Dexter [49] in 1957 once the QE is more than unity. It was experimentally demonstrated in 1974 by Piper et al. [50] in Pr^{3+}-doped YF_3 and LaF_3 fluoride phosphors under vacuum ultraviolet (VUV) excitation of Pr^{3+} ions at 185 nm (6.70 eV). In 1999, Wegh et al. [51] demonstrated this process with QE of 200% (two photons emitted per every photon that is absorbed) by using both Gd^{3+} and Eu^{3+} ions in $Eu:LiGdF_4$ phosphors under VUV excitation of Gd^{3+} ions at 254 nm (4.88 eV). Since their invention, it has been examined in diverse materials for the development of photonics devices [52]. It was assumed that the conversion efficiency could reach as high as 38.6% by integrating Si solar cell and a DC layer [47] under the illumination of feeble sunlight.

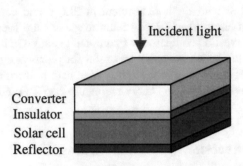

Fig. 11.2 Schematic design of the DC system. The DC is placed on the top of a solar cell having the bandgap energy, E_g. Photons with high-energy $hv > 2E_g$ are captured by the DC and efficiently transformed into two photons of low energy with $hv > E_g$, which can both be absorbed effectively by the solar cell. In an alternative design, the DC is placed on the rear surface of the solar cell. In both cases the solar cell and the DC are isolated from each other, and a perfect reflector is located on the back surface of the system. (Reproduced from Ref. [47], copyright 2002, American Institute of Physics)

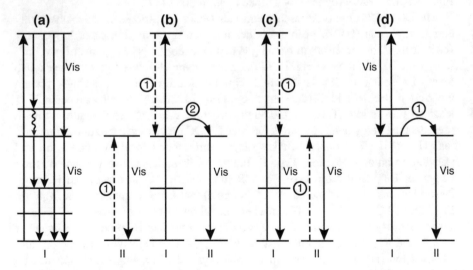

Fig. 11.3 Energy level diagrams of DC for two types of Ln^{3+} ions (I and II). Type I is an ion for which emission occurs from a high-energy level. Type II is an ion to which ET takes place. (**a**) QC from a single ion I by the sequential emission of two visible photons. (**b**) QC by a two-step ET. In the first step (indicated by "1" in circle), a part of the excitation energy is transferred from ion I to ion II by CR. Ion II returns to the ground state by emitting a photon in the visible region. Ion I is still in an excited state and can transfer the remaining energy to a second ion of type II (indicated by "2" in circle), which also emits a photon in the same region, giving a QE of 200%. (**c, d**). The remaining two possibilities involve only one ET step from ion I to ion II. This is sufficient to obtain visible QC if one of two visible photons can be emitted by ion I. (Reproduced from Ref. [45] copyright, 1999, Elsevier)

This value is very significant compared to 30.9% (limiting efficiency) for the same solar cell in the absence of DC layer. Moreover, NIR QC in Ln^{3+}/Yb^{3+} (Ln^{3+} = Tb, Tm, and Pr) co-doped oxide phosphors [53–56] have to be reported under the excitation of Ln^{3+} at around blue photon. The emission of two NIR photons at 980 nm of Yb^{3+} ions by ET from Ln^{3+} to Yb^{3+}, with an ideal QE close to 200%. Hence, NIR QC has attracted significantly for its potential applications as the Si solar cells highly sensitive to the radiation in wavelength range of 900–1100 nm (1.37–1.12 eV) for releasing of charge carriers in c-Si solar cells.

Initially, research on QC has limited to single Ln^{3+} ions which are capable of cascade emission. The Ln^{3+} ion should fulfill two requirements for efficient DC emission from single Ln^{3+} ion [57]. These are (i) the bandgap energy between the ground and excited levels that should be as large as possible to avoid multiphonon relaxation; and (ii) the branching ratio of visible photon emission essentially as high as possible. The energy level structure of Pr^{3+} ion favors the QC emission is shown in Fig. 11.4. The VUV photons absorbed by Pr^{3+} ion from its ground state ($^3H_4$) and promote to the 4f–5d levels located are above the $^1S_0$ level of Pr^{3+}, the excitation decays to the $^1S_0$ level situated at around 47,000 cm^{-1} (5.82 eV). The $^1S_0$ is de-populated through two-step $4f^2 \rightarrow 4f^2$ transitions, linking the transitions $^1S_0 \rightarrow$ $^1I_6$, ^3P_J (~400 nm, 3.09 eV) followed by $^3P_0 \rightarrow {}^3F_J$, ^3H_J (480–700 nm, 2.58–1.58 eV). The possibility of QC phenomenon in other Ln^{3+} ions such as Gd^{3+} [57], Tm^{3+} [58], and Er^{3+} [59] have also been examined. It is noticed from the literature that the QE more than 100% is not possible with a single Ln^{3+} ion, due to the differing emissions in the UV or NIR regions simultaneously.

In order to achieve an efficient visible and NIR QC emissions, i.e. QE must be more than unity, a special focus has been made on the pair of Ln^{3+} ions. Since these emissions could be possible through ET which allows the excited ion transfer its energy partially to the two acceptor ions, and each ion emits a visible photon. The QC in pairs of Ln^{3+} ion-, Ce^{3+}/Yb^{3+}-, Pr^{3+}/Yb^{3+}-, Nd^{3+}/Yb^{3+}-, Eu^{3+}/Yb^{3+}-, Tb^{3+}/Yb^{3+}-, Ho^{3+}/Yb^{3+}-, Er^{3+}/Yb^{3+}-, and Tm^{3+}/Yb^{3+}-doped systems can be understood by using their energy level schemes, as shown in Figs. 11.5 and 11.6, and their QE is compared in Table 11.1. To realize QC, the Ln^{3+}-doped materials should possess a low phonon energy to suppress the multiphonon relaxations among the energy levels of Ln^{3+} ions [52]. In addition, the Ln^{3+}-doped materials should exhibit high emission lifetime, high external quantum efficiency (EQE), good stability, and negligible reflection to couple the light into the excited levels of Ln^{3+} ions [60]. Moreover, this QC phenomenon has also investigated in Ln^{3+}-doped crystals and ceramics, but they involve a lot of complexity during the fabrication compared to oxide glasses. For these reasons, the tellurite glass is used as the host that combines the peculiar property of low phonon energy, 700 cm^{-1} (0.08 eV), as it is compared to silicates (1100 cm^{-1}, 0.14 eV) and fluoride glasses (500 cm^{-1}, 0.06 eV) [61, 62].

The NIR QC emission of Yb^{3+} from Ce^{3+}/Yb^{3+} co-doped $8SiO_2$–58CaO–$27Al_2O_3$–$7MgO$–$0.2Ce_2O_3$–xYb_2O_3 (SCAM), where x = 0.2, 0.4 and 0.6 mol % glasses, has been evidenced under the excitation of 405 nm (3.06 eV) of Ce^{3+} ions [63]. The 4f-5d transition is allowed. The lowest 5d level makes the ET efficiently from Ce^{3+} to Yb^{3+}, as shown in Fig. 11.5a. The QE is also high about 154% and also

Fig. 11.4 Energy level
diagram of Pr^{3+} ion
presenting the QC emission
under VUV excitation.
(Reused from Ref. [52],
copyright, 2010, Elsevier)

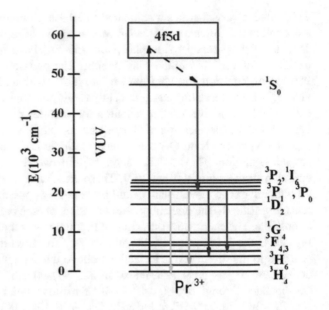

compared in Table 11.1. Furthermore, the emission of Yb^{3+} is equal to the bandgap
of Si solar cells. The Ce^{3+}/Yb^{3+}-doped SCAM glass sample can be used as a DC
layer for solar cells due to broad excitation band, high coefficient of absorption, and
exceptional mechanical, thermal, and chemical stability [60]. Boccolini et al. [64]
reported that an optimal thickness of the sample (0.83 mm) and an optimal doping
concentration of the Yb^{3+} ions (1.0 mol%) contribute to the DC process. These are
essential to minimize the self-absorption of Yb^{3+} ions and enhance the PLQY of the
borate glasses ($70B_2O_3$–$7BaO$–$8CaO$–$14La_2O_3$–$0.5Ce_2O_3$–$1.0Yb_2O_3$). On the other
hand, Wang et al. [65] studied the NIR QC emission in $Ce^{3+}/Er^{3+}/Yb^{3+}$ tri-doped
oxyfluoride GCs exhibiting broader absorption bands compared to Er^{3+}/Yb^{3+}
co-doped GCs. The effective ET processes occur from Ce^{3+} to Er^{3+}, Er^{3+} to Yb^{3+},
and Ce^{3+} to Yb^{3+} simultaneously. The addition of Ce^{3+} with Er^{3+} and Yb^{3+} ions
could enhance the absorption in the UV and visible regions which leads to obtain the
NIR emission in the range of 960–1040 nm (1.29–1.19 eV) with enhanced QE. This
reveals that the tri-doped $Ce^{3+}/Er^{3+}/Yb^{3+}$ materials are promising for improving the
efficiency of Si solar cells through spectral correction.

The NIR QC emission of two NIR photons per every blue photon has been
realized in $0.1Pr^{3+}/xYb^{3+}$-doped SiO_2–Al_2O_3–NaF–YF_3–PrF_3–YbF_3 ($x = 0.1, 0.2,$
0.5, 1.0, and 1.5 mol%) silicate glasses and GCs containing β-YF_3 nanocrystals
[61]. Under 482 nm (2.57 eV) excitation, the NIR emission of Yb^{3+} increases
significantly with increasing Yb^{3+} ion concentration from 0 to 1.0 mol% and
decreases slightly for 1.5 mol% Yb^{3+} ion concentration. The former one is due to
cooperative ET from Pr^{3+} to Yb^{3+} [66] (as shown in Fig. 11.5b), and the latter one is
due to concentration quenching. The QE found to be as high as 194% for a glass with
0.1Pr/1.5Yb. Further studies have to be planned to measure the efficiency of solar
cell on integration of DC material.

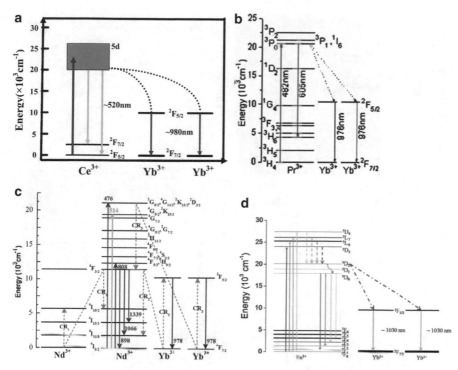

Fig. 11.5 Energy level diagrams: (**a**) Ce^{3+} and Yb^{3+} ions in Ce^{3+}/Yb^{3+} co-doped silicate glass showing the NIR QC emission under 405 nm (3.06 eV) excitation. (Liu et al. [63], copyright, 2014, Elsevier). (**b**) Pr^{3+} and Yb^{3+} ions in Pr^{3+}/Yb^{3+} co-doped oxyfluoride glass-ceramics, exhibiting the NIR QC emission through cooperative ET mechanism under excitation at 482 nm (2.57 eV). (Chen et al. [66] copyright, 2008, Elsevier). (**c**) Nd^{3+} and Yb^{3+} ions in Nd^{3+}/Yb^{3+} co-doped tellurite glasses under 476 nm (2.60 eV) excitation, indicating NIR QC through CR between Nd^{3+} ions (CR1) and Nd^{3+} and Yb^{3+} ions (CR2 and CR3). (Costa et al. [67] copyright, Wiley). (**d**) Eu^{3+} and Yb^{3+} ions in Eu^{3+}/Yb^{3+} co-doped YAG phosphor showing QC DC under excitation at 394 nm (3.15 eV). (Lau et al. [69] copyright, 2012, Elsevier)

Costa et al. [67] investigated DC mechanism and ET efficiency of Nd^{3+}/Yb^{3+} co-doped TeO_2-WO_3 tellurite glasses at different Yb^{3+} ion concentration (0.5, 1.0, 2.0, and 4.0 mol%) and a fixed Nd^{3+} ion concentration (1.0 mol%). It is found that the NIR emission of Nd^{3+} disappears and emission of Yb^{3+} enhances significantly with increasing Yb^{3+} ion concentration as high 4.0 mol%. It is evidenced that the non-radiative relaxations are negligible for high Yb^{3+} concentration which boosts the ET efficiency. The ET mechanism, CR channels, excitation, and de-excitation processes were visualized in Fig. 11.5c. The ET efficiency through CR of high Yb^{3+}-doped sample found to be as high as 96%. Due to these excellent properties, the TeO_2-WO_3 glass system could be considered as a potential candidate to integrate with c-Si solar cells for enhancing its efficiency.

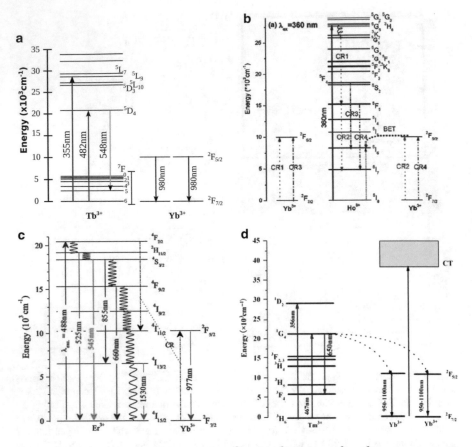

Fig. 11.6 Partial energy level diagrams: (**a**) Tb^{3+} and Yb^{3+} ions in Tb^{3+}/Yb^{3+} co-doped tellurite glasses showing the NIR QC under 355 (3.49 eV) and 482 nm (2.57 eV) excitations. (Luciano et al. [71], copyright, 2016, Elsevier); (**b**) Ho^{3+} and Yb^{3+} ions in Ho^{3+}/Yb^{3+} tellurite glass-ceramics showing the NIR QC mechanism under 360 nm (3.44 eV) excitation, CR (cross-relaxation), BET (back energy transfer). (Zhou et al. [72], copyright, 2014, Wiley); (**c**) Er^{3+} and Yb^{3+} ions in Er^{3+}/Yb^{3+} co-doped tellurite glasses showing the NIR QC under 488 nm (2.54 eV) excitation. (Figueiredo et al. [73], copyright, 2015, Elsevier); (**d**) Tm^{3+} and Yb^{3+} ions in Tm^{3+}/Yb^{3+} co-doped germinate glass showing the NIR QC emission under 356 (3.48 eV) and 467 nm (2.65 eV) excitations. (Zhang et al. [75], copyright, 2010, Wiley)

The NIR QC emission of Yb^{3+} ion has been explored by Smedskjaer et al. in Eu^{3+}/Yb^{3+} co-doped silicate glass and GCs [68], Lau et al. [69] in YAG phosphors (3.0 mol % of Eu^{3+} and 5.0 mol% of Yb^{3+}), and Qiao et al. in LiYb(MoO$_4$)$_2$ phosphors [70] for their possibility to integrate with solar cells. Under 394 nm (3.14 eV) excitation, there is an intense NIR emission at 1030 nm (1.20 eV) of Yb^{3+}:^{2}F$_{5/2}$ → ^{2}F$_{7/2}$ due to cooperative ET (as shown in Fig. 11.5d) from Eu^{3+} to Yb^{3+} in YAG phosphors. The QE found to be as high as 144% [69]. The NIR emission of Yb^{3+} increases initially (0.01 to 8.0 mol% of Yb^{3+}) and starts to decease (8.0 to 10.0 mol% of Yb^{3+}) with increasing Yb^{3+} ion concentration under 266 nm (4.66 eV) excitation of Eu^{3+} ion

Table 11.1 Comparison of quantum efficiency of the co-doped downconverting Ln^{3+}/Yb^{3+}: systems

S. No.	Host matrix	Ln^{3+} ions	Excitation wavelength (nm)	Quantum efficiency (%)	Ref.
1	Borate, $70B_2O_3$–$7BaO$–$8CaO$–$14La_2O_3$–$0.5Ce_2O_3$–$1.0\ Yb_2O_3$	Ce^{3+}/Yb^{3+}	305 (4.06 eV)	150	[64]
	Silicate, $8SiO_2$–$58CaO$–$27Al_2O_3$–$7MgO$–$0.2Ce_2O_3$–$0.6\ Yb_2O_3$		405 (3.06 eV)	154.3	[60]
2	Oxyfluorotellurite, $48.5TeO_2$–$30ZnO$–$10YF_3$–$10NaF$–$0.5Pr_2O_3$–$1.0Yb_2O_3$	Pr^{3+}/Yb^{3+}	440 (2.82 eV)	166	[78]
3	Tellurite, $95(0.8TeO_2 + 0.2WO_3)$–$1.0Nd_2O_3$–$4.0Yb_2O_3$	Nd^{3+}/Yb^{3+}	532 (2.33 eV)	196	[67]
4	Yttrium aluminum garnet ($Y_3Al_5O_{12}$, YAG): 3.0 Mol% Eu_2O_3 and 5.0 Yb_2O_3 Mol%,	Eu^{3+}/Yb^{3+}	394 (3.14 eV)	144	[69]
5	Tellurite, $85.0TeO_2$–$15.0ZnO$–Tb_4O_7 (1.0 wt%) and Yb_2O_3(5.0 wt%)	Tb^{3+}/Yb^{3+}	355 (3.49 eV), 482 (2.57 eV)	113	[71]
6	Tellurite, $75TeO_2$–$20ZnO$–$5Na_2O$–$0.5Ho_2O_3$–$10Yb_2O_3$	Ho^{3+}/Yb^{3+}	360 (3.44 eV), 449 (2.76 eV)	160, 165	[72]
7	Oxyfluorosilicate, $32SiO_2$–$31.5CdF_2$–$18.5PbF_2$–$9AlO_{1.5}$–$5.5ZnF_2$–$3.5(ErF_3,YbF_3)$	Er^{3+}/Yb^{3+}	378 (3.28 eV)	199.4	[79]
	Tellurite, $95.5(0.8TeO_2 + 0.1Li_2O + 1.0TiO_2)$–$0.5Er_2O_3$–$4Yb_2O_3$		488 (2.54 eV)	156	[73]
8	Germanate, $Tm_{0.05}Yb_{0.5}La_{0.94}AlGe_2O_7$	Tm^{3+}/Yb^{3+}	356 (3.48 eV), 467 (2.65 eV)	159.9	[75]
	Transparent ceramics, $(Tm_{0.01}\ Y_{0.94}Yb_{0.05})O_3$		464 (2.67 eV)	136	[80]

due to cooperative ET [69]. In addition to the variation of Ln^{3+} ion concentration, the heat-treated (500, 540, 560, 580, and 600 °C for 8 h under H_2/N_2 atmosphere) Eu^{3+} (1.0 mol%)/Yb^{3+} (1.0 mol%) silicate glasses exhibit high NIR emission of Yb^{3+} under 325 nm (3.81 eV) [68].

Recently, Luciano et al. [71] have explored the DC properties of Tb^{3+}- and Yb^{3+}- doped tellurite glasses ($85.0TeO_2$–$15.0ZnO$–Tb_4O_7 (1 and 2 wt%)–Yb_2O_3 (5 and 7 wt%), (TZTY)). The emission intensity of Yb^{3+} increased for a glass with $5.0Yb^{3+}$/ $1.0Tb^{3+}$ ion concentration and decreased for $7.0Yb^{3+}/1.0Tb^{3+}$ in Tb^{3+}/Yb^{3+}-doped tellurite glasses under 355 nm excitation. This is due to ET between Tb^{3+} and Yb^{3+}

ions and it is pictured in Fig. 11.6a. The emission intensity that decreases with increasing Yb^{3+} ion concentration is due to concentration quenching. As can be seen from Table 11.1, the QE is found to be 113% for $5.0Yb^{3+}/1.0Tb^{3+}$-doped glasses. The single and co-doped Tb^{3+}/Yb^{3+} tellurite glasses have been placed on the top of the commercial Si and gallium phosphide (GaP, $E_g = 2.26$ eV) solar cells. An enhancement of efficiency was observed with a dependence on the Ln^{3+} ion concentration due to the modification of the incident radiation in the IR region. The efficiency of the Si and GaP solar cells found to be 7.46% and 0.69%, respectively. In addition to glasses, Zhang et al. [54] reported the NIR QC phenomenon from one-dimensional (1D) Tb^{3+}/Yb^{3+}-doped $Yb_xGd_{1-x}Al_3(BO_3)_4$ nanorods. The results reveal that the QE of 1D $Yb_xGd_{1-x}Al_3(BO_3)_4$ nanorods found to be as high as 196% in the NIR region under 485 nm (2.55 eV) excitation. In principle, efficient DC material, in an ideal case, converts photons of visible region into two or more photons at NIR region. These Tb^{3+}/Yb^{3+}-doped $Yb_xGd_{1-x}Al_3(BO_3)_4$ (1D) nanorods have a possible potentiality in realizing high-efficiency Si solar cells by DC of the green-to-UV part of the solar spectrum to 1.0 μm (1.23 eV) photons with almost twice the number of photons.

The tellurite glasses with composition, $75TeO_2–20ZnO–5Na_2O–0.5Ho_2O_3–xYb_2O_3$ ($x = 0, 0.7, 1, 3, 5, 10, 20$), have been explored by Zhou et al. [72] for the NIR QC emission at 977 nm (1.27 eV, $^2F_{5/2} \rightarrow {}^2F_{7/2}$) from Yb^{3+} and 981 nm (1.26 eV, $^5F_5 \rightarrow {}^5I_7$) and 1020 nm (1.21 eV, $^5S_2 \rightarrow {}^5I_6$) from Ho^{3+} under both excitations at 449 nm (2.76 eV) and 360 nm (3.44 eV) of Ho^{3+} ions. An enhancement of Ho^{3+} NIR emission at 1193 nm (1.04 eV) due to $Ho^{3+}:^5I_6 \rightarrow {}^5I_8$ transition has been explained by back ET from Yb^{3+} to Ho^{3+}. The ET mechanism between Ho^{3+} and Yb^{3+} ions is visualized in Fig. 11.6b. The QE of NIR emission found to be as high as 166% for a glass with 20 mol% of Yb^{3+} ion concentration. It is observed that these three NIR emissions could be absorbed by Si and improve the efficiency of the silicon-based solar cell.

Figueiredo et al. [73] have observed the DC/QC emission in Er^{3+}/Yb^{3+} co-doped $TeO_2–Li_2O–TiO_2$ glasses at different Yb^{3+} ion concentration and a fixed Er^{3+} ion concentration (0.5 mol%). The NIR QC emission at 980 nm (1.26 eV) of Yb^{3+} increased with increasing Yb^{3+} ion concentration under 488 nm (2.54 eV) excitation due to ET from Er^{3+} to Yb^{3+}. This ET between Er^{3+} and Yb^{3+} ions is visualized in Fig. 11.6c. This ET efficiency found to be as high as 156%. On the other hand, NIR QC emission at 977 nm (1.26 eV) of Yb^{3+} has been observed in $1.0Er^{3+}/1.0Yb^{3+}$-doped $80TeO_2–20WO_3$ by Pandey et al. [74] under 325 nm (3.81 eV) excitation. The emission intensity of co-doped glass is enhanced around 14 times compared to the single ion-doped glass.

The QC emission at 1.0 μm (1.23 eV) of Yb^{3+} has been investigated by Zhang et al. [75] in Tm^{3+}/Yb^{3+}-doped $La_2O_3–Al_2O_3–GeO_2$ germinate glasses under 467 nm (2.65 eV) excitation and found that the QE is as high as 159.9%. At the similar excitation wavelength, the QC emission in Tm^{3+}/Yb^{3+}-doped $SiO_2–Al_2O_3–LiF–GdF_3–TmF_3–YbF_3$ oxyfluorosilicate glasses has been observed by Lakshminarayana et al. [76] with a QE of 187% for $0.5Tm^{3+}/30Yb^{3+}$. The GCs containing nanocrystals exhibit better QE compared to their parent glass [77]. The

ET mechanism from Tm^{3+} to Yb^{3+} ions is envisioned in Fig. 11.6d. However, Tm^{3+}-Yb^{3+}-doped tellurite glasses have to be explored in this direction.

Recently, Chen et al. [81] have been applied the different strategies by selecting the two activator ions, Er^{3+} and Tm^{3+}, due to large radiative rates, high QEs, and excellent prospects for different applications. They observed the QC phenomenon in Er^{3+} and Tm^{3+} co-doped $70TeO_2$–$25ZnO$–$5La_2O_3$–$8Er_2O_3$–$0.5Tm_2O_3$ (TZLET) tellurite glasses and tested their feasibility for Ge solar cells. It is interesting to note that the luminescence intensity at 1800 nm (0.68 eV) enhanced (around 20 times) significantly for the $Er^{3+}(8\%)$/$Tm^{3+}(0.5\%)$ co-doped TZLET tellurite glasses compared to those of the single $Tm^{3+}(0.5\%)$ glass and 5.0 times more than that of single $Er^{3+}(0.5\%)$-doped tellurite glasses. Significant multiphoton NIR QC phenomenon from novel Er^{3+}-Tm^{3+} ion pairs has been perceived. This can facilitate for the development of next-generation environmentally friendly Ge solar cells. In addition, these glasses could be used to construct NIR light sources by exciting with GaN LEDs.

To evaluate the actual performance of the Ln^{3+}-doped QC materials, which should exhibit the QE close to 200%, on the photoelectric conversion efficiency, these materials can be used as DC layer which can be placed on the front surface of solar cell.

11.3   Upconversion

Upconversion (UC) is a process of converting two or more photons of low energy into one photon of high energy. This process has been reported for the first time by Auzel [82] and later on evidenced in a wide variety of Ln^{3+}-doped tellurite glasses [83–92]. The researchers have been concentrated extensively on Tm^{3+}-, Er^{3+}-, and Er^{3+}/Yb^{3+}-doped materials for their feasibility to integrate them to the Si solar cells. There are different types of UC mechanism that are proposed in Ln^{3+} ions which include excited state absorption (ESA) [93], energy transfer upconversion (ETU) or Addition de Photon par Transferts d'Energie (APTE) [94], cooperative UC [95], and photon avalanche (PA) [96].

The transmission of radiation is another loss mechanism in Si solar cells. These losses can be eliminated by adopting the impurity photovoltaic (IPV) effect which was proposed by Wolf [97] in 1960. He suggested that the doping of impurities with bandgap energies more or less similar to the bandgap of solar cell material may reduce the transmission losses. Another way is the use of light convertor. This converts the photons at low energy that are transmitted from the solar spectrum to the photons at higher energy, which can be utilized by the solar cell. This was experimentally evidenced by Gibart et al. [98] in 1996 by using Ln^{3+}-doped GC at the rare face of the substrate-free GaAs ($E_g = 1.42$ eV) solar cell. The efficiency of the cell has been enhanced to 2.5% under the excitation at a laser power of 1.0 W of 891.3 nm (1.39 eV) laser through APTE process.

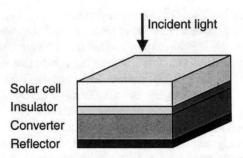

Fig. 11.7 An upconverter-based solar cell design consists of an upconverter and a solar cell. The photons of sub-bandgap transmitted from the solar cell are partially converted into photons of high energy, which are consequently absorbed by the solar cell. An insulator is placed between the convertor and solar cell. A reflector is placed backside of the upconverter. (Reproduced from Ref. [39], copyright 2002, American Institute of Physics)

Trupke and co-workers [39] in 2002 have been proposed a PV system design that consists of a bifacial single junction solar cell and an upconvertor as shown in Fig. 11.7. The solar cell is made up of a martial having the bandgap energy, E_g. The upconvertor transforms photons of low energy transmitted by the solar cell into photons of high energy. The overall efficiency of c-Si solar cell with the presence of an upconverter boosts from 30% to 37.4% under one-sun illumination. In the Er^{3+}/Yb^{3+} ion pair, the Yb^{3+} provides a high absorption cross-section at 980 nm (1.26 eV), and the Er^{3+} offers the emission at blue, green, and red wavelengths. The other solar cell technologies with large bandgap energy were tested with this system. In 2005, Shalav et al. [39] applied the $Yb^{3+}/Er^{3+}:NaYF_4$ UC phosphor to the back surface of the bifacial Si solar cell and studied its responsivity. The PLQY and external quantum efficiency (EQE) of $NaYF_4:Yb^{3+}/Er^{3+}$ were found to be 3.8% and 2.5%, respectively, under the excitation at 5.10 mW of 1523 nm (0.82 eV) laser. The UC mechanism among the Er^{3+} ions is shown in Fig. 11.8. In 2010, Wild et al. [99] have applied $NaYF_4:Yb^{3+}/Er^{3+}$ nanophosphor to a-Si solar cells and illuminated under 980 nm (1.26 eV). However, the efforts are under way to develop more stable and efficient upconvertors.

Recently, Venkata Krishnaiah et al. [25] have been investigated the Er^{3+}-doped $74TeO_2–(18–x)WO_3–8ZrO_2–xEr_2O_3$ (x = 0.01, 0.5 and 3.0 mol%), named as TWZE, tellurite glasses at different Er^{3+} ion concentration for solar energy applications. The UC emission spectra of Er^{3+}-doped TWZE glasses acquired under 1500 nm (0.83 eV) laser excitation are shown in Fig. 11.9, and their emission is displayed in the inset of Fig. 11.9. It is observed that a high UC emission intensity at 975 nm (1.27 eV) has been observed for 3.0 mol% Er_2O_3-doped TWZE glass compared to 0.01 and 0.5 mol% Er_2O_3-doped TWZE glasses. The ETU is confirmed from the temporal evolution studies that are responsible for observed UC emissions, which is in good agreement with the proposed simple rate-equation model. The optimum sample is 3.0 mol% Er_2O_3-doped TWZE tellurite glass which was placed on the top of the Si solar cell to study the photocurrent of the cell. It is found that the

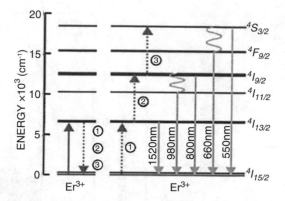

Fig. 11.8 Energy level showing the three-step UC mechanism of Er^{3+} ion. The ET from Er^{3+} ion (the sensitizer) to a neighboring Er^{3+} ion (activator) leads to emission of photons at higher energy. Absorption (up) and emission (down) of photons represented with solid arrows, ET represented with dotted lines, and curved lines represented the non-radiative channels. In the two-step UC process, photons with energies higher than the bandgap of Si are emitted. (Reproduced from Ref. [39], copyright 2005, Elsevier)

Fig. 11.9 Upconversion (UC) emission of Er^{3+} doped TWZE glasses with different Er^{3+} ion concentration under 1500 nm (0.83 eV) laser excitation. The TWZE glass is placed on the top of the solar cell that emits bright green UC emission which is shown in the inset. The excitation was performed right angle to the surface of sample. (Reproduced from Ref. [25], copyright 2017, Elsevier)

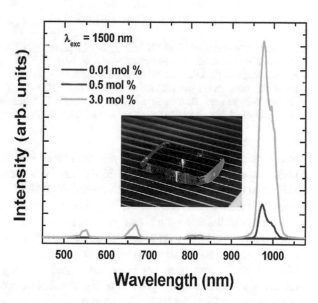

photocurrent enhanced significantly with the presence of TWZE glass. No photocurrent was observed in the absence of glass sample under 1500 nm (0.83 eV) excitation.

Recently, Balaji et al. [26] investigated the ET dynamics of Er^{3+}/Yb^{3+}-doped $75TeO_2-15BaF_2-5AlF_3-(5-x)LaF_3-xErF_3-1.0YbF_3$ ($x = 0.5, 1.0, 2.0,$ and 3.0 mol %), labeled as TBALFEY, oxyfluorotellurite glasses under the excitation of 1550 nm (0.79 eV) for improving the efficiency of c-Si solar cell. With increase of Er^{3+} ion concentration, the photocurrent enhanced significantly, and the optimum Er_2O_3

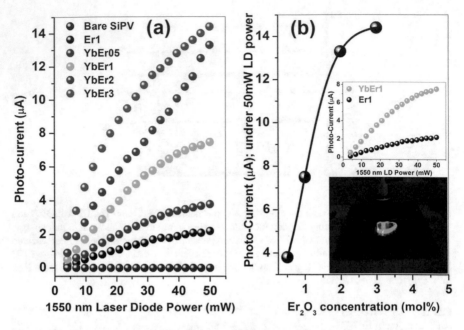

Fig. 11.10 Photocurrent of Si-PV cell in combination with TBALFEY tellurite glasses upon 1550 nm laser excitation. (**a**) Effect of power of laser at 1550 nm (0.79 eV) for different Er^{3+} ion concentration. (**b**) Effect of Er^{3+} ion concentration under excitation at 50 mW constant laser power. Top inset shows the photocurrent of Er- and Er/Yb-doped TBALFEY tellurite glasses at different laser powers. Bottom inset shows the bright UC emission of TBALFEY glass under 1550 nm laser (0.79 eV) excitation. (Reproduced from Ref. [26], copyright 2016, Elsevier)

concentration is about 3.0 mol%. It also increases with increase in the power of the incident laser. The photocurrent of Si-PV has been enhanced significantly around fourfold in the presence of Er^{3+}/Yb^{3+} co-doped TBALFEY glass compared to Er^{3+}-doped TBALFEY glass and a bare PV cell, as shown in Fig. 11.10. The photocurrent enhanced significantly which can be understood on the basis of ET mechanism between Er^{3+} and Yb^{3+} ions upon excitation at 1550 nm (0.79 eV). The influence of Er^{3+} ion concentration and pump power on the ESA and ETU mechanisms and their effect on the UC emission has been elaborated. The significance of ESA or ETU processes were described by considering the de-excitation dynamics of NIR UC emission under 1550 nm (0.79 eV) excitation.

The Er^{3+}-Yb^{3+} co-doped $50TeO_2$–$(49.4-x)PbF_2$–$0.6Er_2O_3$–xYb_2O_3 (where $x = 0$, 0.6, 1.2, 1.8, 2.4, 3.0, and 4.2 mol%), named as TPFEY, oxyfluoride tellurite glasses with different Yb^{3+} ion concentration have been investigated by Yang and co-workers [24] for their feasibility to couple them with PV cells. An intense UC emission at 527 nm (2.35 eV), 544 nm (2.27 eV), and 657 nm (1.88 eV) corresponds to the Er^{3+} transitions, and the maximum was obtained for 1.8 mol% of Yb^{3+} and 0.6 mol% of Er^{3+}. High intensity of UC emission has been noticed for 2.4 mol% of Yb^{3+}-doped glass upon excitation at 100 mW laser power of 980 nm (1.26 eV) laser.

This glass has been placed at the back face of a-Si solar cells in combination with a rear reflector. It is worth mentioning that a significant enhancement of efficiency of 0.45% has been obtained under both the presence of AM1.5 and 400 mW of 980 nm (1.26 eV) laser radiation. The EQE of a solar cell is defined as the ratio of the number of photo-induced e^-–h^+ pairs to the number of incident infrared photons, which increases with increasing the power of incident laser for TPEEY glass with 0.6 mol% of Er^{3+} and 0.6 mol% of Yb^{3+} ions. The PLQY and maximum EQE found to be 1.35% and 0.27%, respectively, were achieved at 300 mW of 980 nm laser excitation. The J-V properties of a-Si solar cells co-excited with AM1.5, and UC emission radiation are presented in Fig. 11.11, where the incident light is perpendicular to the solar cell and a glass. From the results, it is understood that the open-circuit voltage (V_{OC}) remains almost the same, but the short-circuit current density (J_{SC}) and fill factor (FF) initially increase and then reach saturation with increasing the power of laser. As a result, the conversion efficiency increases from 4.25% to 4.70%.

The UC broadband emission at 800 nm (1.55 eV) of Tm^{3+}-doped fluoride glasses (with molar composition in mol%, $37.5InF_3$–$20SrF_2$–$20BaF_2$–$20ZnF_2$–$2.5TmF_3$) has been utilized for solar cell applications. The Tm^{3+} ions absorb photons at the NIR region and transfer their energy through UC process from the $^3F_4$ level to the $^3H_4$ excited level followed by the emission of photons at 800 nm (1.55 eV), which correspond to the $^3H_4 \rightarrow {}^3H_6$ transition. These photons can be absorbed by a Si solar cell which leads for the creation of additional e^+–h^- pairs and then enhance the overall photocurrent [100].

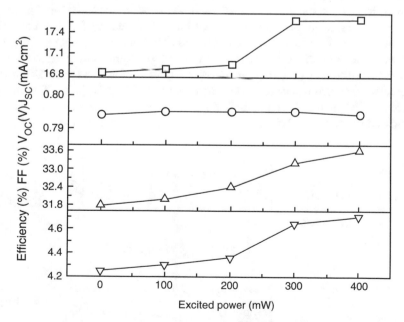

Fig. 11.11 J-V curves of 0.6 Er^{3+}/0.6 Yb^{3+}-doped TPFEY glasses in combination of a-Si solar cells with and without laser excitation. (Reproduced from Ref. [24]. Copyright 2014, Elsevier)

The prime importance of the grouping of solar cell and upconvertor is to enhance the efficiency significantly without any alteration of the solar cell itself. The PLQY of the UC tellurite glasses is to be optimized by doping different Ln^{3+} ion pairs to enhance the efficiency of Si solar cells. In this direction, there is plenty of work required to make use of Ln^{3+}-doped tellurite glass as an upconvertor.

11.4   Photocatalytic Activity

The other way to harvest the solar energy is photocatalysis process in which the water is fragmented into hydrogen (H_2) and oxygen (O_2) in the presence of photocatalyst. The mechanism of photocatalysis [101] is shown in Fig. 11.12. The photocatalytic properties of photocatalyst were derived from the formation of photo-induced charge carriers (h^+ and e^-) which arises under the absorption of ultraviolet (UV)/visible light. The photo-induced h^+ in the valence band (VB) diffused to the surface of photocatalyst and respond to water (H_2O) molecules that are adsorbed. This leads to produce hydroxyl (OH^-) radicals. The photo-induced h^+ and the OH^- ions oxidize nearby organic molecules on the surface of photocatalyst. In the meantime, the e^- in the conduction band (CB) usually participates in the reduction and reacts with O_2 in the air to produce superoxide radical anions ($O_2^{\bullet-}$). This type of mechanism can be utilized extensively in the fields of environmental applications such as water decomposition, microbe destruction, toxin removal, water decontamination, hazardous waste remediation, and water purification [102–106]. The H_2 energy is considered as renewable, ideal, and clean energy for the future. The decomposition of water into H_2 and O_2 is an environmentally friendly approach that can be realized under UV [107] and visible light illumination [108].

Titanium dioxide (TiO_2) is an excellent photocatalyst and demonstrated its photocatalytic performance as it can split water into H_2 and O_2 [109]. The bandgap energy of TiO_2 is 3.2 eV. Photocatalytic response of amorphous TiO_2 is negligible compared to crystalline TiO_2 as the crystalline structure reduces the recombination

Fig. 11.12 Formation of photo-induced charge carriers (h^+ and e^-) under the influence of UV and visible light. (Reproduced from Ref. [101], copyright 2012, Elsevier)

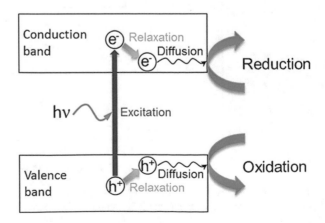

of photo-induced e^- and h^+ pairs [110]. Since its invention, several semiconductor photocatalytic candidates were proposed and investigated that include Fe_2O_3 (2.0–2.2 eV) [111], CdS (2.4 eV) [112], ZnS (2.5 eV) [113], $BiTiO_3$ (2.5 eV) [114], WO_3 (2.8–3.0 eV) [115], ZnO (3.2 eV) [116], SnO_2 (3.7 eV) [117], and ZrO_2 (3.25–5.1 eV) [118] for harvesting H_2 from water. These catalysts cannot only be used to transform less intense solar energy to high-density storable H_2 but also for mineralizing the environmental contaminants. However, the efficiencies of these processes are not up to the mark to become economically viable.

It is desirable to develop an innovative and efficient method for pollutant removal and energy production. In this direction, photocatalysis-based wastewater/pollution treatment is one of the passionate topics among the researchers. In this scenario, Ln^{3+}-doped glasses are coated with photocatalytic materials, such as TiO_2, to achieve photocatalytic activities on the surface of glass including self-cleaning properties. However, these coatings have maintenance issues due to leaching problems. The powder-based semiconductor photocatalyst has a lot of difficulties including separation, dispersion, and collection of photocatalyst for industrial applications. Powder catalyst also induces impurities in water during catalysis, and a separate process is required to extract the particles from the water. The fabrication of a photocatalyst using a photoactive glass plates could resolve these issues and is easier to apply for large-scale operations. The self-cleaning property will widely extend the applications of these kinds of multifunctional glasses in the architecture and other fields.

Even though various photocatalysts are available in the recent times, TiO_2 has several advantages such as high photocatalytic activity, strong oxidizing ability, good chemical stability, long durability, safeness, nontoxicity, transparency to visible light, and low costs [98, 119]. However, its wide bandgap allows it to absorb the UV light (around 380 nm, 3.26 eV), leading to a least utilization of visible light [120]. To improve the absorption ability of TiO_2 in the visible region and enhance its photocatalytic activity, several strategies have been employed including the doping with nonmetals, transition metals, and Ln^{3+} ions [121–123]. One of the promising approach to improve the visible light absorption is doping of photocatalysts with Ln^{3+} ions, which have shown tremendous potential as dopants not only shift the absorption region but also improve the photocatalytic activity of TiO_2 [123]. Usually, Ln^{3+}-doped TiO_2 exhibits excellent luminescent properties. Thus, beside visible emission under UV excitation, these materials can also present UC luminescence. This process results in transformation of the light from NIR and visible spectral range into the UV wavelengths, which is the reason for increasing interest and genuine requirement of TiO_2 photocatalysts. Particularly, those dopants are Ln^{3+} ions which have broad absorption in the UV, visible and NIR regions through DC and UC processes. In the recent times, the Ln^{3+}-doped TiO_2 has been employed extensively as a photocatalyst in photocatalysis process. The best photocatalytic performance has been evidenced for the platinum-coated 0.5 mol% Gd_2O_3-, Eu_2O_3-, Yb_2O_3-, and Ho_2O_3-doped TiO_2 particles [124].

Visible light-induced photocatalytic activity of tellurite glasses has been explored for the production of H_2, self-cleaning, and estrogenic pharmaceutical pollutants. Transparent glasses of composition $(100 - x)$ $TeO_2 - xCaCu_3Ti_4O_{12}$, where

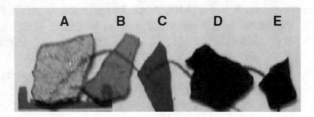

Fig. 11.13 Images of xCCTO–(100–x)TeO$_2$ glasses with photocatalytic crystals concentration (*A*) *x* = 0.25 mol%, (*B*) *x* = 0.5 mol%, (*C*) *x* = 1.0 mol%, (*D*) *x* = 2.0 mol%, and (*E*) *x* = 3.0 mol% (Reprinted with from Ref. [125], copyright, 2017, SPIE)

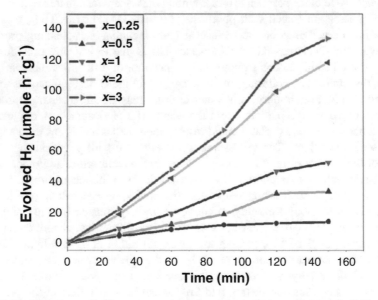

Fig. 11.14 Evolution of H$_2$ energy through photocatalytic activity of (100 − *x*) TeO$_2$ − xCaCu$_3$Ti$_4$O$_{12}$, nanocrystalline tellurite GCs which was heat treated at 600 °C for 2 h under visible light excitation. (Reprinted from Ref. [125], copyright, 2017, SPIE)

x = 0.25–3 mol%, have been synthesized by the melt-quenching technique [125]. An appropriate amount of crystalline phase of CaCu$_3$Ti$_4$O$_{12}$ was first synthesized by using oxalate route [126] and was added to the TeO$_2$ and melted at 850 °C for 30 min in a platinum crucible. The photographs of as-synthesized tellurite glass samples are shown in Fig. 11.13. The glass samples were annealed at 500 °C, 550 °C, and 600 °C for 2 h to induce the photocatalytic crystals in tellurite glass matrix. The size of these crystals was found to be 10 nm. A higher photocatalytic H$_2$ production rate has been observed as high as 135 μmole h^{-1}g^{-1} for the crystalline phase of 3CaCu$_3$Ti$_4$O$_{12}$-97TeO$_2$ glass sample under visible light excitation. The evolution of H$_2$ with respect to time for the GC plates of CaCu$_3$Ti$_4$O$_{12}$-TeO$_2$ with different amounts of CaCu$_3$Ti$_4$O$_{12}$ for the decomposition of H$_2$O into H$_2$ is shown in Fig. 11.14. The linear behavior of graph clearly shows the stable evolution of H$_2$, not observed without a catalyst glass plate or in the dark. In addition, self-cleaning performance of these nanocrystalline tellurite glasses has also been observed using contact angle measurements for water under dark and light conditions. These visible light active nanocrystalline GCs are a cost-effective sustainable solution for water treatment, H_2 production, and self-cleaning applications [125].

11.5 Conclusions

The state of the art of Ln^{3+}-doped transparent glasses and glass ceramics, in particular, tellurite glasses, for solar energy harvesting by downconversion, upconversion, and photocatalysis has been presented. Quantum cutting through downconversion process in Ln^{3+}-doped materials has been employed in different Ln^{3+} pairs including Ce^{3+}/Yb $^{3+}$, Pr^{3+}/Yb^{3+}, Nd^{3+}/Yb^{3+}, Eu^{3+}/Yb^{3+}, Gd^{3+}/Eu^{3+}, Tb^{3+}/Yb^{3+}, Ho^{3+}/Yb^{3+}, Er^{3+}/Yb^{3+}, and Tm^{3+}/Yb^{3+} to overcome the thermalization losses in c-Si solar cells. Among these, Nd^{3+}/Yb^{3+}, Gd^{3+}/Eu^{3+}, and Er^{3+}/Yb^{3+} co-doped materials show high quantum efficiency (QE) close to 200%. Such high QE materials at 1.0 μm (1.23 eV) wavelength can be integrated with c-Si solar cells ($E_g = 1.1$ eV) to further enhance its conversion efficiency. For these, Er^{3+}- and Yb^{3+}-doped systems are highly recommended as their emission matches the bandgap energy of Si.

A variety of upconverting materials have been studied with an aim to reduce the transmission losses of solar cell, among these Er^{3+} and Yb^{3+} ion-doped low phonon energy tellurite glasses, are promising. Such upconvertors can be integrated to solar cells having the bandgap energies 1.1 eV (c-Si), 1.8 eV (a-Si), and GaAs (1.48 eV) for the enhancement of their overall efficiency by reducing the transmission losses. Still efforts are under way to find the best suitable materials for these purposes and may be resolved in the near future.

The Ln^{3+}-doped materials can also be extended to harvest solar energy by photocatalysis process. There are different photocatalysts including Fe_2O_3 (2.0–2.2 eV), CdS (2.4 eV), ZnS (2.5 eV), $BiTiO_3$ (2.5 eV), WO_3 (2.8–3.0 eV), ZnO (3.2 eV), TiO_2 (3.2 eV), SnO_2 (3.7 eV), and ZrO_2 (3.25–5.1 eV); their performance was investigated and it was concluded that TiO_2 could be the best photocatalyst. However, the TiO_2 has been extensively studied due to their superior properties. The aim of using the Ln^{3+}-doped photocatalytic materials is to enhance the absorption of sunlight through upconversion and downconversion/quantum cutting other than the region of photocatalyst. The nanocrystalline-doped Ln^{3+} : tellurite systems may find potential for improving the efficiency of solar cells and photocatalytic activity by utilizing the solar radiation effectively. The other possibility is the purification (i.e., elimination of the unwanted transition metal impurities) of the starting reagents that could enhance the QE, which leads to improve the overall efficiency of the PV and photocatalytic devices.

Acknowledgments Dr. K.V. Krishnaiah is thankful to RGMCET for providing the necessary facilities. One of the authors Dr. C.K.J is grateful to DAE-BRNS for the sanction of mega research project (NO.2009/34/36/BRNS/3174) under MoU between SVU, Tirupati; RRCAT, Indore; and

BARC, Mumbai. Dr. Venkatramu is indebted to DST, New Delhi, for the sanction of India-Portugal bilateral research project (No. INT/PORTUGAL/P-04/2017) under scientific and technological cooperation.

References

1. E. Snitzer, Phy. Rev. Lett. **7**, 444 (1961)
2. G. Boulon, J. Alloys Compounds **451**, 1 (2008)
3. F.J. Duarte, *Tunable laser applications* (CRC, New York, 2009)
4. I. Iparraguirre, J. Azkargorta, R. Balda, K. Venkata Krishnaiah, C.K. Jayasankar, M. Al-Saleh, J. Fernández, Opt. Express **19**, 19440 (2011)
5. V.D. Del Cacho, D.M. da Silva, T.A.A. de Assumpção, L.R.P. Kassab, M.I. Alayo, E.G. Melo, Opt. Mater. **38**, 198 (2014)
6. T. Sun, Z.Y. Zhang, K.T.V. Grattam, A.W. Paimer, Rev. Sci. Instruments **69**(12), 4179 (1998)
7. J. Fernandeza, A. Mendioroz, A.J. Garcia, R. Balda, J.L. Adam, J. Alloys Compound. **323–324**, 239 (2001)
8. K. Venkata Krishnaiah, E.S. de Lima Filho, Y. Ledemi, G. Nemova, Y. Messaddeq, R. Kashyap, Sci. Rep. **6**, 21905 (2016)
9. C.E. Mungan, J. Opt. Soc. Am. B **20**, 1075 (2003)
10. J. Wang, X. Zhang, Q. Su, Rare earth solar spectral convertor for Si solar cells, in *Phosphors, Up Conversion Nano Particles, Quantum Dots and their Applications*, ed. by R. S. Liu, (Springer, Singapore, 2016)
11. X. Liu, J. Lin, Solid State Sci. **11**, 2030 (2009)
12. N.Q. Wang, X. Zhao, C.M. Li, E.Y.B. Pun, H. Lin, J. Lumin. **130**, 1044 (2010)
13. S. Shen, A. Jha, X. Liu, M. Naftaly, K. Bindra, H.J. Bookey, A.K. Kar, J. Am. Ceram. Soc. **85** (6), 1391 (2002)
14. W.J. Chung, J. Heo, Appl. Phys. Lett. **79**, 326 (2001)
15. A. Mori, IEEE J. Lightwave Technol. **LT-20**, 822 (2002)
16. J.S. Wang, E.M. Vogel, E. Snitzer, Opt. Mater. (Amsterdam, Neth.) **3**, 187 (1994)
17. P. Joshi, S. Shen, A. Jha, J. Appl. Phys. **103**, 083543 (2008)
18. K. Vu, S. Madden, Opt. Express **18**, 19192 (2010)
19. R. Josea, Y. Ohishi, Appl. Phys. Lett. **90**, 211104 (2007)
20. L. Gomes, J. Lousteau, D. Milanese, E. Mura, S.D. Jackson, J. Opt. Soc. Am. B. **31**, 429 (2014)
21. S.F. León-Luis, U.R. Rodríguez-Mendoza, E. Lalla, V. Lavín, Sensors Actuators B **158**, 208 (2011)
22. V.K. Rai, D.K. Rai, S.B. Rai, Sensors Actuators A **128**, 14 (2006)
23. Z.-X. Jia, L. Liu, C.-F. Yao, G.-S. Qin, Y. Ohishi, W.-P. Qin, J. Appl. Phys. **115**, 063106 (2014)
24. F. Yang, C. Liu, D. Wei, Y. Chen, Jingxiao Lu, Shi-e Yang. Opt. Mater. **36**, 1040 (2014)
25. K. Venkata Krishnaiah, P. Venkatalakshmamma, C. Basavapoornima, I.R. Martín, K. Soler-Carracedo, M.A. Hernandez-Rodríguez, V. Venkatramu, C.K. Jayasankar, Mater. Chem. Phys. **199**, 67 (2017)
26. S. Balaji, D. Ghosh, K. Biswas, A.R. Allu, G. Gupta, K. Annapurna, J. Lumin. **187**, 441 (2017)
27. M.A. Green, Prog. Photovolt. Res. Appl. **13**, 447 (2005)
28. K. Sasaki, T. Agui, K. Nakaido, N. Takahashi, R. Onitsuka, T. Takamoto, in *Proceedings of 9th International Conference on Concentrating Photovoltaics Systems*, Miyazaki, Japan (2013)
29. A. Kojima, K. Teshima, Y. Shirai, T. Miyasaka, Organometal halide perovskites as visible-light sensitizers for photovoltaic cells. J. Am. Chem. Soc. **131**, 6050 (2009)

30. H.J. Snaith, J. Phys. Chem. Lett. **4**, 3623 (2013)
31. V. Aroutiounian, S. Petrosyan, A. Khachatryan, K. Touryan, J. Appl. Phys. **89**, 2268 (2001)
32. A. Hagfeldt, G. Boschloo, L. Sun, L. Kloo, H. Pettersson, Dye-sensitized solar cells. Chem. Rev. **110**, 6595 (2010)
33. G. Gruner, J. Mater. Chem. **16**, 3533 (2006)
34. A.C. Mayer, S.R. Scully, B.E. Hardin, M.W. Rowell, M.D. Mc Gehee, Mater. Today **10**, 28 (2007)
35. T. Saga, NPG Asia Mater. **2**, 96 (2010)
36. W. Shockley, H.J. Queisser, J. Appl. Phys. **32**, 510 (1961)
37. M.A. Green, A. Ho-Baillie, H.J. Snaith, Nat. Photonics **8**, 506 (2014)
38. M.A. Hernández-Rodríguez, M.H. Imanieh, L.L. Martín, I.R. Martín, Sol. Energy Mater. Sol. Cells **116**, 171 (2013)
39. A. Shalav, B.S. Richards, T. Trupke, K.W. Krämer, H.U. Güdel, Appl. Phys. Lett. **86**, 013505 (2005)
40. B.S. Richards, Sol. Energy Mater. Sol. Cells **90**, 2329 (2006)
41. C. Strumpel, M. McCann, G. Beaucarne, V. Arkhipov, A. Slaoui, V. Svrcek, C. del Canizo, I. Tobias, Sol. Energy Mater. Sol. Cells **91**, 238 (2007)
42. D.C. Law, R.R. King, H. Yoon, M.J. Archer, A. Boca, C.M. Fetzer, S. Mesropian, T. Isshiki, M. Haddad, K.M. Edmondson, D. Bhusari, J. Yen, R.A. Sherif, H.A. Atwater, N.H. Karam, Sol. Energy Mater. Sol. Cells **94**, 1314 (2010)
43. J.H. Werner, S. Kolodinski, H.J. Queisser, Phys. Rev. Lett. **72**, 3851 (1994)
44. G. Conibeer, N. Ekins-Daukes, J.-F. Guillemoles, D. Kőnig, E.-C. Cho, C.-W. Jiang, S. Shrestha, M. Green, Sol. Energy Mater. Sol. Cells **93**, 713 (2009)
45. T. Trupke, M.A. Green, P. Wurfel, J. Appl. Phys. **92**, 1668 (2002)
46. D. Ross, E. Klampaftis, J. Fritsche, M. Bauer, B.S. Richards, Sol. Energy Mater. Sol. Cells **103**, 11 (2012)
47. T. Trupke, M.A. Green, P. Wurfel, J. Appl. Phys. **98**, 4117 (2002)
48. R.T. Wegh, H. Donker, K.D. Oskam, A. Meijerink, J. Lumin. **82**, 93 (1999)
49. D.L. Dexter, Phys. Rev. **108**, 630 (1957)
50. W.W. Piper, J.A. Deluca, F.S. Ham, J. Lumin. **8**, 344 (1974)
51. R.T. Wegh, H. Donker, K.D. Oskam, A. Meijerink, Science **283**, 663 (1999)
52. Q.Y. Zhang, X.Y. Huang, Prog. Mater. Sci. **55**, 353 (2010)
53. P. Vergeer, T.J.H. Vlugt, M.H.F. Kox, M.I. Den Hertog, J.P.J.M. van der Eerden, A. Meijerink, Phys. Rev. B **71**, 014119 (2005)
54. Q.Y. Zhang, C.H. Yang, Y.X. Pan, Appl. Phys. Lett. **90**, 021107 (2007)
55. Q.Y. Zhang, C.H. Yang, Z.H. Jiang, X.H. Ji, Appl. Phys. Lett. **90**, 061914 (2007)
56. Q.Y. Zhang, G.F. Yang, Z.H. Jiang, Appl. Phys. Lett. **91**, 051903 (2007)
57. R.T. Wegh, H. Donker, A. Meijerink, R.J. Lamminmaki, J. Hölsä, Phys. Rev. B **56**, 13841 (1997)
58. Z. Yang, J.H. Lin, M.Z. Su, Y. Tao, W. Wang, J. Alloys Compd. **308**, 94 (2000)
59. R.T. Wegh, E.V.D. van Loef, G.W. Burdick, A. Meijerink, Mol. Phys. **101**, 1047 (2003)
60. B.S. Richards, Sol. Energy Mater. Sol. Cells **90**, 1189 (2006)
61. L.G. Hwa, S.L. Hwang, L.C. Liu, J. Non-Cryst. Solids **238**, 193 (1998)
62. Z. Liu, Y. Yu, N. Dai, Q. Chen, L. Yang, J. Li, Y. Qiao, Appl. Phys. A Mater. Sci. Process. **108**, 777 (2012)
63. Z. Liu, J. Li, L. Yang, Q. Chen, Y. Chu, N. Dai, Sol. Energy Mater. Sol. Cells **122**, 46 (2014)
64. A. Boccolini, J. Marques-Hueso, D. Chen, Y. Wang, B.S. Richards, Sol. Energy Mater. Sol. Cells **122**, 8 (2014)
65. W. Wang, X. Lei, H. Go, Y. Mao, Opt. Mater. **47**, 270 (2015)
66. D. Chen, Y. Wang, Y. Yu, P. Huang, F. Weng, Opt. Lett. **33**, 1884 (2008)
67. F.B. Costa, K. Yukimitu, L.A.O. Nunes, M.S. Figueiredo, J.R. Silva, L.H.C. Andrade, S.M. Lima, J.C.S. Moraes, J. Am. Ceram. Soc. **100**, 1956 (2017)
68. M.M. Smedskjaer, J. Qiu, J. Wang, Y. Yue, Appl. Phys. Lett. **98**, 071911 (2011)

69. M.K, Lau and Jian-Hua Hao. Energy Procedia **15**, 129 (2012)
70. X. Qiao, T. Tsuboi, H.J. Seo, J. Alloys Compound. **687**, 179 (2016)
71. L.d.A. Florêncio, L.A. Gómez-Malagón, B.C. Lima, A.S.L. Gomes, J.A.M. Garcia, L.R.P. Kassab, Sol. Energy Mater. Sol. Cells **157**, 468 (2016)
72. X. Zhou, Y. Wang, X. Zhao, L. Li, Z. Wang, Q. Li, J. Am. Ceram. Soc. **97**, 179 (2014)
73. M.S. Figueiredo, F.A. Santos, K. Yukimitu, J.C.S. Moraes, L.A.O. Nunes, L.H.C. Andrade, S.M. Lima, J. Lumin. **157**, 365 (2015)
74. A. Pandey, R.E. Kroon, V. Kumar, H.C. Swart, J. Alloys Compound. **657**, 32 (2016)
75. Q. Zhang, B. Zhu, Y. Zhuang, G. Chen, X. Liu, G. Zhang, J. Qiu, D. Chen, J. Am. Ceram. Soc. **93**, 654 (2010)
76. G. Lakshminarayana, H. Yang, S. Ye, Y. Liu, J. Qiu, J. Phys. D. Appl. Phys. **41**, 175111 (2008)
77. S. Ye, B. Zhu, J. Luo, J. Chen, G. Lakshminarayana, J. Qiu, Opt. Express **16**, 8989 (2008)
78. D. Rajesh, M. Reza Dousti, R.J. Amjad, A.S.S. de Camargo, J. Non-Cryst. Solids **450**, 149 (2016)
79. V.D. Rodrígueza, V.K. Tikhomirov, J. Me'ndez-Ramos, A.C. Yanes, V.V. Moshchalkov, Sol. Energy Mater. Sol. Cells **94**, 1612 (2010)
80. H. Lin, S. Zhou, X. Hou, W. Li, Y. Li, H. Teng, T. Jia, IEEE Photon. Tech. Lett. **22**, 866 (2010)
81. X. Chen, S. Li, L. Hu, K. Wang, G. Zhao, L. He, J. Liu, C. Yu, J. Tao, W. Lin, G. Yang, G.J. Salamo, Sci. Rep. **7**, 1976 (2017)
82. F. Auzel, C. R. Acad. Sci. Paris **262**, 1016 (1966)
83. V.K. Rai, K. Kumar, S.B. Rai, Opt. Mater. **29**, 873 (2007)
84. I. Iparraguirre, J. Azkargorta, J.M. Fernández-Navarro, M. Al-Saleh, J. Fernandez, R. Balda, J. Non-Cryst. Solids **353**, 990 (2007)
85. V.K. Rai, S.B. Rai, D.K. Rai, Opt. Commun. **257**, 112 (2006)
86. M. Reza Dousti, R.J. Amjad, R. Hosseinian, M. Salehi, M.R. Sahar, J. Lumin. **159**, 100 (2015)
87. G. Poirier, F.C. Cassanjes, C.B. de Araújo, V.A. Jerez, S.J.L. Ribeiro, Y. Messaddeq, M. Poulain, J. Appl. Phys. **93**, 3259 (2003)
88. M. Kochanowicz, J. Zmojda, D. Dorosz, P. Miluski, J. Dorosz, Proc. SPIE **9228**, 92280B-1 (2014)
89. A. Pandey, V. Kumar, R.E. Kroon, H.C. Swart, J. Lumin. **192**, 757 (2017)
90. H. Lin, K. Liu, E.Y.B. Pun, T.C. Ma, X. Peng, Q.D. An, J.Y. Yu, S.B. Jiang, Chem. Phys. Lett. **398**, 146 (2004)
91. Y. Luo, J. Zhang, X. Zhang, X. Wang, J. Appl. Phys. **103**, 063107 (2008)
92. K. Venkata Krishnaiah, J. Marques-Hueso, K. Suresh, G. Venkataiah, B.S. Richards, C.K. Jayasankar, J. Lumin. **169**, 270 (2016)
93. N. Bloembergen, Phys. Rev. Lett. **2**, 84 (1959)
94. J. Wright, Up-conversion and excited-state energy transfer in rare-earth doped materials, in *Radiationless Processes in Molecules and Condensed Phases*, Topics in Applied Physics, ed. by F.K. Fong, vol 15, (Springer, New York, 1976), p. 239
95. E. Nakazawa, S. Shionoya, Phys. Rev. Lett. **25**, 1710 (1970)
96. J.S. Chivian, W.E. Case, D.D. Eden, Appl. Phys. Lett. **35**, 124 (1979)
97. M. Wolf, Proc. IRE **48**, 1246 (1960)
98. P. Gibart, F. Auzel, J.C. Guillaume, K. Zahraman, Jpn. J. Appl. Phys. Part **1**(35), 4401 (1996)
99. J. de Wild, J.K. Rath, A. Meijerink, W.G.J.H.M. van Sark, R.E.I. Schropp, Sol. Energy Mater. Sol. Cell **94**, 2395 (2010)
100. H. Rodríguez-Rodríguez, M.H. Imanieh, F. Lahoz, I.R. Martín Sol, Energy Mater. Sol. Cells **144**, 29 (2016)
101. K. Nakata, A. Fujishima, J Photochem Photobiol C: Photochem Rev **13**, 169 (2012)
102. P.K.J. Robertson, J.M.C. Robertson, D.W. Bahnemann, J. Hazardous Mater. **211-212**, 161 (2012)
103. M. Adams, I. Campbell, C. McCullagh, D. Russell, D.W. Bahnemann, P.K.J. Robertson, Int. J. Chem. React. Eng. **11**, 621 (2013)

104. C. McCullagh, P.K.J. Robertson, M. Adams, P.M. Pollard, A. Mohammed, J. Photochem. Photobiol. A **211**, 42 (2010)
105. D.W. Bahnemann, L.A. Lawton, P.K.J. Robertson, The application of semiconductor photocatalysis for the removal of cyanotoxins from water and design concepts for solar photocatalytic reactors for large scale water treatment, in *New and Future Developments in Catalysis*, ed. by S.L. Suib, 1st edn., (Elsevier, Amsterdam, 2013), pp. 395–415
106. C. Kim, M. Choi, J. Jang, Catal. Commun. **11**, 378 (2010)
107. H. Fan, G. Li, F. Yang, L. Yang, S. Zhang, J. Chem. Technol. Biotechnol. **86**, 1107 (2011)
108. R. Brahimi, Y. Bessekhouad, A. Bouguelia, M. Trari, J. Photochem. Photobiol. A Chem. **186**, 242 (2007)
109. A. Fujishima, K. Honda, Nature **238**, 37 (1972)
110. G. Tomandl, F.D. Gnanam, *Sol-Gel Processing of Advanced Ceramics* (Oxford and IBH Publishing Co. Pvt. Ltd, New Delhi, 1996)
111. T. Vincent, M. Gross, H. Dotan, A. Rothschild, Int. J. Hydrog. Energy **37**, 8102 (2012)
112. Q. Wang, J. Li, Y. Bai, J. Lian, H. Huang, Z. Li, Z. Leia, W. Shangguan, Green Chem. **16**, 2728 (2014)
113. J. Zhang, Y. Wang, J. Zhang, Z. Lin, F. Huang, J. Yu, ACS Appl. Mater. Interfaces **5**, 1031 (2013)
114. Y. Huo, Y. Jin, Y. Zhang, J. Mol. Catal. A Chem. **331**, 15 (2010)
115. J. Cao, B. Luo, H. Lin, B. Xu, S. Chen, Appl. Catal. B Environ. **111–112**, 288 (2012)
116. N. Serpone, D. Lawless, J. Disdier, J.-M. Herrmann, Langmuir **10**, 643 (1994)
117. Q. Luo, L. Wang, D. Wang, J. Environ. Chem. Eng. **3**, 622 (2015)
118. J.A. Navio, M.C. Hidalgo, G. Colon, S.G. Botta, M.I. Litter, Langmuir **17**, 202 (2001)
119. K. Hashimoto, H. Irie, A. Fujishima, Jpn. J. Appl. Phys. **44**, 8269 (2005)
120. V. Etacheri, M.K. Seery, S.J. Hinder, S.C. Pillai, Adv. Funct. Mater. **21**, 3744 (2011)
121. D. Dvoranová, V. Brezová, M. Mazúr, M.A. Malati, Appl. Catal. B Environ. **37**, 91 (2002)
122. W. Choi, A. Termin, M. Hoffmann, J. Phys. Chem. **84**, 13669 (1994)
123. J. Reszczynska, T. Grzybb, J.W. Sobczakc, W. Lisowski, M. Gazda, B. Ohtani, A. Zaleska, Appl. Catal. B Environ. **163**, 40 (2015)
124. M. Zalas, M. Laniecki, Sol. Energy Mater. Sol. Cells **89**, 287 (2005)
125. H.S. Kushwaha, P. Thomas, R. Vaish, J. Photon. Energy **7**, 016502 (2017)
126. H. Kushwaha, N.A. Madhar, B. Ilahi, P. Thomas, A. Halder, R. Vaish, Sci. Rep. **6**, 18557 (2016)

Chapter 12
Development of Bioactive Tellurite-Lanthanide Ions–Reinforced Hydroxyapatite Composites for Biomedical and Luminescence Applications

S. H. Nandyala, P. S. Gomes, G. Hungerford, L. Grenho, M. H. Fernandes, and A. Stamboulis

Abstract Human skeletal bone loss is a major health concern in the twenty-first century, with massive socioeconomic implications. The objective of the current work is to develop and characterize bioactive tellurite glasses for biomedical applications. As so, tellurium oxide- (TeO_2) and lanthanide (Ln^{3+})-doped borate host systems have been developed and incorporated in a hydroxyapatite (HA) matrix, being adequately characterized regarding solid-state parameters and for in vitro biological response. In the proposed work, the following scientific questions will be addressed: Will the reported tellurite-lanthanide ($Te-Ln^{3+}$) host glass-reinforced hydroxyapatite (HA) ceramic materials influence the cell behavior, such as proliferation and differentiation? Does this Te-Ln material show any luminescence response? Further, the research on lanthanide-based materials is promising, with potential application in prospective medical applications. Consequently, investigation into the role of $Te-Ln^{3+}$-HA host scaffold materials for bone repair is a relatively new approach that deserves a special attention.

S. H. Nandyala (✉) · A. Stamboulis
School of Metallurgy and Materials, University of Birmingham, Edgbaston, Birmingham, UK
e-mail: s.h.nandyala@bham.ac.uk

P. S. Gomes · L. Grenho · M. H. Fernandes
Laboratory for Bone Metabolism and Regeneration, Faculty of Dental Medicine,
University of Porto, Rua Dr. Manuel Pereira da Silva, Porto, Portugal

REQUIMTE/LAQV – University of Porto, Porto, Portugal

G. Hungerford
HORIBA Jobin Yvon IBH Ltd, Glasgow, UK

© Springer International Publishing AG, part of Springer Nature 2018
R. El-Mallawany (ed.), *Tellurite Glass Smart Materials*,
https://doi.org/10.1007/978-3-319-76568-6_12

12.1 Introduction

Human skeletal bone loss and infection are the major health concern in the twenty-first century with massive socioeconomic implications. In Europe, it is estimated that around 25,000 patients die annually because of infections caused by resistant bacteria translating into an estimated costs of EUR 1.5 billion per annum. Globally it was estimated that around 700,000 deaths result from antimicrobial resistance. Unless action is taken, this is projected to rise to 10 million deaths each year by 2050, with the added impact of a cumulative $100 trillion of economic output at risk due to the rise of drug-resistant infections [1–7].

Tellurium oxide (TeO_2) glass is a smart material, and it plays a multi-role in the composition of innovative materials, being potentially useful in the enhancement of luminescence and biological properties. In a preliminary work, the tellurium oxide (TeO_2) and lanthanide (Ln^{3+}) ion-doped glass host systems have been developed and incorporated in the hydroxyapatite (HA) matrix for their in vitro biological studies. The research on lanthanides materials for medical applications is promising but requires further investigation. The role of Te-Ln^{3+}-HA host scaffold materials for bone repair is a relatively innovative approach that deserves a special attention, as shown in the schematic diagram Fig. 12.1. The objective is to develop bioactive-based Te-Ln^{3+} ion-doped host glasses for biomedical and luminescence applications.

Tellurium (symbol Te and atomic number 52) exhibits great biological potential. According to Ba and Cunha [8, 9], the usage and potential of tellurium in biological applications is at its beginning. It has long appeared as a nearly "forgotten" element in biology, but over the last decade, several discoveries have led to an increased interest in this element. The toxicology of tellurium has received less attention than that of selenium, and historically it has found application in the treatment of

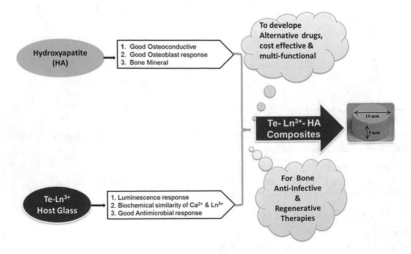

Fig. 12.1 A schematic representation of Te-Ln^{3+}-HA composites

microbial infections. Prior to the discovery of antibiotics, it was used in the treatment of syphilis and leprosy. Another potential application of tellurium is that of an antisickling agent of red blood cells in the treatment of sickle cell anemia and in immunomodulating drugs [8, 9]. There were no reports of acute intoxication caused by this element. It has also found usage in the metallurgic industry to improve the mechanical properties of steels and other ferrous alloys [10]. Tellurium was discovered in 1782, by Franz Müller von Reichenstein, although he was unable to identify it. The metal was first isolated in 1798 by Klaproth, who was interested in the work of Müller. He also suggested its name, which derives of the Latin "tellus," meaning earth [10]. In the context of the work presented here, tellurium dioxide (TeO_2) is a conditional glass former.

Hydroxyapatite (HA), $Ca_{10}(PO_4)_6(OH)_2$, has been widely used as a bone substitute due to its chemical similarity to the inorganic matrix of natural bone, excellent osteoconductivity, and bioactivity. The major drawback of HA is its poor mechanical properties, especially brittleness and low fracture toughness. Therefore, it cannot be used in load-bearing applications. To overcome these disadvantages, HA has to be combined with other materials [11–13].

Lanthanide ions (Ln^{3+}), also known as rare earths [14], are a group of elements from lanthanum to lutetium ($Z = 57–71$), plus scandium ($Z = 21$), and yttrium ($Z = 39$). Lanthanides have been previously shown to display an effective antibacterial activity and to further modulate bone metabolism, broadly due to their affinity to calcium interaction sites [15]. Webster et al. reported that HA doped with divalent and trivalent cations exhibited enhanced bone regenerative properties [16]. Sooraj et al. have also studied lanthanide-doped HA composites for bone tissue applications and reported that osteoblasts response to these materials was improved [17]. It has been also been reported that bioactive, luminescent, and mesoporous europium-doped hydroxyapatite has interesting properties as drug carrier [18]. Thus, the incorporation of Ln^{3+} into hydroxyapatite may represent a great advantage for innovative drug development. Therefore, in the present study, we used lanthanum oxide (La_2O_3).

12.2 Fabrication of Te-Ln³⁺-HA Composites

In literature, there are no studies to our knowledge relating to tellurium-doped hydroxyapatite. However, in a recent study by Mohd Shkir et al. [19], nanorods and nanosheets of pure and Te-doped HAp with different tellurium concentrations have been fabricated by microwave-assisted technique at low temperature. An improved mechanical and antimicrobial effect against some pathogenic gram-negative bacteria, gram-positive bacteria and yeast, has been observed [19]. In another study, tellurite-containing glass ceramics showed a better bioactivity during the in vitro test than that of the silicate [20]. Since last two decades, Sooraj et al. have been working on different tellurium, borosilicate, borophosphate, and zinc borosilicate host glassy materials. The research team also reported $CaO-P_2O_5$

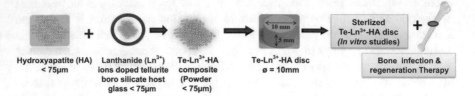

Fig. 12.2 A schematic fabrication design of a Te-Ln-HA composite

glass-reinforced hydroxyapatite composites for bone regenerative applications [11, 12, 21]. In a similar way, the present Te-Ln^{3+}-HA composites have been prepared. The fabrication procedure has two parts: the first part is the glass preparation (Ln^{3+} ion-doped tellurite borosilicate) in the following composition of LaTe = 100-x (B$_2$O$_3$-SiO$_2$-Na$_2$O-CaO-TeO$_2$- xLn^{3+}) (where x = 0.5 & 2 mol% of Ln^{3+} = La$_2$O$_3$) by a melting-water quenching technique at 1000 °C, using a porcelain crucible, for an hour. The melt was then quenched in water. The obtained glass frit was dried at 80 °C overnight prior to crushing and then sieved to less than 75 microns. After the glass preparation, the Te-La^{3+} glass-reinforced hydroxyapatite composites (second part) were obtained by the mixture of 5 wt% of LaTe glass with 95 wt% of hydroxyapatite (HA) powder. A commercial HA powder (Urodelia SA, France) with a particle size of 75 microns was used in the composite preparation. In order to ensure an effective mixing, the HA and glass powders were mixed for 3 h, in dry conditions in a WAB T2F turbula. The composite powders were used to prepare disks by uniaxial pressure at 10 MPa, and sintered in a furnace, at 1300 °C for 1 h, using a 2 °C/min heating rate followed by natural cooling inside the furnace. A schematic fabrication design of the composite is shown in Fig. 12.2. The developed composites are abbreviated as GRHA-0.5(Te-0.5La-HA) and GRHA-2 (Te-2La-HA).

12.3   Luminescence Performance of Ln^{3+} Ion-Doped Tellurite Host Glasses

A schematic indicating different techniques that can be used for the characterization of Ln^{3+}-doped tellurite host glass is shown in Fig. 12.3. There are many reports in the literature showing the good glass-forming abilities for different lanthanide ions as dopants in glass hosts such as silicates, borates, phosphates, tellurite's, etc., enabling a wide variety of applications [21]. Among them, TeO$_2$-based glasses are reported to have good optical transparency, stability, and a good moisture resistance [21]. TeO$_2$ hosts have become more attractive and important for different potential applications, such as in the development of laser glasses and in optical communications [22–26]. In fact, the TeO$_2$ belongs to an intermediate class of glass-forming oxides; it does not readily form a glass but can do so when it is mixed with certain other oxides such as B$_2$O$_3$, P$_2$O$_5$, and SiO$_2$. It has been reported that pure borate glasses have

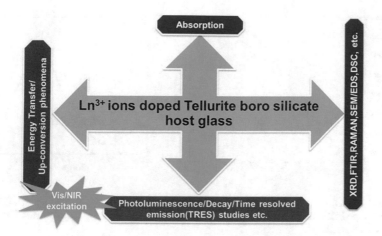

Fig. 12.3 A schematic different characterization techniques for Ln^{3+} ion-doped tellurite borosilicate host glass

phonon energies in the order of 1300–1500 cm^{-1}, tellurite (~700 cm^{-1}), germanate (~900 cm^{-1}), silicate (~1100 cm^{-1}), and phosphate glasses (~1300 cm^{-1}). The glasses with low phonon energy are more suitable as a host material and enable higher quantum efficiency of photoluminescence performance along with longer-lived excited states. Therefore, the addition of glass former or modifier reduces phonon energy significantly. The distinguishing feature of these tellurite-based glasses is that the atomic network appears more open than in other glasses, as the Te-O bond in tellurite glasses is easily broken, and thus it is advantageous for use with lanthanide ion dopants [27, 28].

Because of the selective absorption of light in the visible spectral region by the lanthanide ions, solutions or solids containing lanthanide ions have a specific color. Binnemans and Gorller-Walrand [14, 29] have considered in detail the color caused by the 4f-4f transitions of lanthanide ions. The color of each lanthanide ion was predicated based on the position and the intensity of the absorption bands. They also reported that only intraconfigurational 4f-4f transitions are responsible for the color of trivalent lanthanide ions [14]. Further, colors are not identified with a unique wavelength but by a wavelength interval. The color of lanthanide ion even changes from host to host and depends also on the light source used. The yellow-to-blue luminescence intensity ratios and color chromaticity coordinates of the Dy^{3+}-doped lithium tellurofluoroborate glasses have also been estimated to evaluate the white-light emission as a function of Dy^{3+} ion concentration [30] and luminescence investigations of Dy^{3+}-doped yttrium calcium silicoborate glasses for cool white LED applications [31].

Sooraj et al. have reported the energy level structure, bonding, band intensities, and radiative properties of the Nd^{3+}-doped borotellurite glasses [32]. The effects of change in the alkali cations (Li, Na, and K) in the glass chemical composition were verified. In

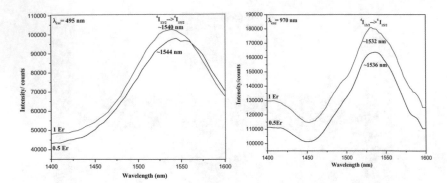

Fig. 12.4 NIR transition ($^4I_{13/2} \to {}^4I_{15/2}$) emission spectra for (20TeO$_2$–0.5Er$_2$O$_3$)- and (20TeO$_2$-1Er$_2$O$_3$)-doped tellurium borate glasses. Left (λ_{ex} = 495 nm) and right (λ_{ex} = 970 nm). (Data were taken from Ref. [33])

another work, an IR-transmitting circular glass window of Nd^{3+}-doped borophosphate tellurite system has been reported [32]. The results shown were very encouraging, and this glass is an ideal optical material for development of the intense lasing action at 1.06 microns due to the $^4F_{3/2} \to {}^4I_{11/2}$ transition. In another work, erbium oxide (0.5–4 mol %)-doped lithium borotellurite glasses have been reported and characterized for NIR emission band at 1.53 μm. By pumping with LED excitation sources at 495 nm and 970 nm, a broader emission transition ($^4I_{13/2} \to {}^4I_{15/2}$) and lifetimes ($\tau_{mea}$) of NIR luminescence transitions $^4I_{13/2} \to {}^4I_{15/2}$ have shown [33] in Figs. 12.4 and 12.5.

12.4 Time-Resolved Decay Measurements of Different Glasses

It should be noted that luminescent transitions can present different lifetimes. There is one problem associated with steady-state luminescence (SSL) techniques, where only an average of the individual contribution of different environments or luminescent centers is observed and hence valuable information can be missed. Time-resolved emission spectroscopy (TRES) can be used to help elucidate the distribution and different centers in the samples [34, 35]. Time-dependent spectral shifts are usually studied by measurement of the time-resolved emission spectroscopy (TRES).This technique is a very sensitive analytical method for structure investigation in more detail. It depends strongly on the kind of active ion, its local site, and site distribution in the given host matrix. Moreover, overlapping bands can often be discriminated by time-resolved spectroscopy provided the emitting levels have different decay times [36–38].

Measurements were done through a HORIBA Scientific FluoroCube, again with DeltaDiode excitation, and a H10330-75 detection module close coupled to the

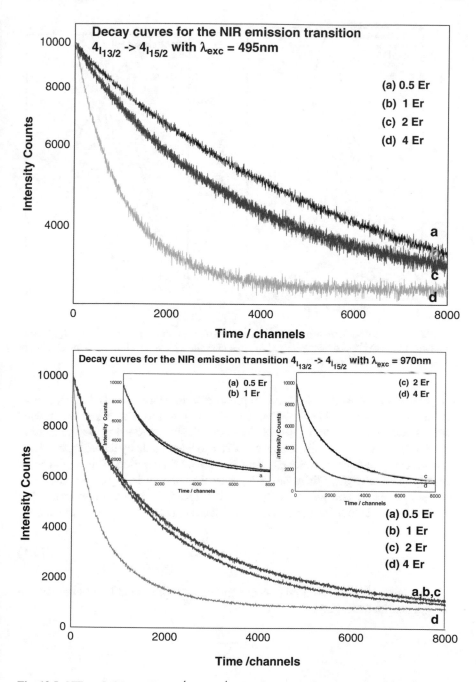

Fig. 12.5 NIR emission transition ($^4I_{13/2} \rightarrow {}^4I_{15/2}$) decay trends for erbium oxide (0.5–4 mol %) doped tellurium borate glasses. Left (λ_{ex} = 495 nm) and right (λ_{ex} = 970 nm). (Data were taken from Ref. [33])

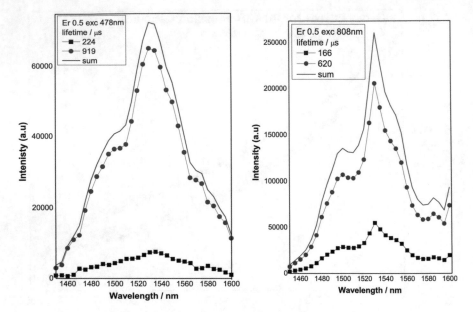

Fig. 12.6 Time-resolved decay-associated spectra of NIR emission transition ($^4I_{13/2} \rightarrow {}^4I_{15/2}$) decay trends for erbium oxide (0.5 mol %)-doped calcium phosphate glass. Left ($\lambda_{ex} = 478$ nm) and right ($\lambda_{ex} = 808$ nm) (unpublished data)

timing electronics using a CFD-2G amplifier/discriminator. The DeltaDiode laser excitation sources (emitting at 478 nm and 808 nm) were operated in "burst mode" with the 100 MHz excitation pulse chain automatically gated using the DataStation software. Time-resolved emission spectra (TRES) were made by recording the time-resolved decay at equal wavelength steps for a fixed time period. This produced an intensity – wavelength – time, i.e., 3-D, dataset as shown in Figs. 12.6 and 12.7. Both simple decay and global analyses of the TRES data were performed using DAS6 software. Decays were fitted as a sum of exponential components of the form

$$I(t) = \sum_{i}^{n} \alpha_i \exp(-t/\tau_i) \tag{12.1}$$

with the "amount" of each luminescent component represented by the normalized pre-exponential factor, i.e.,

$$\alpha_i = \frac{\alpha_i}{\sum_{i=1}^{n} \alpha_i} \tag{12.2}$$

Goodness of fit was assessed by evaluation of the reduced chi-squared value and weighted residuals. The outcomes of the global analysis of the TRES data were then

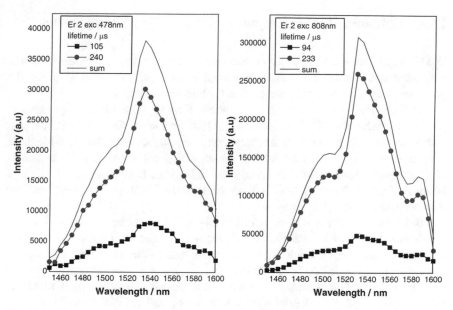

Fig. 12.7 Time-resolved decay-associated spectra of NIR emission transition ($^4I_{13/2} \rightarrow {}^4I_{15/2}$) decay trends for erbium oxide (2 mol %)-doped calcium phosphate glass. Left ($\lambda_{ex} = 478$ nm) and right ($\lambda_{ex} = 808$ nm) (unpublished data)

treated to produce decay-associated spectra by plotting (using Origin software) the pre-exponential components weighed by their lifetime, against wavelength.

Er^{3+} emission spectra (0.5 mol%) acquired at different periods of time are shown in Fig. 12.6 with two different pumping sources (478 nm and 808 nm). Exciting the 0.5 Er sample at 478 nm, the NIR emission transition ($^4I_{13/2} \rightarrow {}^4I_{15/2}$) seen in the steady-state spectrum, can be described by two component spectra. It is dominated by a component spectrum associated with a lifetime of 919 μs, with a smaller contribution from a spectrum associated with a lifetime of 224 μs. Similarly for 2 mol% Er^{3+}-doped glass, presented in Fig. 12.7, the transition exhibits a dominant longer-lived ($\tau = 240$ μs) decay. Another feature that can deduced from these spectra is the bands shifting. It is not very pronounced, but it takes place. Also a shoulder, near the peak, on the shorter-lived spectra is not properly observed but is evident in the longer-lived decay-associated spectra. Moreover, the bands' broadening is not constant for all decay-associated spectra. Thus, these results reinforce the idea that the microenvironment of ion sites changes their spectroscopic properties. Furthermore, the time-resolved decay-associated spectra of different lanthanide (Eu^{3+} and Tm^{3+}) ion-doped tellurite borosilicate host glasses also show the same trends (date not shown).

12.5   Osteoblastic Cell Response of Te-Ln^{3+}-HA Composites

The biological response to developed glass-reinforced hydroxyapatite materials was conducted with human bone marrow-derived stromal cells (hBMSCs), given the prospective application of the composites for bone tissue applications.

hBMSCs (Stemcell Technologies, Grenoble, France), with a immunophenotypic profile of $\geq$90% for CD73, CD90, and CD105 and <5% for CD14, CD34, and CD45, assessed by flow cytometry, were cultured in α-Minimal Essential Medium, supplemented with 10% fetal bovine serum, 50 μg/ml ascorbic acid, penicillin (10 units/ml)/streptomycin (2.5 μg/ml), and 2.5 μg/ml fungizone, at 37 °C, in a humidified atmosphere of 5% CO_2 in air. Cells of the fourth passage were used for the composites' biological evaluation.

Sterile material samples were placed in 24-well tissue culture plates and seeded with a cell density of 10^5 cells/cm^2. Cultures were grown for 7 days. HA samples, with similar microtopographic surface characteristics to the developed composites, were used as a control material. Seeded Te-Ln^{3+} glass-reinforced composites and HA samples were characterized by confocal laser scanning microscopy (CLSM) for the assessment of the cell/biomaterial interaction, cell proliferation (total DNA content), and alkaline phosphatase (ALP) activity as shown in Figs. 12.8 and 12.9.

12.5.1   Assessment of the Cell/Biomaterial Interaction: CLSM Analysis

For CLSM observation, live cells were stained with MitoSpy Red CMXRos (Biolegend, Germany), at 150 nM, for 15 min. The cells were then fixed (3.7% paraformaldehyde, 15 min), permeabilized (0.1% Triton in PBS, 5 min), and incubated with a solution of 10 mg/ml of bovine serum albumin with 1 μg/ml RNAse in

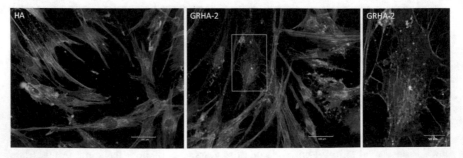

Fig. 12.8 Representative CLSM images of hBMSCs grown for 3 days on the surface of HA and GRHA-2. Cells were stained for mitochondrial localization (red), F-actin cytoskeleton (green), and nucleus counterstaining (blue). Inset corresponds to a high magnification image of the GRHA-2 micrograph, delimited by the white box. Scale bar: 100 μm (low) and 50 μm (high) magnification images

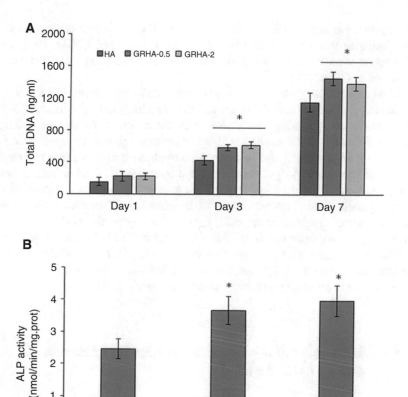

Fig. 12.9 Total DNA content (**a**) and ALP activity (**b**) of hBMSCs grown on HA and GRHA. ALP activity was determined at day 7 of culture. Statistical analysis of data was performed using the Kruskal-Wallis test, computed with the software SigmaStat 3.5. * – significantly different from HA, $p \leq 0.05$

PBS (1 h). Cell cytoskeleton filamentous actin (F-actin) was stained by treating the cells with 5 U/ml Flash Phalloidin™ Green 488 (Biolegend, Germany) (20 min). Cultures were counterstained with DAPI (0.1 µg/ml), mounted in Vectashield, and visualized on a Leica SP2 AOBS microscope. Representative images are shown for 3-day cultures grown over HA and GRHA-2 as shown in Fig. 12.8.

Cells grown on HA presented an active mitochondrial activity, as evidenced by the red staining of the used mitochondrial probe, particularly focused on the perinuclear region. Comparatively, cells grown on GRHA-2 presented a more intense staining, further polarized around the nucleus. The perinuclear arrangement of the mitochondrial staining has been previously associated, on cells growing over a material surface, with a cytocompatible profile, as opposing to a scattered mitochondrial arrangement broadly associated to cell toxicity [39]. On both HA and GRHA, the perinuclear mitochondrial arrangement was evident, sustaining the adequacy of both substrates to endure cell functionality. Of additional relevance, a higher probe signal intensity was attained on cells growing on developed GRHA composites, which seems to correlate with a higher mitochondrial membrane potential and higher cellular metabolic activity [40].

CLSM further allowed the assessment of cell morphology. Phalloidin staining for cytoskeletal F-actin evidenced, on seeded HA, an elongated cell structure, with a high cytoplasmic spreading and dense actin stress fibers. At the cells' border, abundant filopodia – thin, actin-rich plasma-membrane protrusions that extend out from the cell edge – could also be easily identified. A similar structure and organization could be acknowledged on cells grown over GRHA. Furthermore, it is also evident, particularly at the high magnification inset, the organized filopodial processes expanding isotropically. Filopodial structures were shown to contribute to microenvironment probing of the cells, further assisting on the modification of the surrounding extracellular matrix (ECM) and also facilitating the intercellular communication. Furthermore, these adhesion structures appear to consolidate the adhesive interaction with the substrates, modulating cellular cascades that regulate cell behavior, further enhancing cell survival and proliferation [41].

12.5.2 Assessment of the Cell Proliferation and Functional Activity (ALP Activity)

Cell proliferation was estimated by total DNA quantification, and functional activity of hBMSCs was determined by the assessment of ALP activity. For the DNA assay, cultured cells were lysed with 0.1% Triton X-100, and DNA content was determined by the fluorimetric PicoGreen DNA quantification assay (Quant-iT™ PicoGreen® dsDNA Assay Kit, Molecular Probes Inc., Eugene). Fluorescence was measured at 480 and 520 nm, excitation and emission, respectively, in a Synergy HT, Biotek, ELISA system.

ALP activity was assessed in cell lysates by the incubation with p-nitrophenyl phosphate (in an alkaline buffer pH 10.3; 30 min, 37 °C) and measuring the conversion to p-nitrophenol at $\lambda = 400$ nm, in a Synergy HT, Biotek, ELISA system. ALP levels were normalized to total protein content, previously quantified by the Bradford's method.

On cultures grown over HA, total DNA content, a measure of cell proliferation, increased throughout the assayed culture period. Comparatively, cultures established on the two formulations of GRHA (GRHA-0.5 and GRHA-2) presented higher levels and were found to attain statistical significance at day 3 and 7 of culture, suggesting an increased cell proliferation on cultures established on the composites' surface as shown in Fig. 12.9.

ALP activity was also found to be increased on cultures grown on both composites' surface, at day 7. Within the osteogenic differentiation process, ALP is synthesized during early extracellular matrix organization, assisting on the hydrolysis of organic phosphates and ensuring the microenvironmental conditions for the mineral deposition and matrix mineralization [42]. It is further commonly used as an early marker of the in vitro osteogenic differentiation process [42].

Despite the absence of literature addressing the biological behavior of tellurite-containing biomaterials for bone tissue applications, attained data is in line with previous observations of cell culture studies conducted on other lanthanides-containing composites. For instance, lanthanum-doped HA was found to enhance the proliferation and osteoblastic differentiation of osteoblastic cells [16], whether samarium-reinforced HA promoted the adhesion and proliferation of human osteoblastic-like populations [43]. Also cerium-containing substrates were found to promote the adhesion and proliferation of osteoblastic cells [44]. Taking it all together, a strong body of evidence converges to the idea that HA doping with lanthanides enhances cell response, improving adhesion, proliferation, and differentiation processes. Lanthanides may contribute to modulate surface chemical properties of the composites, influencing protein adsorption, which has a great impact on determining the subsequent cell activity [45]. In addition, lanthanides, being unable to cross cell membranes, may also interfere with membrane-bound signalling mechanisms, particularly those regulated by ionic calcium, as the Ca-sensing receptor [46].

Acknowledgment The authors (SHN and AS) would like to thank the European Union's Horizon 2020 research and innovation program under the Marie Sklodowska-Curie grant agreement no. 753636.

References

1. See, http://www.nhs.uk/conditions/osteomyelitis
2. See, http://www.medicinenet.com/osteomyelitis/article.htm
3. R. Berendt et al., *Osteomyelitis. Oxford Textbook of Medicine*, 5th edn. (Oxford University Press, Oxford, UK, 2010)
4. P. Bejon et al., Medicine **41**(12), 719 (2013)
5. See, http://ec.europa.eu/dgs/health_food-safety/docs/communication_amr_2011_748_en.pdf
6. See, http://ec.europa.eu/health/antimicrobial_resistance/docs/2015_amr_progress_report_en.pdf
7. See, http://www.jpiamr.eu/wp-content/uploads/2014/12/AMR-Review-Paper-Tackling-a-crisis-for the -health-and-wealth-of-nations_1-2.pdf
8. L.A. Ba, M. Döring, V. Jamier, C. Jacob, Org. Biomol. Chem. **8**(19), 4203 (2010)
9. R.L.O.R. Cunha et al., An. Acad. Bras. Ciênc. **81**(3), 393 (2009)
10. See, http://nautilus.fis.uc.pt/st2.5/scenes-e/elem/e05200.html
11. N. Sooraj Hussain, J. D. Santos (eds.), *Biomaterials for Bone Regenerative Medicine* (Trans Tech Publishers, Switzerland, 2010). ISBN- 13: 978-0-87849-153-7
12. N. Sooraj Hussain, et al., Chapter 13. Bonelike® graft for Bone regenerative applications, in Surface Engineered Surgical Tools and Medical Devices, ed. by M.J. Jaccson, W. Ahmad, (Springer Publications, Boston, 2007), pp. 477–512
13. D. Milovaca, et.al. Mater. Sci. Eng. C **42**, 264 (2014)

14. K. A. Gschneidner Jr., L. Eyring (eds.), *Handbook on the 'Physics and Chemistry of Rare Earths'*, vol 25 (Elsevier, New York, 1998)
15. S.P. Fricker, Chem. Soc. Rev. **35**(6), 524 (2006)
16. J.T. Webster et al., Biomaterials **25**, 2111 (2004)
17. J. Coelho, N. Sooraj Hussain, et al. *Current Trends on Glass and Ceramics Materials* (Bentham Science Publishers, Sharjah, 2012), https://doi.org/10.2174/97816080545271130101, eISBN: 978-0-87849-153-7, ISBN: 978-0-87849-153-7, pp. 87–115
18. P. Yang et, al. Biomaterials **29**, 4341 (2008)
19. I.S. Yahia, M. Shkir, et al., Mater. Sci. Eng. C **72**, 472 (2017)
20. G. El-Damrawi, H. Doweidar, H. Kamal, SILICON **9**(4), 503 (2017)
21. N. Sooraj Hussain, J. D. Santos (eds.), *Physics and Chemistry of Rare-Earth Ions Doped Glasses*, Reinhardtstrasse 18, CH-8008 Zurich, Switzerland (Trans Tech Publishers, Enfield, 2008)
22. G. Senthil Murugan, Y. Ohishi, J. Non-Cryst. Solids **341**(1–3), 86 (2004)
23. V.C. Veeranna Gowda, C. Narayana Reddy, K.C. Radha, R.V. Anavekar, J. Etourneau, K.J. Rao, J. Non-Cryst. Solids **353**(11–12), 1150 (2007)
24. V.O. Sokolov, V.G. Plotnichenko, et al., J. Non-Cryst. Solids **352**(52–54), 5618 (2006)
25. R. El-Mallawany, *Tellurite Glasses Handbook, Physical Properties and Data* (CRC Press, Boca Raton, 2002)
26. R. El-Mallawany et al., Opt. Mater. **26**(3), 267 (2004)
27. N.V.V. Prasad, K. Annapurna, N.S. Hussain, et al., Mater. Lett. **57**(13–14), 2071 (2003)
28. T.V.R. Rao, R.R. Reddy, et al., Infrared Phys. Technol. **41**(4), 247 (2000)
29. K. Binnemans, C. Gorller-Walrand, Chem. Phys. Lett. **235**, 163 (1995)
30. U.K. Maheshvaran, K. Marimuthu, G. Muralidharan, J. Lumin. **176**, 15 (2016)
31. C.R. Kesavulu et al., J. Alloys Comp. **726**, 1062 (2017)
32. N. Sooraj Hussain, K. Annapurna, S. Buddhudu, Phys. Chem. Glasses **38**(1), 51 (1997)
33. J. Coelho, J. Azevedo, G. Hungerford, N. Sooraj Hussain, Opt. Mater. **33**, 1167 (2011)
34. N. Akiyama, S. Muramatsu, et al., J. Lumin. **87**, 568 (2000)
35. B. Marmodée, J.S. De Klerk, et al., Anal. Chim. Acta **652**, 285 (2009)
36. J. R. Lakowicz (ed.), *Principles of Fluorescence Spectroscopy*, 2nd edn. (Kluwer Academic/Plenum Publishers, New York, 1999)
37. A. Herrmann, D. Ehrt, J. Non-Cryst. Solids **354**, 916 (2008)
38. P. Bernard et al., J. Eur. Ceram. Soc. **16**, 195 (1996)
39. D.K. Patel et al., Polymer **106**, 109 (2016)
40. K. Mitra, J. Lippincott-Schwartz, *Analysis of Mitochondrial Dynamics and Functions Using Imaging Approaches*, Curr. Protoc. Cell Biol. **46**: 4.25.1–4.25.21 (2010), https://doi.org/10.1002/0471143030.cb0425s46
41. G. Jacquemet, H. Hamidi et.al. Curr. Opin. Cell Biol. **36**, 23 (2015)
42. E. Birmingham et al., Eur. Cells Mater. **23**, 13 (2012)
43. D.S. Morais, J. Coelho, M.P. Ferraz, P.S. Gomes, et al., J. Mater. Chem. B **2**, 5872 (2014)
44. Y. Goh, A. Alshemary, et al., Ceram. Int. **40**, 729 (2014)
45. C.J. Wilson et al., Tissue Eng. **11**, 1 (2005)
46. J. Zhang et al., Mini. Rev. Med. Chem. **11**, 678 (2011)

Index

A
Absorption
    coefficient, 118, 119, 123, 124
    cross section, Sm^{3+}, 114
    edge, 117, 119, 122, 123
    emission wavelengths, 112
    UV-Vis-NIR spectroscopy (*see* UV-Vis-NIR absorption spectroscopy)
Absorption spectrum fitting (ASF) method, 72
Adatoms, 107
Addition de Photon par Transferts d'Energie (APTE), 261
Alkaline phosphatase (ALP), 284, 286–287
Amorphization, 106
Amorphous, 111, 118, 119, 122–124
Applications of tellurite glasses
    amorphous silicon solar cells, 8
    biomedical, 13
    DC electrical conductivity, 8
    gold nanoparticles, 11
    NIR emission properties, 12
    noncrystalline solid, 10
    optical, 11
    photonic, 9
    radiation shielding, 9
    solar cell, laser and luminescent display, 12
    solar energy harvesting, lanthanide-doped tellurite glasses, 12, 13
    solar energy technology and laser devices, 9
    structural and luminescence properties, 9, 11
    thermoluminescence (TL) mechanism, 8
    thermometry, 11
    Yb^{3+} concentrations, 8

Artificial neural network (ANN) technique, 1
Asymmetrical distribution, electrons, 130
Avogadro's number, 121

B
Band gap
    energy, 125
    optical, 123
Band tail, 119, 120, 124, 127
Barium fluorotellurite (BFT) glasses, 172
Beer-Lambert's law, 52, 118
Binary glasses synthesis, 108
Blue-red spectrum, 227, 231–234
Bonding parameter and nephelauxetic ratio, 129–132
Bone marrow, 284
Bradford's method, 286
Branching ratios (β_R), host glass network, 57–59, 63
Broadening mechanisms, 110

C
Carbon nanotubes, 251
CdS nanocrystals, 43, 44
Chalcogenide, 107
Chemical reaction, manufacturing of glass material, 106
Cluster, RE ion, 108, 112
Coherent oscillations, NPs, 113
Conduction band, 115, 117, 119, 124, 135
Cooperative upconversion (CUC), 157
Corrosion resistance, 106
Covalent bond, 130

Printed in the United States
By Bookmasters